T0039166

# Foundations of Software Engineering

# Foundations of Software Engineering

**Ashfaque Ahmed**
*SCM Consulting, Bhilai, Chattisgarh, India*

**Bhanu Prasad**
*Florida A&M University, Tallahassee, USA*

CRC Press is an imprint of the
Taylor & Francis Group, an **informa** business
AN AUERBACH BOOK

CRC Press
Taylor & Francis Group
6000 Broken Sound Parkway NW, Suite 300
Boca Raton, FL 33487-2742



Printed on acid-free paper by CPI Group (UK) Ltd, Croydon, CR0 4YY
Version Date: 20160127

International Standard Book Number-13: 978-1-4987-3759-3 (Hardback)



**Library of Congress Cataloging-in-Publication Data**

Names: Ahmed, Ashfaque, author. | Prasad, Bhanu, author.
Title: Foundations of software engineering / Ashfaque Ahmed and Bhanu Prasad.
Description: Boca Raton : CRC Press, 2015. | Includes bibliographical
references and index.
Identifiers: LCCN 2015045650 | ISBN 9781498737593 (acid-free paper)
Subjects: LCSH: Software engineering.
Classification: LCC QA76.758 .A398 2015 | DDC 005.1--dc23
LC record available at http://lccn.loc.gov/2015045650

**Visit the Taylor & Francis Web site at**
**http://www.taylorandfrancis.com**

**and the CRC Press Web site at**
**http://www.crcpress.com**

# Contents

# Preface

Software engineering is a vast field. It has the following core areas: requirements management, software architecture, software design and construction, and software testing. All of these areas are vast and are mostly separated from each other. The core area software design and construction is further divided into user interface design and construction, business logic design and construction, and database design and construction.

The authors have several years of experience working with and teaching all areas of software engineering. The lead author has also been writing university-level textbooks in areas related to software engineering for the last 10 years.

This book provides in-depth coverage of the areas of software engineering that are essential to learning this field. Advanced areas such as commercial-grade security mechanisms, and performance management techniques for high-end software products, are not discussed in this book. Discussing these areas may distract the reader from the primary goal of learning software engineering.

This book was written after receiving feedback from several professors and software engineers from the United States and India. A book on software engineering was thought to be necessary, and it not only should cover the theory of software engineering but also should have content for practice. Without practical examples to work on, it is difficult to learn any subject. This is especially true for a subject such as software engineering, which is used to build large and complex software products. At the same time, it is also true that many universities have a curriculum in software engineering that includes building a software product by undertaking a software project. This book covers both theory and practical examples of building software products.

As a result of extensive research conducted by the authors on what content to include in the book and how to present that content, the material provided in this book is relevant for teaching software engineering to students. Rather than using UML or other formal notations, the content is explained in plain (and easy-to-understand) English. Basic programming knowledge using an object-oriented language is helpful to understand the contents of this book. We have provided a complete case study using Java. The knowledge gained from this book can be readily used in other relevant courses or in a typical software development environment of a company.

The best way to learn software engineering is by understanding its core and peripheral areas. In this book, a complete chapter is devoted to each of the core and peripheral areas. Requirements management is covered by explaining requirement specifications and use cases. Software high-level design aspects are covered including software architecture patterns, software architectures, component design, component diagrams, and data flow diagrams. Software detailed design and construction for business logic development covers classes, class diagrams, objects, object diagrams, statechart diagrams, sequence diagrams, database programming, web-based programming, refactoring, and so on. User interface design includes Model–View–Controller architecture, HTML, client-side scripts, and AJAX, among others. Database design covers schemas, relational database concepts, referential and entity integrity rules, primary keys, secondary keys, normalization, and so on. Software testing includes unit testing, integration testing, system testing, user acceptance testing, and testing strategy, to name a few.

The book also covers peripheral areas of feasibility study, software engineering methodologies, project management, configuration management, and software maintenance.

Some of the salient features of the book include the following:

- Complete coverage of all important areas of software engineering.
- Strong foundation for almost all areas of software engineering. Each software engineering concept is backed by solid examples.
- Complete alignment with university-level courses in software engineering offered at several universities in the United States and India.
- Complete case study with source code (source code is available at the course website).
- Object-oriented design and programming with complete concepts in modeling and design using classes and objects.
- Database design with entity-relationship concept.

The case studies that have been provided for all core areas of software engineering are a useful feature of this book. They provide an understanding of software engineering as well as how things work while building a software product. The following artifacts are provided for a case study that deals with an online banking system called OBAAS:

- Complete requirements specifications along with the use cases
- Software high-level design including software architecture, software component diagrams, and data flow diagrams
- Detailed design for developing the business logic including class diagrams, object diagrams, statechart diagrams, and sequence diagrams
- Software construction including complete source code and technical documentation
- Database design (entity-relationship diagrams and table structure)
- User interface design (mockup screens)
- Testing strategy including testing samples

This case study is provided at the end of each of the core chapters under a separate section. While all the design documents and requirements specifications are provided in the book itself, the source code for this software product is provided at the course website: http://ahmedashfaque.wordpress.com. People can easily download the source code as well as the database scripts from this website. In case this website is down for any reason, the details of the new website will be posted at http://www.bhanuprasad.org/fse.html. There are two sets of source code: OBAAS Version 1.1 and OBAAS Version 1.2. We have used Java technology to help readers learning how to apply software engineering concepts develop an object-oriented project. The other case study deals with an order management system for a restaurant. It is provided within the chapters.

The case study on OBAAS can also be used for doing additional exercises. We have provided some ideas about enhancing OBAAS at many places in the case study. For example, we are currently using hard-coded values to populate the service types in a drop-down list in the source code. This functionality can be extended so that instead of hard-coded values these values can come from a database. You can find a list of future enhancements in Chapters 4 and 8, which can be taken as a practical exercise or assignment in the classroom. These exercises can be in any area of software engineering. For example, a feasibility study can be done to find out if any of the enhancements listed there are feasible.

All in all, this book has everything one needs to learn software engineering easily. There are also myriad code samples, comparisons, and examples provided throughout the book on all the topics covered. They will help the reader clearly understand the concepts.

This book has an appendix that contains answers to all the questions at the end of each chapter. Instructor material is available at the course website for instructors who adopt this book as a textbook.

The book will be very useful for students taking a course in software engineering or its equivalent at the undergraduate or graduate level. People working in the software industry and those who want to bridge their knowledge gap in any area related to software engineering may also find this book useful.

We invite you to e-mail us any comments concerning this book at the addresses provided below. We will use them to improve future editions of this book. We also invite you to e-mail us any errors that you notice. The errata will be posted at the course website. Hope you enjoy reading this book!

**Ashfaque Ahmed**
*ashfaque.a@gmail.com*

**Bhanu Prasad**
*bhanupvsr@gmail.com*

# Acknowledgments

This book was prepared after getting feedback from a large number of professors who teach courses in software engineering or its equivalent at several universities in the United States and India. We are thankful to all these people for their invaluable feedback.

I thank Dr. Nadimpalli Mahadev, professor, Department of Computer Science, Fitchburg State University, Massachusetts; Dr. Victoria Hilford, assistant professor, Department of Computer Science, University of Houston, Texas; Dr. Guowei Yang, assistant professor, Department of Computer Science, Texas State University, San Marcos; and Dr. Scott Stoller, professor, Computer Science Department, Stony Brook University, New York. We are also thankful to many other professors, whose invaluable advice and feedback helped in creating the structure and content of this book.

Writing a large textbook is very demanding and labor-intensive. While I was working on writing the book, I was fully supported by my family. Without their support, I could not have done it. Credit goes to my father, mother, wife, and brother Javed, who have always stood by me and my commitment to writing this book. Javed also helped in writing the source code for the case studies. During breaks in writing, my family's kids, Arisha, Shija, Aashi, Jasim, and Ismail, enlivened me. Playing with kids is so much fun and relieves my stress. After playing with kids, I get recharged for the next load of hard work.

Sadly, during the time that I was (too engrossed) writing this book, my youngest brother, Aslam, passed away because of heart failure. It jolted me so much that for quite some time I was not able to concentrate on anything. Somehow, I mustered enough courage and strength to finish writing this book.

**Ashfaque Ahmed**

I am thankful to many people, including my colleagues, students, relatives, and friends, who helped me in writing this book and for providing valuable input from time to time. I am thankful to Florida A&M University for providing an environment where I was able to work on this book. I am thankful to Dr. Bobby Granville, Dr. Zoran Majkic, Samuel, John, Sreedevi, Anjali, Ram, and Brown for their valuable feedback and help.

I am thankful to India, the country where I was born and brought up and from which I received my academic education. I am also thankful to all my teachers and professors. Finally and most importantly, I am thankful to my parents for all their support and encouragement in pursuing my academic education.

**Bhanu Prasad**

# Authors

**Ashfaque Ahmed** is a seasoned software engineering professional with more than 27 years of experience. He has worked as a programmer analyst, software engineer, project manager, test engineer, test manager, and business analyst in his career. He has work experience in all areas of software engineering including requirements management, software design, software construction, software testing, and database design. He has worked for leading multinational companies in his career, in countries including the United States, Canada, United Arab Emirates, India, Libya, and Nigeria.

Ahmed is also a well-known author. He has written *Software Testing as a Service* (2009, CRC Press, USA); *Software Project Management: A Process-Driven Approach* (2011, CRC Press, USA); *The SAP Materials Management Handbook* (2014, CRC Press, USA); and two smaller books. His books are recommended for reading at large universities in countries including the United States, India, Canada, South Korea, Thailand, and Indonesia.

Ahmed is passionate about helping mankind by using technology and spreading awareness to combat climate change and global warming. Using renewable energy and finding better ways to reduce our carbon footprint is one area that is close to his heart.

Ashfaque Ahmed earned his bachelor of engineering degree from the National Institute of Technology, Raipur, India, in 1988 and master of business administration degree from Indira Gandhi National Open University, India, in 1997.

**Bhanu Prasad** earned his master of technology and PhD degrees, both in computer science, from Andhra University and the Indian Institute of Technology Madras, respectively. He is currently serving as an associate professor in the Department of Computer and Information Sciences at Florida A&M University, Tallahassee, where he was also the chairman of the department. He served as an assistant professor of computer science at Georgia Southwestern State University, Americus. He worked with Hyperion Solutions Corporation (now part of Oracle) in the United States and Infosys Technologies and Future Software (now part of Aricent) in India.

Dr. Prasad's current research interests are in the areas of artificial intelligence and software engineering. He has authored or co-authored more than 70 research papers in various international conference proceedings, journals, or books in computer science. Some of the courses he has taught at the graduate level include software engineering, software development and maintenance, advanced system design principles, human–computer interaction and interface design, internetworking architecture and protocols, programming languages, topics in advanced artificial intelligence, operating systems, and theory of computation. He has also taught software engineering and approximately 20 other courses at the undergraduate level. Dr. Prasad received several certificates including the Rational Rose certificate and Rational Unified Process certificate from the Rational Software Corporation (now part of IBM). He has supervised graduate students to successfully complete their thesis work and served as an external reviewer or committee member for graduate theses or dissertations from several universities around the world. He is or was a reviewer, program committee member, organizing committee member, co-chair, chair, keynote speaker, editorial board member, guest editor, co-editor, editor, and editor in chief for more than a hundred international conferences, journals, or books in computer science.

# 1

# Introduction to Software Engineering

**In Chapter 1, we will learn**

- **What computers are**
- **What software engineering is**
- **Why software engineering is important**
- **Challenges in building software products**
- **Software engineering careers**

## 1.1 Introduction

Human civilization has evolved from cave dwelling to modern-day city living. During the course of evolution, humans learned trades, improved their skills, and refined their living standards. A key ingredient for this evolution is the understanding of science and applying the scientific findings to improve the living standards of humans. On their way to early evolution, humans learned how to farm, use wheels, make tools, and so on, which permanently changed the way early humans lived. Since, new discoveries in science continue to change the lives of humans.

Hundreds of years ago, during the Industrial Revolution, engineering started to shape human lives. Engineering helped create large man-made structures using applied science. Humans were able to build large bridges, factories, buildings, and so on, which paved the way for making modern-day infrastructures.

Sometime around the 1940s, a new revolution started. This revolution was the introduction of computers. Computers are able to perform large and complex calculations quickly. With the increase in knowledge, humans are able to use computers to do many things in areas such as telecommunications, business, entertainment, technology, medicine, education, and research. Today, we use devices such as mobile phones, the Internet, and automobiles. Most of these are driven or controlled by computers.

## 1.2 Components of a Computer

Computers consist of hardware and software components. The hardware components are physical components that you can touch, smell, and taste. The main hardware components include the input devices, output devices, processor or CPU, memory, storage devices, and communication devices. A keyboard, screen, RAM, C drive or hard drive, and modem are examples of an input device, output device, memory, storage device, and communication device, respectively.

A computer system also contains many types of software. Software or a program is an instruction or sequence of instructions that tells the computer what tasks it needs to perform and how to perform them. There are two types of software: application software and system software. Application software is intended to operate one or more applications such as a database program, a word processing program, a video editing program, or a web browser program. System software deals with operating or maintaining either the computers or the devices connected to the computers. In other words, system software works as an interface between the hardware of the computer, the computer user, and the application software of the computer. System software can be categorized as operating systems and utility software. Microsoft Windows, Mac, and UNIX are some examples of operating systems. Antivirus programs, disk cleaning programs, and screen savers are some examples of utility software. To use an application software on a computer, that computer should have an operating system compatible with (or that supports) that application software. For example, the Mac operating system is compatible with the Apple Safari browser and vice versa.

While computer engineers, better known as hardware engineers, design and create hardware parts, it is the software engineers who develop the software products.

## 1.3 Building the Software Products

Almost all software products or application programs running on your computer have been developed by a team of people. A team is required to develop a software product because of the sheer size of such products or application programs. For example, the word processor that you use is actually a very large application that was developed by a team of hundreds of software engineers at Microsoft Corporation. Websites such as ebay.com or amazon.com are also very large. A large team of software engineers develops and maintains such websites. A very small number of software products can be developed by just a bunch of people.

Any computer program, small or big, is qualified as a software product. However, in this book, a software product refers to a large program (or a set of large programs) that needs at least a group of people to develop it for several weeks or longer. The terms *software*, *product*, and *software product* are used synonymously in this book.

Consider the task of constructing a huge apartment (i.e., an apartment building). Construction of an apartment involves several things. In a sample scenario, first, the owner (or developer) of the proposed apartment building notices that he or she can construct an apartment to make rental income. The motive for this project is depicted in Figure 1.1. He or she engages an architect who makes the design for the apartment. The owner then engages a construction company and it brings in people, machines,

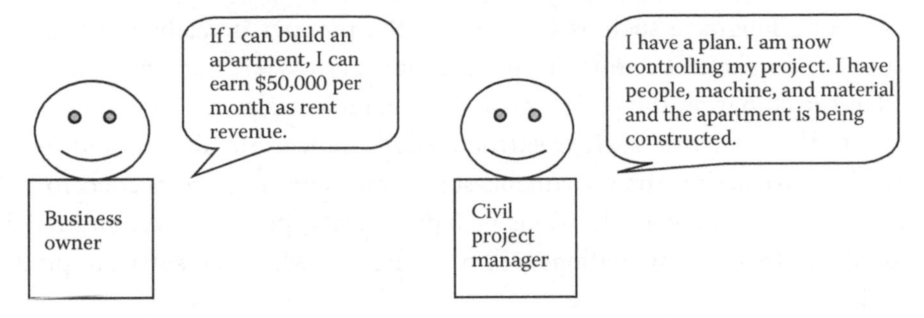


**Figure 1.1** Project for building an apartment initiated by a business owner and managed by a civil project manager.

and materials to construct the apartment building. The civil engineer who is in charge of the construction (i.e., the civil project manager) performs several tasks. On the basis of the design of the apartment, he or she makes estimates for the items such as the material, machines, and manpower that are needed for the construction. He or she also makes an estimate of the time required to finish the construction. In essence, the civil engineer makes a project plan. When the project starts, the civil engineer tracks the project to ensure that everything is going as per the project plan. Once the construction of the apartment is completed, the project ends.

An apartment also requires electricity, cable, sewerage, water, and so on. The apartment should meet some standards prescribed by the government. These standards include safety measures such as earthquake resistance, fire alarms, fire exits, fire extinguishers, emergency exits, and open spaces around the building.

Building a software product is an activity similar to the construction of an apartment. In a sample scenario, a software company notices that a software product needs to be developed because there is a market for that product. The motive for this project is depicted in Figure 1.2. The company then initiates a project. A project team is formed. A software project manager is appointed to create a project plan. The project plan includes details such as how much time and how many people are needed. Business analysts create a list of product features (i.e., the features that the proposed product should have). The software designers prepare the architecture of the product. Software developers then construct the product (i.e., write the source code using

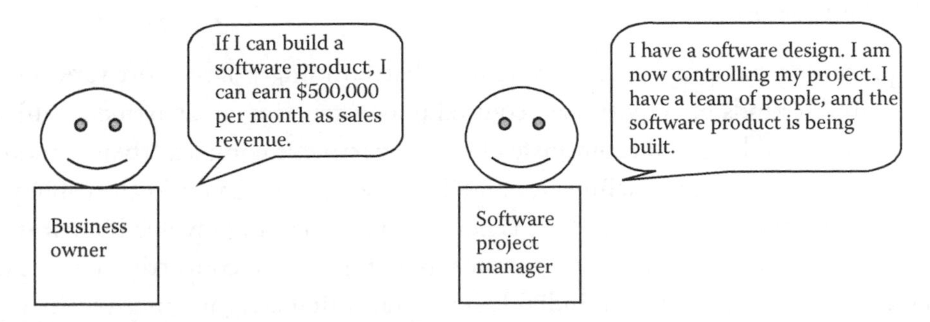


**Figure 1.2** A business owner initiates a project for building a software product and the software project manager plans and controls the software development project.

a programming language such as C++). Once the product is ready and tested, it is shipped to the market. The software project manager oversees all these activities to ensure that the things on the project go as per the project plan.

Similar to the measures (such as earthquake resistance, fire alarms, and fire exits) taken while constructing the apartment, some considerations are needed to build a software product. These considerations include security, performance, and reliability of the product. These considerations will be addressed when the software product is developed.

### 1.4 What Is Software Engineering?

Software engineering can be defined as the study and application of engineering to the design, development, and maintenance of software products. The keyword here is *engineering*. Engineering always connotes two things: designing a product in detail and a methodical approach to carrying out the design and production so that the product is produced with the desired qualities and a large number of people can use it. Software engineering thus annotates building a software product with a detailed design, and at the same time, the tasks of designing and building will involve a methodical approach at each and every step. The software product thus built can be used by a large number of people.

Nowadays, a modular design is more popular for designing new products. What this means is that the complete product is divided into parts that can be easily assembled into a complete and working product once they are developed.

As a note, software engineering is not only limited to creating new products. Existing software products need maintenance. Maintenance includes activities such as correcting the errors in a product, enhancing the capabilities of a product, and removing the outdated features from a product. As a comparison, consider, again, the apartment example. An apartment needs maintenance such as fixing water leaks and installation of a security system. Software maintenance is also a part of software engineering.

### 1.5 Why Software Engineering?

In the preceding paragraphs, it is explained that building a large software product involves similar kinds of activities as required in constructing an apartment building. If you need to build a small cabin instead of an apartment complex, then you do not need a design, manpower, machinery, or civil engineer. You can build the cabin yourself with a handful of materials. There is not much engineering involved in constructing the cabin. Similarly, small programs such as a program for converting temperature from Celsius to Fahrenheit or an individual programming assignment given in a typical classroom do not need any software engineering. However, building large software products does require software engineering.

Computers are used in almost all aspects of our lives. Accordingly, more and more software products are being developed to fulfill these needs, and this trend is accelerating. Definitely, software engineering techniques make it possible to create such useful software products.

If you need to build a software product, then you need to take into consideration the cost economics, time factors, and quality factors. If a software product is built with excessive cost, then the company that is paying to build the product will incur losses. If a software product cannot be built within the accepted time limits, then the product cannot be launched into the market at the right time, and thus the company that owns the product may not make the expected profits. If a software product fails to meet the quality standards, then it cannot be used effectively by users. Maintaining a poor-quality software product is also costly. The company owning such a software product may incur heavy support costs in terms of time spent by the support personnel in fixing and addressing customer complaints. In fact, some software defects can pose serious risks to people as well. For example, a faulty software system deployed at a hospital may give false diagnosis reports to the doctors, and thus patients may get wrong treatments.

A software product can be built with sufficient quality, within the time limits, and within the budget if proper software engineering practices are adopted. Software engineering is thus extremely important.

### *1.5.1 Reduction of Development Costs*

If you look at most industries, you will notice that the cost of production of goods continues to decline year after year. How is this trend achieved? It is achieved by using more productive processes, reducing waste, and so on. The same is true when building software products. In general, the cost of building software products should decrease continuously year after year because of market dynamics. Software engineering methodologies can help achieve this goal.

Most software products evolve over time by maintaining them or by developing new products based on the experience gained from the existing products. Software engineering helps reduce the cost of development by reusing the existing code, using better tools, and following sound software engineering practices.

### *1.5.2 Reduction of Development Time*

If you look at the time that was needed to build a software product in the past, say 10 years ago, you will notice that you can build the same product in less time today. This is because you have better development tools and are aware of better software engineering practices. Programming languages continue to mature and offer many ready-made libraries that can be used instead of writing some fresh source code to achieve the same functionality. There are some software products and services available in the market that provide the needed functionality immediately; thus, the project team can avoid writing fresh source codes altogether to achieve the same functionality. The end result is that the productivity of the project team continues to improve, which allows the faster development of software products.

### *1.5.3 Increasing the Quality*

Quality is the single most important ingredient in making any product successful. Software engineering plays an important role in building quality software products. By using better software engineering methodologies and software engineering tools, it is possible to build better-quality software products. Making the quality assurance processes an intrinsic part of the software development processes ensures the development of quality software products. Creating awareness about quality is also an effective method to increase the quality of software products.

Most software products evolve over time. To ensure that a software product can be enhanced in the future, careful consideration is needed while designing it. Careful planning for future maintenance of the software product ensures that quality does not deteriorate when software product features are changed or enhanced.

## 1.6 Challenges in Software Engineering

Developing a software product is always a challenging job. The reasons are numerous. The conversion of user requirements into a software product involves a long chain of processes. For example, suppose a user requires a software product that is able to analyze the popularity of his or her personal web page. How are you going to build such a product? First, you need to design a program that will count the number of hits to that web page. A simple design could be "counting the number of unique visitors to the web page within a specified period of time" (say 24 h) and then reporting this count to an e-mail address of the user. Once you design the software, you can write the source code using a programming language. After thorough testing, you can deliver the product to the user.

Building such a simple software product is easy. However, consider that the user's requirements are not only to provide the number of hits to his or her web page but also to categorize such hits according to the following criteria:

- Number of hits for the last year
- Number of visitors by country per day, per week, and per month
- Number of visitors by different browsers used by the visitors
- Number of visitors by the operating system used
- Number of hits and the number of unique visitors

Now, if you try to design the product according to the new requirements of the user, you will realize that it is not easy to develop the required software product this time. As the complexity and size of the user requirements increase, the design and implementation sizes of the product also increase. This increase in complexity and size is the main problem associated with the development of software products. This is the reality that most of software product development projects face.

### 1.7 Project Management and Software Engineering

As discussed in Section 1.6, the total effort required to complete a project becomes bigger when the requirements become large and complex. In this situation, a single person will no longer be able to do everything from creating the requirements to designing, implementing, and testing the product. A team of people is needed to do these tasks. At the same time, as each project task becomes more specific and specialized in nature, you need people with specialized skills to work specifically on each of these tasks. Therefore, you will have a bunch of people who will be creating the requirement specifications. After the requirement specifications are created, a set of people who have expertise in software design will create the design specifications for the user interface, design various parts of the software product, and so on. Then, there will be people who write the source code for the product and another set of people to test it. To maintain an existing product, on the basis of the nature of maintenance, you need one or more groups of people.

All the tasks required for building a software product need to be carried out in the most cost-effective and timely manner. Because of the nature and size of these tasks, they need to be performed under a controlled and managed environment. This is possible only when a project is instituted to develop the specific software product.

What all this means is that you will need project management to build a new software product or to maintain an existing software product. In either case, a project manager will be involved to manage the tasks of the project. As explained in Section 1.5, software engineering is required to create a new product. Thus, both project management and software engineering are needed to develop a new software product or maintain an existing software product.

### 1.8 Costs Involved in Software Development

As explained earlier, building a software product requires people with specialized skills. At the same time, there are many people involved in developing a software product. This translates to high costs in building a software product. Apart from paying for the office space, computers, electricity, and so on, you need to pay all the people who will be working on the project. The salaries for highly skilled people are high (which is always the case with software projects because software engineers are highly skilled people). Therefore, your total costs for the project will be high. For example, suppose you need six people on the project and the project lasts for 7 months. If the average salary for each person on the project is US$5000 per month, then the total cost of the project will be US$5000 × 6 × 7 = US$210,000. Now, imagine if a project lasts for 2 years and there are 100 people involved, then the cost will amount to several millions of dollars.

The cost of software maintenance is also very high. Software maintenance costs include customer support costs as well as costs involved in making changes to the software product. To make changes to an existing software product, a maintenance project is undertaken. There could be many such maintenance projects over the life of a software product.

### 1.9 Methodologies Used for Software Development

A software product can be developed using any of the popular software engineering methodologies (or models or approaches). At one extreme, the entire product is developed under one project using a big project planning model. At the other extreme, you can develop the product incrementally. In between these two extremes, other models of software development exist.

Which model should be used for software development for a specific project boils down to the preference of the sponsors or clients of the project. If the clients prefer to have a product with minimal (or essential) features at the beginning and launch it into the market and, on the basis of the market response, continue adding additional features later on, then the incremental model is preferred. A software product can be incrementally built using any of the many popular agile methodologies such as eXtreme Programming, Scrum, and Rational Unified Process (used by Eclipse platforms). On the other hand, if the clients want the entire product to be delivered in one shot, then the big project planning model is preferred, better known as the Waterfall model.

There are many factors that need to be considered in deciding which model is best suited for a given project. More details on these models are available in Chapter 2.

### 1.10 Some Careers in Software Engineering

Software engineering provides several careers to pursue. A person having skills and interest in finding defects or shortcomings in software products can pursue a career in software testing. Anyone who enjoys programming can become a software developer and write source code. A person good in designing and architecture can become a software designer or architect. A person who likes the task of collection and classification of information can become a business or technical analyst. If a person likes discovering new things or analyzing how things work, he or she can pursue a research career. There are teaching careers for those who wish to educate others. A person who has gained experience in more than one area of software engineering can become a software project manager. People with different skills can also be useful because different projects need different skill sets. The list of careers provided here is not exhaustive. New careers are emerging as software engineering technology emerges.

Still, there are some additional roles required for software projects. Because of the increased use of web-based software products, web designers are needed on such projects. Web designers have specialized skills for creating the user interface for web-based software products. Database administrators are needed to manage the databases.

### 1.11 Software Industry Size

The software industry is a large and fast-growing industry, where millions of people are currently employed.

Billions of people use computers, the Internet, and cell phones, and play games every day. All these items use software systems in one way or another. On the basis of the report of the industry analyst Gartner, the global size of the software industry was US$427.3 billion in 2014, an increase of 4.8% over 2013. Microsoft, Oracle, IBM, and SAP were respectively the four largest software vendors. The size of the software industry in the United States stood at US$131 billion in 2014. There are more than 100,000 software companies registered in the United States that cater to software development, software maintenance, and other software-related services. These companies hire software engineers, software architects, software testers, software project managers, business analysts, and so on. The total number of people employed with these software companies is in excess of 2 million in the United States alone.

### 1.12 Code of Ethics

Many industries influence the lives of people in many ways. For example, the medical industry tries to save the lives of people through medical treatment. The medical practitioners, including doctors and health professionals, abide by a code of ethics to prevent misuse by their profession. The power of medical science can easily be misused to destroy lives instead of saving them.

Software products can be misused to harm people. For example, malware can be created to steal money from bank accounts of other people. Similarly, viruses can be created to infect the computer systems of other people. Thus, a software engineer needs to abide by a code of ethics to shy away from such activities. It is quite possible to create powerful software systems that can harm a large number of people. All software engineers should abide by a code of ethics to work for the betterment of humanity and not to harm others.

### 1.13 Book Organization

Essential parts of software engineering include software engineering methodology, software requirement specifications, software architecture, software implementation, and software testing. Additional parts of software engineering include feasibility study considerations, software release considerations, software configuration management, and software project management. Since each of these parts of software engineering is a complete area of study by itself, a balanced approach has been followed to include all of them in this book. The book is organized into 13 chapters. The title of the chapter and a few lines outlining that chapter are provided below for Chapters 2 through 13.

Chapter 2: Software Engineering Methodologies. This chapter covers major software engineering methodologies including the Waterfall methodology, eXtreme Programming, Scrum, Boehm's Spiral model, and the Rational Unified Process model.

Chapter 3: Feasibility Study. This chapter covers the topics related to the economic and technical feasibility studies including prototyping that are needed for software projects.

Chapter 4: Software Requirements Specifications. This chapter presents software requirement gathering and requirement specifications.

Chapter 5: Software High-Level Design and Modeling. This chapter covers the details of designing a software product at the system level and some other levels.

Chapter 6: Software User Interface Design and Construction. This chapter covers the design and construction aspects of user interfaces.

Chapter 7: Software Middle Layer Design and Construction. Software middle layer contains business logic. This chapter presents the design and construction of software middle layer.

Chapter 8: Database Design and Construction. A database is the place where permanent data are stored for software products. This chapter presents details on databases.

Chapter 9: Software Testing (Verification and Validation). Software testing is an integral part of any software product development process. This chapter presents various testing strategies that are used in the software industry.

Chapter 10: Software Release. Software release is an important activity that occurs once the software product is fully developed. This chapter presents various aspects related to software release.

Chapter 11: Software Maintenance. Software maintenance is an important activity that occurs once the users start using the product. This chapter presents the details of that activity.

Chapter 12: Configuration and Version Management. Configuration and version management tools are widely used in the software industry to organize several artifacts that are being developed as part of a software project. They are discussed in this chapter.

Chapter 13: Software Project Management. Software project management is all about planning, execution, and control of the tasks involved in software product development. This chapter presents the software project management activities.

### 1.14 Chapter Summary

Software engineering is a branch of computer science that is used for building commercial-grade software products. Software engineering methodologies consist of a set of well-defined processes for collecting and preparing the requirements and then designing, building, and testing the software products. The process of building or maintaining a software product is similar to the processes that are used to construct an apartment building. Most software projects are large in nature; thus, they require a team of software professionals to perform specialized tasks such as design, construction (also known as coding or implementation), testing, and maintenance.

Software product development is not an easy task because of the complexity of the products, difficulty in learning the technology, immaturity of available processes, and intangibility of these products. To address these challenges, relatively new and easy-to-handle technologies, such as object-oriented design and multilayered product architecture, are being widely used.

The software industry is large and it employs millions of people worldwide. Software engineers and all types of professionals engaged in building or maintaining the software products are always in demand and are well paid.

## QUESTIONS

1. What is software engineering? Explain with some examples.
2. What is a computer?
3. Explain some advantages of computers.
4. What is a software engineering methodology?
5. Why is the cost of building software products so high?
6. Why is the task of building software products difficult?
7. What are the remedies to mitigate the challenges in building software products?
8. What career options are available to software engineers?
9. What is a software project?

## Recommended Reading

Bernd Bruegge, Allen H. Dutoit (2010), *Object-Oriented Software Engineering: Using UML, Patterns and Java*, 3rd Edition, Prentice Hall, Boston.

Pankaj Jalote (2008), *A Concise Introduction to Software Engineering*, Springer-Verlag, London.

Pontus Johnson, Mathias Ekstedt (2005), *The Grand Unified Theory of Software Engineering*, 2nd Edition, Industrialla informations-och styrsystem, KTH, Stockholm, Sweden.

Roger S. Pressman, (2007), *Software Engineering: A Practitioner's Approach*, 6th Edition, McGraw Hill Publications, Columbus, OH.

Ian Sommerville (2011), *Software Engineering*, 9th Edition, Addison-Wesley, Boston.

Frank Tsui, Orlando Karam, Barbara Bernal (2013), *Essentials of Software Engineering*, 3rd Edition, Jones & Bartlett Learning, Burlington, MA.

# 2
# Software Engineering Methodologies

**In Chapter 1, we learned**

- **What computers are**
- **What software engineering is**
- **Why software engineering is important**
- **Challenges in building software products**
- **Software engineering careers**

**In Chapter 2, we will learn**

- **What a software engineering methodology is**
- **Waterfall methodology, its benefits and drawbacks**
- **Rational Unified Process methodology, its benefits and drawbacks**
- **Scrum methodology, its benefits and drawbacks**
- **eXtreme Programming methodology, its benefits and drawbacks**
- **Boehm's Spiral methodology, its benefits and drawbacks**
- **What incremental software development methodology is**
- **What a process standard is**

## 2.1 Introduction

A methodology is a way of making or adopting a model to perform a task or a set of tasks so that the goal of that task can be achieved as predicted. A methodology can also be defined as a single method or a set of methods through which a goal can be achieved. The second definition is explained using an example from civil engineering. Consider the task of constructing a large apartment building. Depending on the situation and requirements, you have two choices. You can opt for prefabricated structures and transport them on-site and then assemble them to build the apartment building, or you can build each part completely on-site without using any prefabricated structures. In this example, constructing the apartment is the goal, and the construction using prefabricated structures is one of the available methods. The decision of selecting a specific method or methods depends on the requirements of that method (or methods) as well as considerations such as time to build, cost, security, and availability of technology. A software product can be developed using either some well-known methodologies or any methodology that may be available to the developer.

**Usage note:** In this book, the terms *methodology*, *model*, and *method* are used synonymously. Similarly, the terms *process*, *task*, and *phase* are used interchangeably.

## 2.2 Why a Methodology?

When you are building a software product, you implicitly acknowledge that you are building it for commercial or critical use and you are using the most competitive, cost-effective, and sound engineering techniques available. In order to apply the engineering techniques to build a product, a methodology that has been proven successful and effective in terms of cost, time, quality, and so on is needed.

Software development methodologies (also called software engineering methodologies) have evolved over time to build software products. When you undertake a project to build a software product, you determine which software engineering methodology is best suited for the given circumstances. As a comparison, to construct an apartment, the civil engineer may prefer to build each part completely on-site if the transportation cost of the prefabricated structures is too high.

The oldest software development techniques are of the big bang type and they do not contain any methodology. In the big bang type, a project is initiated and people with adequate skills are hired to do the project work. The management in this scenario has no idea how the project will be executed. They just hope that the people will somehow develop the software product. Since no particular methodology is in place, it is the people who will decide and keep working as per their individual plans. This kind of technique was used during the early days of software development because there were no well-established software engineering methodologies available at that time. In this type of software development, the results are unpredictable. Later, software practitioners realized the need for software engineering methodologies. Some methodologies are discussed in the next few sections.

## 2.3 Agile Methodologies

In manufacturing industries, agile technologies such as Kanban are used for responsive supply chains. An aspect of the agile methods used in the manufacturing industry is that you should never keep unnecessary inventory with you but only inventory that is immediately needed. If you keep the entire inventory and the market conditions suddenly change and this inventory is no longer needed, then you will be in a mess. Therefore, it is a lot better to keep the necessary minimum inventory. Then if the market conditions change, you will be able to cope with the changed conditions well. In nonagile software development methodologies such as the Waterfall model, all the requirements are taken together and the software development planning takes place. This scenario is similar to keeping a large and unnecessary inventory in the manufacturing industry. In contrast, in agile software development, you take only a few requirements at a time and develop the software product. If the market receives the product well, then you build the next set of product features and add them to the existing product. This kind of product development ensures that you are always in tune with the markets. Some agile methodologies that have become popular and are discussed later in this chapter include eXtreme Programming (XP), Scrum, and Rational Unified Process (RUP).

## 2.4 Waterfall Model

In the 1970s, a software engineering methodology known as the Waterfall model was developed. In those days, the military needed software products for various programs and the existing software engineering techniques were very time-consuming and costly. Thus, the military was not able to get the needed software products on time. At the same time, the cost of developing these software products was too high. To address this concern, the software practitioners proposed a methodology known as the Waterfall model. It was named as such because the model looks like a waterfall when sketched on a paper. Similar to water falling from a higher to a lower level, in a Waterfall model, the specifications created at a higher (i.e., earlier) level are used at the lower (i.e., next) levels to create more detailed specifications for building the product. The task at a given level will not start unless all the tasks are completed at the level just above the given level. The Waterfall model is one of the oldest models for software development. Therefore, it is considered the traditional methodology for software development. The Waterfall model promotes the linear operation of its process with a well-planned and well-documented method of software development. This is why it is also known as plan-driven methodology.

### *2.4.1 Details of the Waterfall Model*

In the Waterfall model (also known as the Waterfall life cycle), first, a project team is set up to collect all the requirements from the customer (or the users). Once these requirements are collected and the requirement specifications are documented, then the software designing phase starts. Once the software design is completed, the software construction (coding) phase starts. After the construction of the entire software product is completed, the software testing phase starts. Finally, once the software product is tested and found to be working fine, the software release phase starts.

Most Waterfall model variations implement a quality gate system. In Figure 2.1, a Waterfall model with quality gates is shown. After each phase, there is an exit criterion. After the completion of all the activities for a phase, a quality inspection or testing is done on all the artifacts (outcomes) of that phase. Only if all the artifacts in

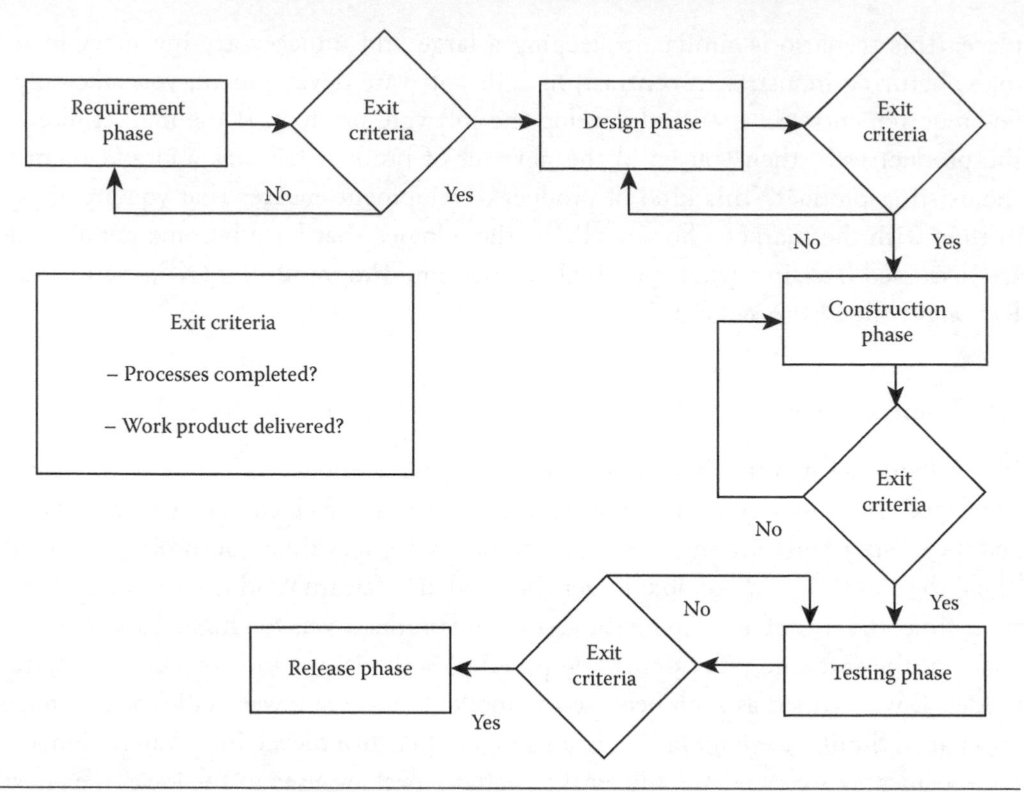


**Figure 2.1** Waterfall model with quality gates.

that phase are found to be of good quality can the project proceed to the next phase of the model. Otherwise, the faulty artifacts are reworked and tested again for quality. This whole process is depicted in Figure 2.1.

The quality gate system has one more purpose. When a software development life cycle phase reaches a quality gate, a check is made to ensure that all the required artifacts have been created during the phase just before that quality gate. Also, in the same way, all the artifacts are checked for completeness.

The benefit of such a process model is that the quality is maintained throughout the development life cycle. Remember the saying "garbage in, garbage out!" If your input artifact for any phase contains errors or defects, then the resulting artifacts will contain errors or defects. However, if your input artifact is defect-free, then the output artifact will likely be defect-free.

### *2.4.2 Salient Features of the Waterfall Model*

Some of the salient features of the Waterfall model are as follows:

- It supports plan-driven product development: When the project is initiated, the entire project is planned even before the software development processes actually start. There are a baseline, milestones, and deadlines against which the project is tracked during its execution. A plan-driven development leads to better control over all the processes in the project because of clear visibility in all the processes of the development cycle. Project visibility is explained next.

- It supports project visibility: Since the entire project along with the milestones and deadlines is planned in advance, at any given point in time, any stakeholder of the project can get a status report (i.e., progress report) of the project. The status reports ensure that the stakeholders can make decisions about any changes to ensure that the project progresses smoothly. Project status reports are of two types: external and internal. The external reports are for the stakeholders. Stakeholders can view these reports, and if the reports are not satisfactory, then they may ask for action to be taken. The internal status reports are for the project team to evaluate the progress and take remedial actions if needed. For example, if it is found that a module of software design slipped by 10 days and subsequently the construction of this module got delayed by 10 days, then the project manager must take action. The project manager can ask the project team to work overtime or assign extra resources to finish the ongoing tasks in time. Generally, the stakeholders get the external project reports periodically. These reports contain a summary of the project progress. Hence, the stakeholders do not get all the minute details about the project. Generally, the stakeholders do not get any information related to small problems on the project because the external project reports will not include that information. The project managers deal with these small problems on a day-to-day basis without any interference from the stakeholders.
- It supports quality gates: After each phase, the artifacts produced during that phase are tested and only when the test results are satisfactory will that phase be completed. For instance, the software design phase should not start until all the requirement specifications have been verified and the problems, if any, are fixed. Quality gates ensure that no quality defects in the produced artifacts of the current phase are passed on to the next phase of product development.
- It supports the development of large software products: If you need to develop a large software product within a short period, then the Waterfall model is a suitable option. The Waterfall model allows a project team to be of any size, from a few people to hundreds of people. Therefore, if the software product to be developed is large in size, then deploying a large project team can result in developing the product within a short period. For instance, suppose the estimated size of the software product is 1,000,000 lines of source code (i.e., the programs written using a programming language such as C++). It is required that this software product be developed in 1 year (365 days). Assume that the productivity of the project team (consisting of five people)

is 20 lines of source code per day. In this scenario, we will need 1,000,000/(20 × 300) = 166 people for developing the source code (assuming that there are 300 working days in a year and one person in the project team is writing the source code). Note that in a project team, there are business analysts, software testers, software designers, and software developers. These people are doing different tasks, such as design, source code development, and testing related to the project. When the estimates are made for the productivity of a project team, the average speed of each person (doing different things on the project) is considered. The final productivity of the team is derived after noting the productivity of each team member.

#### *2.4.3 Drawbacks of the Waterfall Model*

The Waterfall model has several drawbacks. It has a large management overhead because the product is developed as a whole but not incrementally. When a complete software product needs to be developed as a whole, a large number of people need to be deployed on the project at the same time. Managing a large number of people on the project necessitates appointing them in a management capacity. This results in a large management overhead. Most of the time, the customers are not computer professionals; therefore, the customer requirements (that are provided during the beginning of the project) may contain ambiguities. Proper estimation of some parameters (such as the required time and cost) to develop the product is difficult because of these ambiguities. Such faulty estimations lead to difficulties during the software product development.

In the Waterfall model, all the requirements need to be frozen at the beginning of a project because the entire project is planned in advance, on the basis of the requirements. In reality, freezing the software requirements at the beginning of the project is not possible. Some changes to the requirements (from the customer) are bound to happen after the project starts. For example, if the project is currently at the testing phase and a change is needed at the design phase (because of a changed requirement from the customer), then the testing phase and all its previous phases, namely, construction phase and design phase, need to be changed accordingly. Such changes will affect the entire project plan. In many projects, the requirements need to be fine-tuned based on the discussions between the customer and the developer while the project is progressing. There is no way of doing this fine-tuning in the Waterfall model if there are any changes in the requirements (from the customer) once the project starts. For these reasons, the Waterfall model is not suitable to develop projects in which the customers change their requirements once the project starts.

In reality, in most projects, the customers change their requirements once it starts. This is one of the major reasons why the Waterfall model is not widely used nowadays.

### *2.4.4 When to Use the Waterfall Model*

The Waterfall model is appropriate to use in some scenarios. Most of the time, large projects fall behind their schedule because of faulty project plans. If a project needs to be completed as per an agreed-upon schedule, then the Waterfall model is suitable because the entire project is planned in advance. The Waterfall model emphasizes documentation at each and every phase. Therefore, if a person leaves the team, then the replacement person can read the documentation of the person who left and be productive quickly. If the requirements are clear and fixed, then the Waterfall model is less risky to develop a product.

## 2.5 Rational Unified Process

As explained in Section 2.4.3, the Waterfall model is well suited for simple and risk-free software projects. However, most of the projects involve risks such as changes to the requirements (from the customers) once the projects started.

Ivar Jacobson, Grady Booch, and James Rumbaugh have proposed what is known as the Rational Unified Process (RUP) model, which better suits risky projects. They proposed that any design or specification cannot be perfect when it is first developed. Any software project goes through evolution, and this factor must be incorporated into the process model. RUP is an iterative process for software development. In RUP, a software design can be iterated to incorporate any requirement changes as well as any changes to the design itself. A similar type of iteration can occur among various process steps during the entire software project.

In Figure 2.2, the RUP model is depicted. The $x$-axis shows the progress of each process (project phases) over time as well as the iterations (inception phase, elaboration phase, construction phase, and transition phase) within each project phase. If you look closely at the graph, you will notice that the iterations do not start in the inception phase. This is because, in the inception phase, no actual software development process takes place. The inception phase is mostly for project management and environment work. Environment work is related to process tailoring. Process tailoring requires studying the project requirements and finding the best-fit software life cycle processes. For example, if you find that you will need at least two iterations over the requirement analysis, then you will create your process map accordingly.

From the elaboration phase, the actual software development processes start. In Figure 2.2, there are three iterations (Iterations 1–3) for the elaboration phase, two iterations (Iterations 1 and 2) for the construction phase, and two iterations for the transition phase. On the $y$-axis, workflows are defined. There are two kinds of workflows: main workflows and supporting workflows. The main workflows include business modeling, requirements, analysis and design, implementation, test, and deployment. The supporting workflows include configuration management, project management, and environment.

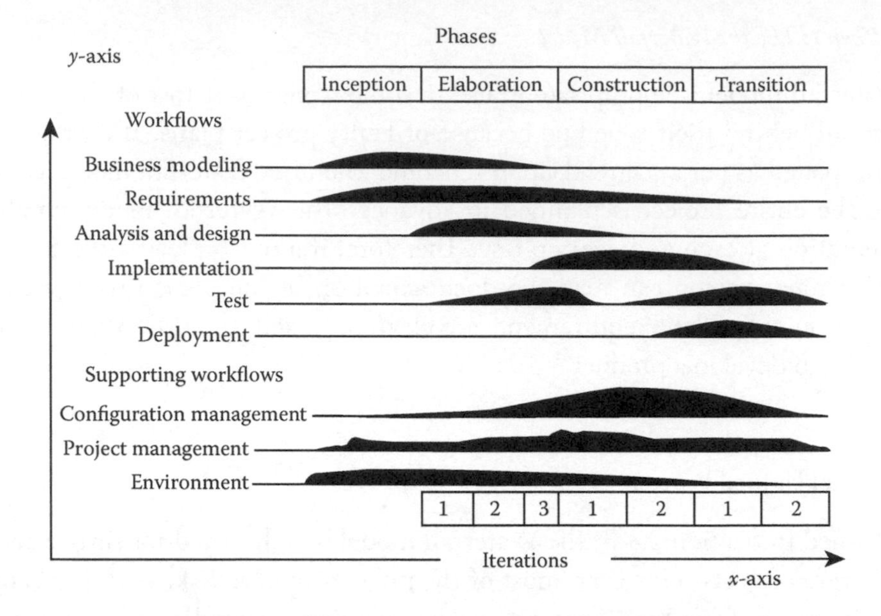


**Figure 2.2** Rational Unified Process.

During the elaboration phase, the requirements are elicited from the end users. These requirements are refined over several iterations. This is the most important aspect about RUP because this methodology believes that the complete requirements cannot be gathered in one stroke. You need to elaborate each of the requirements over several iterations; only then will the requirements become perfect and free of any ambiguities. The construction phase is related to the design and construction activities. The transition phase is related to testing and final preparation to implement the software product at the customer premises.

The workflows shown on the *y*-axis in Figure 2.2 are self-explanatory. Business modeling workflow is all about creating a software model or prototype. The requirements workflow is all about the requirement management. Analysis and design workflow is for software analysis and design work. Implementation, test, and deployment workflows are for software implementation, software testing, and software deployment, respectively. The supporting workflow of project management is for all the work related to managing the project. The configuration management workflow is related to all the work involving software build, version control, and configuration management. We have already seen the use of environment management workflow for setting a number of iterations in each project phase.

The black heaps on the graph show the amount of work to be done for a workflow over a period. For example, let us consider the black heap for the business modeling workflow. It shows a large amount of work at the beginning of the project. This means the business modeling work is done early in the project. A major part of business modeling is done in the inception phase. It also shows that work is spread over Iterations 1 and 2 for the elaboration phase. If you see the requirements workflow, then you will notice that the work related to this workflow is spread over a fairly large area spanning from the inception phase to the construction phase. You will also notice that the amount of work is not intensive throughout this large period. What it means is that the requirements have been changing even during the construction phase or they are evolving during this entire period. If you see the graph (Figure 2.2) for analysis and design workflow, you will notice that it is spread only over Iterations 1 and 2 of the elaboration phase.

The supporting workflows are depicted similar to how the main workflows are depicted. If you observe the progress work for configuration management workflow, you will notice that the black heap is shown to be concentrated over two iterations for the construction phase. Some work is also being shown over the elaboration and transition phases. As the source code is built during the construction phase, more work is required to be done in the configuration management area.

### *2.5.1 Salient Features of RUP*

RUP allows the concept of workflows. Even the same workflow (e.g., requirements) runs during several iterations over a single phase (e.g., elaboration) of the project. Similarly, even when a phase changes, through the use of a workflow, some work can be carried out in that other phase. For example, the requirements workflow can be carried out even during the construction phase of the project. Thus, overlapping of work is possible over the phases of the project. This kind of functionality is not available in the traditional Waterfall model.

### *2.5.2 Drawbacks of RUP*

RUP brings an overhead in terms of maintaining the project phases as well as workflows. This kind of functionality is not needed on projects where the requirements are not expected to change. For these projects, usage of RUP is overkill.

### *2.5.3 When to Use RUP*

RUP is a process model for software development where the refinement of an artifact is possible through several iterations. This means that this model is best suited for projects where the requirements from the end users are ambiguous at the beginning because the end users themselves may not be clear as to what they expect in the software product. When they go over their own requirements over time, it becomes clear to them as to what they look for in the software product that will be built. RUP allows the refinement of the requirements over several iterations.

RUP has some great advantages over the traditional Waterfall model. In the traditional model, change requests create a lot of problems in the software development process. Since RUP allows these changes to be incorporated in the development process easily, it is a good model for projects where the change requests are expected.

RUP is definitely suited for software projects in which changes occur in the design or requirements. RUP takes care of these changes by incorporating them during iterations.

### 2.6 Spiral Model

Many people realize that the software development process does not follow a linear path. Evolution takes place at any stage during the entire life cycle of software product development. When the product is at the design phase, the requirements may evolve and thus are changed. Similarly, when the product is constructed and is being tested, some part of the design may evolve and need to be changed.

In essence, there is a risk involved at any stage during the entire life cycle of software product development. For example, if a software design does not represent the software requirement specifications properly (or if the source code does not represent the software design properly), then there is a risk involved. A risk mitigation strategy is needed to ensure that the development cycle is not affected by these risks.

Barry Boehm proposed that the development of a software product should take place in a spiral manner, hence he developed the Spiral model. He stipulated that if the requirements are known, fixed, and can be specified correctly and the design can be correctly made for those requirements, then software development can progress in a linear manner and a risk-free Waterfall model can be adopted for that project. However, if the requirements are not fixed or are not fully known at the initiation of the project, then the agile model is a better solution. Choosing the best software engineering methodology for a software project is the outcome of the Spiral model.

The Spiral model is iterative in nature. At the beginning of each iteration, four basic activities must be performed. The first one is to evaluate whether all the stakeholders' success factors are met. For example, you must consider whether the estimated cost of a project justifies the further development of that project. Here, project cost is a success factor. The second factor is to find out the alternatives available. For example, a factor that is often considered before a project is considered is a "build versus buy" decision. Choosing between building a custom product from scratch and purchasing an available off-the-shelf or "canned" product is a decision factor during the feasibility assessment for a project. The next consideration is to find out all possible risk factors for each alternative. For example, building a product from scratch may be more expensive than buying an off-the-shelf product. However, it may be difficult to add new features to an off-the-shelf product. If a required feature is not available in the off-the-shelf product and it is not possible to develop (and add) that required feature to the off-the-shelf product, then the product is of no use. In that case, a new product may need to be developed from scratch even if that option is costly. Finally, when all the alternatives are considered and the best course of action is selected, approval from all the stakeholders is needed to proceed further.

In the Spiral model, there are provisions for allocating extra efforts for riskier process steps as well as provisions for making more detailed plans for riskier process steps. Building a prototype is also a necessary requirement here. Refining this prototype over many iterations is again an important ingredient in this model. The number of cycles in the spiral shows the amount of risk involved in a project. More number of cycles means the project is riskier. There will be more iterations (spiral cycles) involved here over which many prototypes will be developed before the prototype is good enough to be considered for making the actual software product. The effort to refine the prototypes over several iterations is the essence of the Spiral model. It is hoped that if these detailed plans are followed, then the risks can be handled effectively.

Figure 2.3 depicts the Spiral model. This model consists of two levels of activities. One set of activities is performed in a spiral fashion. The other set of activities can be understood by looking at the four quadrants. Each quadrant depicts the activities

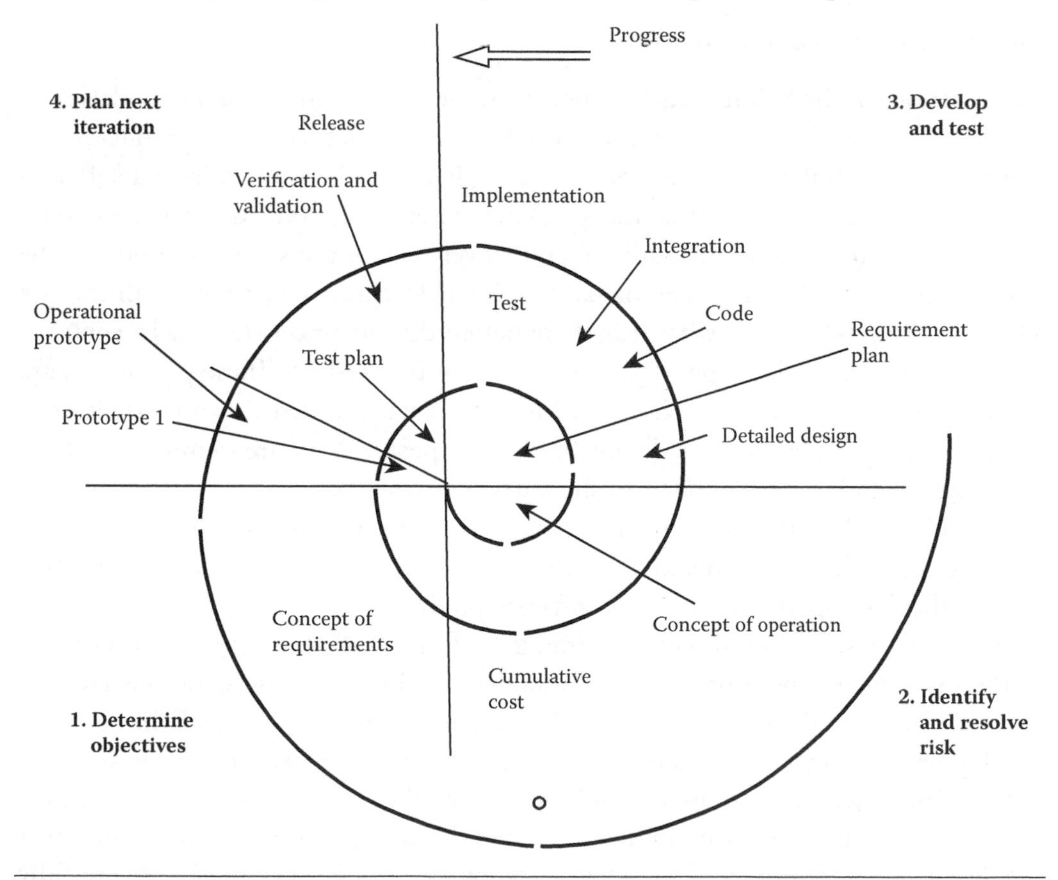


**Figure 2.3** Barry Boehm's Spiral model.

(similar activities are performed inside a quadrant) performed over many iterations in the spiral.

The model is composed of four quadrants. The process starts at the first quadrant in which the objectives (i.e., requirements) are determined. The second quadrant is about identifying and resolving the risks. When the prototypes are ready, they are discussed with the customer. The customers provide feedback, and on the basis of that feedback, all the risk areas are identified and resolved. The third quadrant is about development and testing. When the prototypes are complete and customers are fully satisfied with this prototype, only then can the development work for the software product start. Once the full-fledged design and construction work is done, the product will have to undergo rigorous testing. You can see here all the activities of a typical software development life cycle including a detailed design, code (source code), integration, test, and implementation in this quadrant.

Once an iteration (i.e., one spiral cycle) is completed, you can think about the next iteration. This is done in the fourth quadrant (plan the next iteration). For example, if Prototype 1 is done in the present iteration, then you can think of moving to develop Prototype 2 in the next iteration.

### *2.6.1 Comparison with Other Models*

The iterations in the Spiral model are not the same as those in true agile models such as XP and Scrum. In true agile models, you build a complete software product increment during each iteration. For example, after Iteration 1, you can have a software product Version 1.1 in the agile model. After Iteration 2, you can have a software product Version 1.2. Both are fully functional versions of the software product. The only difference is that, in Iteration 2, additional features (i.e., functionalities) are added to the product over what was there before that iteration started. In contrast, the Spiral model produces prototypes of the software product. These prototypes are not fully functional software products. Instead, they are scaled-down versions of it. These prototypes are used for demonstration purposes. They are demonstrated (by the project team) to the customer to show whether what the project team is building (or planning to build) is correct. Prototypes are a great way of communicating with the customers. Prototypes help reduce the communication gap between the customers and the developers of a proposed software product.

In many ways, the Spiral model is similar to RUP. RUP also has iterations similar to the ones in the Spiral model. The difference lies in how these iterations are carried out. In RUP, there could be several iterations in just one phase of the software development life cycle. For example, there could be three iterations over software design. These iterations are mostly isolated to the other phases (e.g., coding) of software development. The reason for iterating over this design phase is to make sure that the design becomes mature. In the first iteration, the design may not be able to fully capture some aspects of the software requirements. These missing points can be better captured in the next pass during the next iteration. In contrast, in the Spiral model, the iteration takes place over all the phases of the development life cycle including the requirements, design, construction, and testing phases. The assumption here is that the prototype that was developed in the previous cycle may have some missing elements (i.e., product features) and now these elements will be incorporated into the present cycle.

### *2.6.2 What Are Prototypes?*

A prototype is a rapidly developed piece of software, and it may not contain the depth of a software product. For example, a prototype of a menu system may not have all the submenus incorporated as required in the requirement specification. However, the prototype will show (to the customer) how the menus will look and what those menus will do. Prototypes are not tested thoroughly as is the case with actual software products. The purpose of a prototype is not to be used by the end users but to be used only for demonstration purposes. In general, the effort required to develop a prototype is around 5% to 10% of the effort required to build the actual product. Chapter 3 presents more details on prototyping.

### *2.6.3 Salient Features of the Spiral Model*

The Spiral model believes in building a good prototype of the software product first, before building the actual software product. Over the iterations, these prototypes are refined. The customer looks at the more refined prototype and provides his or her input (feedback) as to how the prototype meets his or her requirements. The customer also provides input if there are any missing parts or wrongly assumed product features (by the project team). Only after the customer is fully satisfied with a more refined prototype does the project team start building the actual software product.

### *2.6.4 Drawbacks of the Spiral Model*

The Spiral model carries a lot of overhead in terms of several iterations to develop a software product. Thus, if a software product does not need to be developed using many iterations and, in fact, it can be developed using just one development cycle, then the Spiral model is overkill. For example, a project to develop a software product for transportation management for a retailer may be complex because it needs a large number of complex requirements. In this example, there are bound to be many cases where the prototypes should be developed first, to get better clarity about the requirements. A Spiral model may help in this case. However, a project to develop a software product for a library system for a small library may not require several iterations to understand the requirements. For such projects, other simpler models may be better suited.

*2.6.5 When to Use the Spiral Model*

The Spiral model is in fact a super model. The outcome of a Spiral model is a model that can be an agile or Waterfall model. Therefore, you first run the Spiral model to find a suitable model for your project. In these scenarios, the Spiral model is used for evaluating the risks that may occur in the project. For example, in a Spiral model, after evaluating and mitigating the risks, if it is determined that a Waterfall model is more suitable to the development project, then the Waterfall model can be used.

Generally, the Spiral model can be used for projects where too many risks are involved. The process of considering the risks and the ways to mitigate those risks is costly and lengthy. For these reasons, this model is suitable for high-risk projects only.

## 2.7 Incremental Iteration Model

We have learned about agile methodologies in Section 2.3. An agile methodology mostly depends on building a software product using incremental iterations. It provides many benefits over the traditional approaches of building a software product. The most important of these benefits includes the ability to tap the market at an early date when only a modest product is built quickly using less effort. This approach minimizes the financial risks. Once the modest product is well received by the customers and the preferences of the customers are well known, then additional features can be added to the software product over time. The concept of incremental iteration is explained in Figure 2.4.

As shown in Figure 2.4, the entire project is broken down into several iterations. Each iteration consists of all the phases, including the requirement specification, design, construction, testing, and release of building a software product. Breaking down the entire project into several iterations is done by considering only a few requirements at a time. Using the design, construction, testing, and release phases for these few requirements alone will result in a small software product with only a few product features after the first iteration. During the next iteration, again a few requirements are considered and all the phases of software development are again performed for developing the software features as per these requirements. After completion of each iteration, a partial product is created. This partial product is integrated with the existing product. The process of creating the partial products and integrating them with the main product continues until all the requirements are exhausted.

The beauty of the incremental iterative approach is that the main build of the product is always a working product and it can be marketed while the project team is busy creating additional software product features.

In an incremental iteration model, there is a concept called major and minor releases. Any major or minor release of the software product is a product release that may coincide with the completion of some predefined number of iterations. For example, in Figure 2.5, Major Release 1 of the product coincides with the completion of Iterations

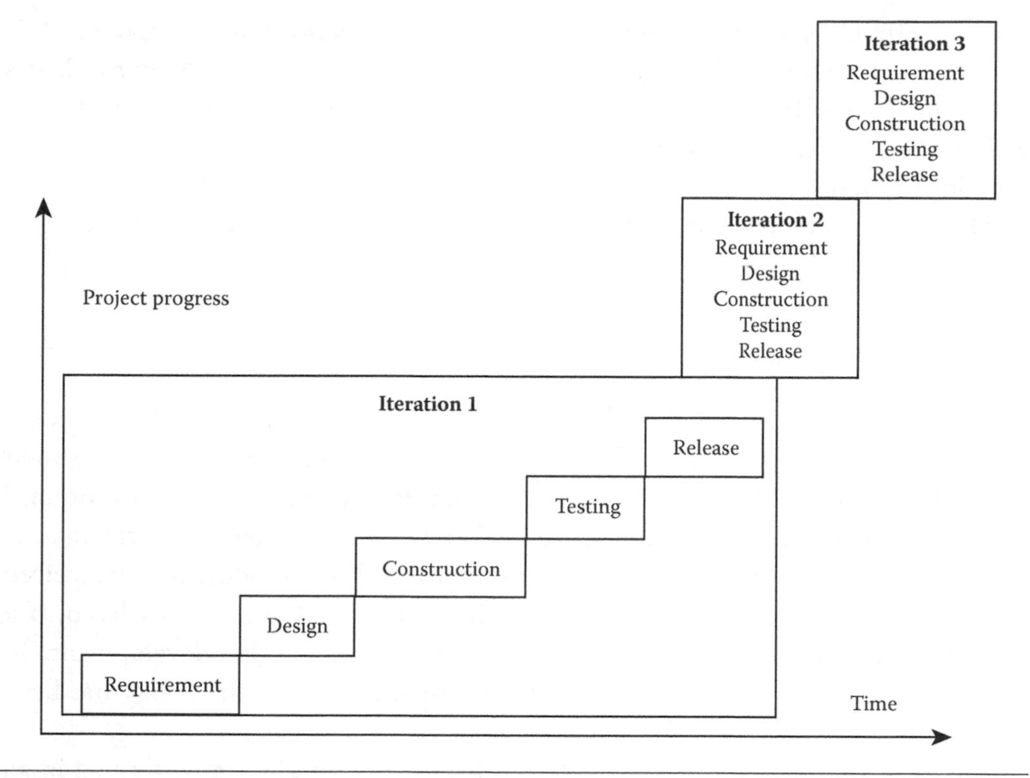


**Figure 2.4** Incrementally building a software product using iterations.

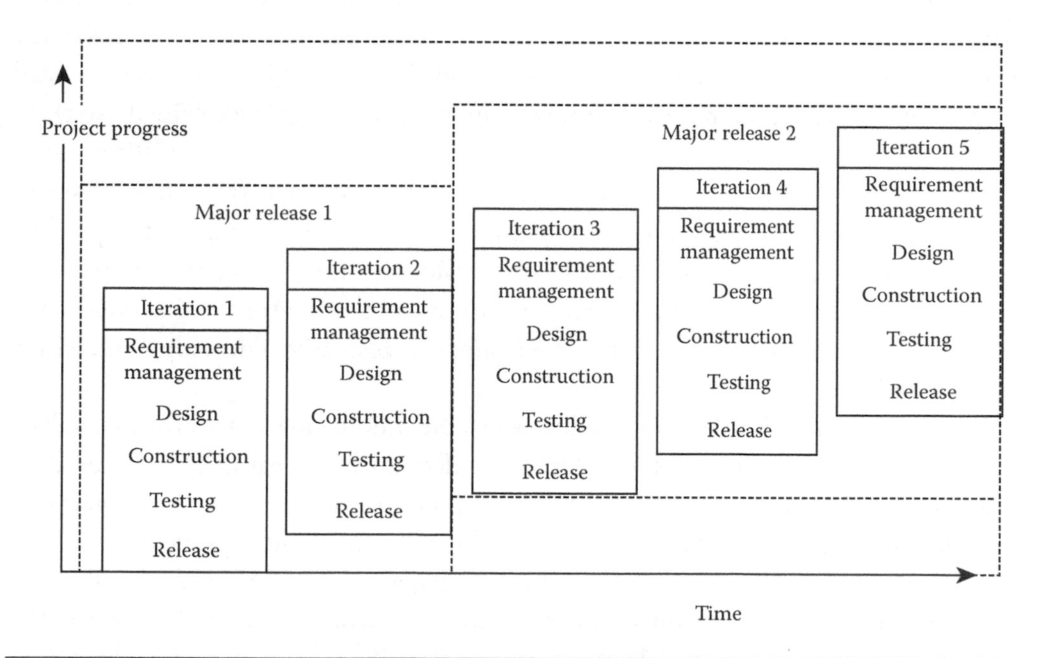


**Figure 2.5** Complete product development plan in the incremental iteration model.

1 and 2. Similarly, Major Release 2 coincides with the completion of Iterations 3–5. There could be Minor Releases 1 and 2 inside Major Release 1, and Minor Releases 3–5 inside Major Release 2. In addition, Minor Releases 1 and 2 could be at the end of Iterations 1 and 2, respectively, and Minor Releases 3–5 could be at the end of Iterations 3–5, respectively.

Most agile methodologies such as XP and Scrum allow incremental building of software products.

## 2.8 eXtreme Programming

The idea behind the Waterfall and RUP models is that the entire software product is developed under one project. However, many companies want to release their products into the market as early as possible, and these companies do not wait until the full products are developed. It is also realized that it is better to test the market with a product containing a small number of features. If the product is well received in the market, then additional features can be developed. On the other hand, if it is not well received, then the product can be stopped from further development. To cater to these kinds of needs, eXtreme Programming (XP) methodology has been developed.

The word *eXtreme* indicates the fact that, in XP, many tasks are performed in an extreme manner when compared to the traditional Waterfall model. In the Waterfall model, the customer is generally not involved in the product development cycle. The customer gives his or her requirements to the project team, and after that, his or her role in the project is no longer required. In the Waterfall model, software testing is not performed until the product is completely developed. In contrast, in XP, the customer is constantly engaged in the project and the product is continuously tested using a concept called pair programming (pair programming is explained later). In fact, in a concept known as test-driven software development, software testing is performed even before the source code is developed. All the artifacts (requirement specifications, product model, untested product, and so on) available during the product development are constantly reviewed by the customer. The customer is responsible for all aspects related to the requirement specifications for the software product. The "eXtreme" is also evident by the fact that XP supports test-driven development, which is explained later.

Since the customer (sometimes also known as the product owner) is always involved in the product development process, he or she decides what features of the product need to be developed during the next iteration. Therefore, incremental iteration cycles are performed. Some product features are developed, tested, and approved by the customer during an iteration. Once the iteration is over, another set of product features is considered and so on, as explained in the incremental iteration model. In essence, the complete software product is developed in multiple short development cycles rather than in a single, long, and complete cycle.

*2.8.1 XP Activities*

XP is based on several activities. The activities include coding, testing, listening, and designing. Using these activities, a set of working product features is developed and added to the existing software product in each iteration cycle.

*2.8.1.1 Coding* The XP methodology believes that the main task of any software product development is to generate a code (using a high-level programming language) that, in turn, will generate a machine-readable code. This machine-readable code (binary code) is also known as the software product. The ultimate goal of all the activities of XP is to generate the software product.

*2.8.1.2 Testing* The XP methodology believes that the best way to achieve product quality is to make the testing procedure the central activity during the entire product development process. Because of this, XP supports test-driven development. In test-driven development, the developer first writes the test cases for each feature of the product and then develops the code that passes these test cases. These tests are known as unit tests because they test each unit of the source code. These tests are intended to test the functional (operational) features of the product. These tests ensure that all the business logic implemented in the source code is free from software defects.

*2.8.1.3 Listening* During the listening activity, the programmers listen to what the customer wants in the product. The customer explains the business logic behind a feature that he or she is expecting in the product. The development team, in turn, understands that business logic and finds ways to implement that logic in the software product.

*2.8.1.4 Designing* In XP, the size of the product developed during the initial iterations is very small. The size keeps growing with the number of iterations. Product design is not a serious activity during the early iterations. However, as the product size increases, the product design becomes a concern. The parts of the product that were developed initially were not designed to be integrated with the features that are added later. This is because, while developing the initial product, it is difficult to predict what additional features the customer will want at the later stages. This problem results in a software product with a shaky performance. To address this issue, the existing code needs to be redesigned so that all parts of the product become more compatible with each other. This kind of designing is known as refactoring.

*2.8.2 XP Concepts*

Let us discuss some concepts that are valid or used in XP.

*2.8.2.1 Customer* A customer in XP is a representative of the company for whom the software product is being developed. The customer provides the list of software features that need to be developed in each iteration. The customer is always present at the location where the software development project team develops the software product. The customer also tests the software product features that are developed during each iteration and ensures that they have no software defects.

*2.8.2.2 Pair Programming* Pair programming was introduced in XP because it was noticed that when one programmer is coding, he or she often fails to implement the business logic correctly. To address this issue, in pair programming, two programmers work on the same computer. While one of them writes the code, the other one reviews each line of that code. The two programmers frequently switch their responsibilities. The programmer who is reviewing the code will pinpoint any errors in the logic.

*2.8.2.3 Planning Game* Planning game is a meeting intended to do the planning for the next release as well as for the next iteration of the product. The meeting is treated more like a game, hence the name *planning game.*

*2.8.2.4 Release Planning* Release planning determines the product features that will be included during the next major release of the product. The customer and the development team are involved in release planning. Release planning consists of three phases: exploration phase, commitment phase, and steering phase. In the exploration phase, the customer provides a brief list of requirements that need to be added to the product during the next release. The development team estimates how much time is needed to convert these requirements into product features and then to develop those features (i.e., add those features to the product). If it is not possible for the development team to properly estimate the time to handle a requirement (provided by the customer), then the customer will subdivide that requirement into two or more requirements so that it can be better understood. In the commitment phase, the development team commits to the customer about the development of some of the requirements out of all the requirements (proposed by the customer) and about the date of the next release. The purpose of the steering phase (while doing the release planning) is to update the overall plan of the project, on the basis of the input (or feedback) provided by the customer as well as the development team so far. During the steering phase, the proposed features can also be modified or removed, or new features can also be added.

*2.8.2.5 Iteration Planning* Iteration planning determines the features that will be included during the next iteration. The development team does the iteration planning to ensure that at least the important features are developed in the present iteration (the customer has already provided the list of features as well as the information on the relative importance among these features). The iteration meeting takes place among the members of the development team alone. The features that are committed to be developed during this meeting are assigned to each team member. Effort estimation (i.e., how much time is needed, etc.) is done for each such feature.

*2.8.2.6 Whole Team* A customer is the person or a group of persons who will be using the software product that is being developed. In XP, the customer is treated as part of the development team. Notice that, in release planning explained earlier, the customer is mentioned separately from the development team for clarity.

*2.8.2.7 Continuous Integration* Continuous integration is the process of integrating the newly developed source code with the existing build of the software product continuously. Whenever a programmer completes his or her code, it is immediately submitted for integration with the existing build. Nowadays, tools are available to facilitate this.

*2.8.2.8 Refactoring* "Refactoring is the process of changing a software system in such a way that it does not alter the external behavior of the code yet improves its internal structure" (Martin Fowler). Object-oriented programming languages are used in all XP projects. The building blocks of object-oriented programming are classes. Using the refactoring techniques, the old classes are modified so that they will have proper interfaces to integrate with the new classes. We will learn more about refactoring in Chapter 7.

*2.8.2.9 Smaller Releases* The incremental product building process has the advantage of the product being marketed quickly even with fewer features. Smaller releases can be developed over a week's time. Most of the products developed these days are deployed over the Internet or can be downloaded from the Internet. Therefore, the customers can easily use or download the new releases of such products whenever they are made available.

### *2.8.3 Salient Features of XP*

XP allows co-located small project teams to work on software development projects. There are no management overheads in XP because project teams self-manage and there is only one project manager involved on such projects. XP allows the incremental building of software products and it is a great way to introduce a software product into the market as early as possible.

### *2.8.4 Drawbacks of XP*

If a large product needs to be built quickly, then it may not be possible to build it by using XP. For example, suppose you want to build a software product that has 100,000 lines of source code. Suppose the speed of your XP team is 20 lines of source code per day (note that 20 lines per day is the general speed at which an XP team consisting of five to six people develops the source code). Then, it will take 5000 days to build this product. Assuming 300 working days per year, it will take around 17 years to build this software product. However, 17 years is a long time to build a software product; thus, it is not a feasible solution. This kind of product can be built using a methodology (e.g., the Waterfall model) that allows large teams to work on the projects. The time taken to develop such a software product using the Waterfall methodology can be drastically reduced by employing a large project team.

### *2.8.5 When to Use XP*

XP is an agile methodology that can be used when there is a need to build a software product incrementally. To start building the product, you do not have to wait until the complete requirements are in your hand. A major risk with software products is that the customers may not find the products (that are developed) suitable because by the time the products are developed, the customers' requirements may have changed. XP reduces this risk because the customer is involved in the incremental addition of new features to the product. This approach makes sure that the developed software product is always as per the customer's requirements. XP is well suited in these scenarios.

XP is most suitable for small co-located teams because of the requirement of frequent meetings and frequent releases of product updates. However, many companies are using XP for building large software products by deploying some large and geographically scattered project teams. This is possible because of the following: the availability of video conferencing, central builds accessible to the teams through the Internet, and the availability of central repositories for electronically keeping the project documents.

## 2.9 Scrum

Scrum is another popular iterative incremental agile methodology for software product development, specifically for developing complex products. Scrum provides a framework for the development teams to handle the difficulty in the development of complex products. Scrum is characterized by different types of project meetings that happen on a daily basis. These meetings deal with feature selection for the product (i.e., the features that need to be developed in an iteration), quality, project progress reporting, and so on.

There are some typical terminologies used in Scrum such as scrum master, product backlog, sprint, and velocity, to name a few. These terminologies are discussed in the following sections.

### *2.9.1 Scrum Roles*

In Scrum, there are three primary roles: scrum master, product owner, and development team. The scrum master is the person who manages the project. The scrum master role differs from the traditional project manager role because the scrum master does not manage a development team. The development team is self-managed. The main role of the scrum master is to ensure that the team can concentrate on its development work. Thus, the role of the scrum master is to provide an environment for the smooth working of the team and to remove any hurdles faced by the team. Sample hurdles include a computer failure, electricity outage, housekeeping, and office maintenance. In other words, the scrum master is the facilitator of the development team.

The product owner is the person who requested the product. His or her responsibility is to manage the product backlog (explained later), announce the product releases on behalf of all the stakeholders of the project, and make sure that the customer requirements are properly converted into features of the software product. The product owner also acts as the interface between the project team and the actual users of the project by ensuring that there are no communication gaps between the project team and the actual users of the product.

The development team is supposed to perform all the development tasks such as design, construction, and testing of the product. The development team gets its input, in the form of product features, from the product owner and makes an effort to develop and deliver the product as per the schedule provided by the product owner.

### *2.9.2 Scrum Events*

Scrum activities are driven through scrum events. These events are discussed in Sections 2.9.2.1 through 2.9.2.4.

*2.9.2.1 Sprint* A sprint is an iteration cycle during which a set of product features are implemented. A sprint is a time-bound event. Before a sprint starts, a sprint planning meeting (explained later) is held between the scrum master, product owner, and the development team. During that meeting, the product owner requests the work that needs to be completed during that sprint. The development team has to approve the amount of work, and the product owner has to approve the evaluation criteria (e.g., the number of errors in each hundred lines of source code) for the completed work. The scrum master determines the duration of the sprint. The sprint starts once a consensus is reached between all the parties. Once the sprint is completed, the product owner evaluates the developed work using the criteria that he or she set up during the sprint planning meeting. The developed work may be accepted or rejected. After a sprint is completed, the next sprint starts. While planning for a sprint, care is taken to only consider (i.e., plan for) the development work that can be finished by the end of that sprint.

*2.9.2.2 Meetings* There are different kinds of meetings for each sprint. There is a sprint planning meeting before the start of each sprint. In this meeting, planning is done to select the product features that need to be developed during the upcoming sprint, as explained earlier. The sprint planning meeting generally lasts for 8 h. All the parties, namely, the product owner, the development team, and the scrum master, attend the first 4 h of the meeting. In this meeting, the product owner discusses the user stories (product features) with the team to find out if these product features can be implemented in a sprint. The developers/designers get clarification from the product owner on any matter related to the implementation of these product features.

The next 4 h meeting is for the development team. The development team tries to find out all the implementation details for the product features to be developed (as given in the user stories by the product owner) in the present sprint. The development team makes its own sprint plan for developing the product features. Apart from the sprint planning meeting, there are daily scrum meetings during the morning, generally at the same location, to discuss the work completed during the previous day and to plan for work for the current day. The duration of the daily scrum meetings is 15 min to make sure that the discussion is relevant and fast. The scrum master, the product owner, and the development team attend the meeting.

Two more meetings, a sprint review meeting and a sprint retrospective meeting, are organized at the end of each sprint. In a sprint review meeting, all the work that was completed during that sprint is analyzed and the work that was planned but not completed during that sprint is noted down. The completed work is presented to the stakeholders. The time limit for the sprint review meetings is around 4 h. The sprint retrospective meeting is organized generally after the sprint review meeting. The purpose of a sprint retrospective meeting is to achieve continuous improvement to the processes followed during that sprint (that was just finished). The items discussed during the sprint retrospective meeting include what processes worked well, what processes created problems, and how to improve or rectify the problematic processes in future sprints.

*2.9.2.3 Backlog Refinement* Product backlog needs to be constantly updated because the priorities get changed or some requirements are themselves changed over time (product backlog is explained later). The backlog must be kept up to date to ensure its integrity and to keep the priorities in order.

*2.9.2.4 Scrum of Scrums* Scrum of Scrums is used to scale up the project team size to build larger software products at a faster rate. Scrum of Scrums is actually a project team composed of many smaller project teams. In other words, Scrum of Scrums is a technique in which a large group of development persons is divided into smaller teams. Meetings are held in which all these teams can participate by nominating one person, as an ambassador from each team, to participate in the meetings. These meetings focus on the areas where the product development responsibilities overlap between these teams and on the integration points between different parts (of the product) that are being developed by these teams.

#### *2.9.3 Scrum Artifacts*

*2.9.3.1 Product Backlog* Product backlog is the repository of the requirements that are used to build the software product. Each requirement is assigned a priority, and the higher-priority requirements are always taken up for developing the product before the lower-priority requirements are considered. The requirements can be a product feature, fixing a bug in the product, or a nonfunctional requirement such as security, performance, and usability of the product. Each requirement (also known as a user story) is written like a story and contains the specification as defined by the stakeholders of the product. Optionally, a user story may contain the details on how the development team will implement the specification. The product owner owns and maintains the backlog. The product owner also keeps the information regarding the risk, expected expenses, and estimated time to develop each requirement.

*2.9.3.2 Sprint Backlog* Sprint backlog is a subset of product backlog. It contains the requirements that are used to develop the current sprint.

#### *2.9.4 Salient Features of Scrum*

As is the case with all agile methodologies, Scrum is also used for incremental development of software products. Each iteration in the Scrum process may be used to develop a minor release of the software product. Many of these minor releases can be part of a major release of the software product.

In Scrum, each iteration is known as a sprint. There is a sprint meeting at the beginning of each sprint. In the sprint meeting, product features are selected (from the product backlog) for development. A development plan is made for the sprint in this meeting. Once a sprint starts, the project team will have daily meetings. These daily meetings are for planning the development activities for the day as well as for resolving the issues, if any.

At the end of the sprint, there is a sprint ending meeting. Through this meeting, it is ensured that all the product features have been developed and thoroughly tested before they are released. Any product feature that could not be developed during that sprint will be taken up in the next one.

#### *2.9.5 Drawbacks of Scrum*

Scrum is meant for developing the software products using small co-located teams. It is not suited for large project teams. Projects that require a fixed cost or fixed duration for developing the products can never be adapted to the Scrum environment because the project planning for building the complete software product is not possible in Scrum.

#### *2.9.6 Benefits with Scrum*

Scrum is a truly agile methodology for building software products. The presence of the product owner, as an integral part of the project team, ensures that it is almost impossible to have any communication gaps between the product development teams and the rest of the stakeholders of the product. The daily scrum meetings provide a platform to quickly clarify the issues among all the stakeholders.

Through the technique of Scrum of Scrums, a cluster of project teams can work on the same project; thus, large software products can also be developed using Scrum. The Scrum of Scrums technique supports geographically scattered teams by using video conferencing and other web-based techniques. Therefore, offshore teams can also participate in a Scrum-based project.

Product backlog helps the development of a software product incrementally. The first release of the software product can be launched in the market in a short amount of time, even if the software product does not contain a large number of product features. Later, additional features can be added to the product.

### 2.10 Methodology for Implementing SaaS Products

In the Software as a Service (SaaS) business model, a vendor develops the software product and installs it on a centralized server. The customers, who are either individual persons or business organizations, can use that product through web browsers by paying a subscription fee to the vendor. The customers do not need to buy or install the product on their own (local) computers. Therefore, the customers are freed from the hardware, software, and maintenance responsibilities of the product because these are carried out by the vendor.

When an existing software product lacks some features that are needed by the customer, then these features need to be developed. For developing these new features, fresh source code needs to be written. These newly developed product features will then be integrated with the existing software product. For example, customers want a software solution for reading the bar codes that are placed on each of their products. If this software product feature of reading the bar codes is not available in the SaaS product of the vendor, then it has to be developed for the customer. This feature will then be integrated with the SaaS product.

In some cases, there could be some existing product features that fulfill the customer needs. For example, there could be an existing product feature that calculates the transportation cost. It does this on the basis of the distance traveled by a vehicle. However, the customer may want to calculate the transportation cost on the basis of the weight of the goods carried. Since this method of cost calculation is not available in the existing product, it needs to be developed by writing some fresh source codes and then integrated with the software product. This process is known as customization (of the existing software product).

In some cases, creating the customer-specific data to an existing software product (i.e., to the database of an existing software product) is enough in order to make that software product suitable to that customer's needs. Once the customer-specific data are created, the end users of that customer will be able to use the software product. Data can be classified as master data and transaction data. Master data and transaction data are explained in Chapter 8.

Thus, such software implementation projects may involve customization, data creation, or addition of new features to an existing product.

There are many vendors who have developed SaaS products. Some examples of SaaS products include the following:

- A web-based Customer Relationship Management application developed by SalesForce.com. Any customer who wants to use this application can register with this company, which also runs this web-based application. After user registration, the users (who belong to the customer) can log into this application and do their transactions.
- A web-based warehouse and transportation management application developed by OneNetwork.com. It has an arrangement similar to that of SalesForce.com for creating the data for the customers' companies and their users.

In Figure 2.6, the process model for SaaS implementation is depicted. There are four phases in a SaaS implementation project: initiation, blueprinting, realization, and go live. These phases are explained below.

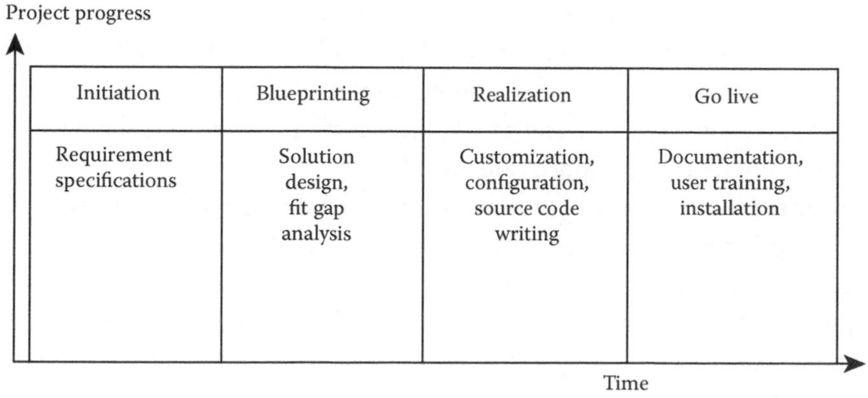

Project progress

| Initiation | Blueprinting | Realization | Go live |
|---|---|---|---|
| Requirement specifications | Solution design, fit gap analysis | Customization, configuration, source code writing | Documentation, user training, installation |

Time

**Figure 2.6** The SaaS implementation process model.

**Initiation:** In the initiation phase, the end user requirements are gathered and a requirement specification document is prepared.

**Blueprinting:** Blueprinting is essentially the process of mapping the end user requirements to the software product features that are available in the software product. A fit gap analysis is also performed during the blueprinting phase. In fit gap analysis, the software product features (that already exist in the product) are mapped to the end user requirements. If any requirements of the end user are met (through any existing product features), then they are mentioned in the "fit" part of the blueprinting document. If any requirements are not met (by any existing software product feature), then they are mentioned in the "gap" part of the blueprinting document. There could also be some requirements that can be met through customizing the software product.

**Realization:** Once the blueprint is ready, then the implementation work can be started. This implementation will result in the creation of the customer-specific version of the SaaS product. This phase is known as realization. During realization, master data and transaction data are also created and fed into the database. The customization process involves both performing the proper feature settings and writing the source code where required. Feature setting is done for the product feature options that are already developed (i.e., those already available in that SaaS product). In feature setting, the required product feature is selected out of all the available features (this is like selecting an option from a drop-down list on the user interface). Fresh source code is written for the features that need to be developed. Once all these activities are completed, the implemented system is ready for testing.

Most features of a SaaS software product have options to perform the things in many different (i.e., alternative) ways by using the features that are already available in that product. For example, the transportation costs (in the example provided earlier) can be calculated in many ways such as by finding the distance and applying the per-mile rate, on a fixed rate (such as monthly rate), or based on the weight of the goods carried. All of these different kinds of transportation cost calculations can be already defined in the SaaS product. During the customization of the product for a specific customer, the settings need to be changed (according to what the customer wants) in the SaaS product to calculate the transportation cost.

To meet the customer requirements, which have been classified as "gaps" in the blueprinting document, some fresh source code may have to be written. The gaps indicate that the software product features matching the customer requirements have never been developed by the software vendor; thus, those features need to be developed to meet customer requirements. Generally, on most software implementation projects, the source code required to be written to bridge the gaps does not comprise more than 5% of the total source code that is already there in that software product.

**Go live:** In this phase, all the documents such as user manuals and technical manuals related to the customized product are created. Some employees from the customer are trained on how to use the system. The go-live phase also includes testing the software product by the end users. During testing, the completely implemented system is tested against the customer requirements. Once testing is successfully completed, the product is hosted at the server of the vendor of the SaaS product and is available to the customer.

*Example:* When a customer wants to use a SaaS product (e.g., the application offered by OneNetwork.com), a project team from the vendor (i.e., OneNetwork.com) will gather the customer (or user) requirements. Then, a blueprint of the required SaaS product is created by matching the customer requirements with the SaaS product features. Some of the SaaS product features will exactly match some of those requirements. Some other requirements may be fulfilled by changing some existing product feature settings, without doing any additional programming (e.g., replacing the existing logo on the SaaS product with the logo of the customer or changing the color of the web page of the existing SaaS product with a new color that the customer wants). Lastly, some other requirements may not have any equivalent features in the existing SaaS product. In such cases, some fresh source codes are written to provide these additional features. All these activities together are part of fit gap analysis and realization.

### *2.10.1 What Is a Customer-Specific Version?*

A SaaS vendor can develop many versions of its software product. For example, the vendor OneNetwork.com has developed many versions of their software product for transportation management solution. These versions have been developed for different needs of the customers. For example, OneNetwork.com has a version of their product that is suitable for third-party logistics providers and another version for large retailers. These versions of the SaaS product are standard.

While each standard version meets the needs of many customers, they may not meet all the requirements of all the customers. For this reason, customer-specific versions of these SaaS products are created. During the implementation of a SaaS product for a customer, the SaaS vendor selects a standard version and then creates some additional product features that are not available in that standard version. These additional product features are integrated with that standard version and then the modified version is installed for that customer. This resultant (or modified) version of the SaaS product is known as a customer-specific version.

## 2.11 Methodology for Implementing COTS Products

Commercial Off-the-Shelf (COTS) software products are built (by software vendors) to be sold in the market as ready-made software products. These software products provide valuable service to the customers. Customers can use them as productivity tools, business applications, or entertainment tools. Some examples of COTS products include spreadsheets, document processing tools, and games.

Business application COTS tools/products include Enterprise Resource Planning packages, customer relationship management packages, and supply chain management packages. Most of these software products are very large. They cannot be used immediately after buying them. If a customer buys a business application software product, he or she has to meet certain requirements before using that product.

The first requirement is that the software product should be populated with some customer-specific data. These data could be related to the customer location information, the types of goods the customer sells, and so on. All of these data are known as master data. Business applications also store transaction data. Examples of transaction data include sales data, purchase data, and production data. Master and transaction data are explained in detail in Chapter 8.

The second requirement is that all the features available in a COTS product may not fit well for the requirements of a specific customer. For example, a purchasing process used by a customer may require a sanction (or permission) from a sanctioning authority as per the customer requirement. This functionality may not be available in the COTS product, in which case it needs to be created in the COTS product. In some other cases, some functionality may already be available in the COTS product but work in a different way. We have already seen an example on transportation cost calculation in the previous section on SaaS. A similar kind of scenario can happen with a COTS product when the existing functionality of a software product cannot meet a specific requirement of a customer. Let us look at an example. A warehouse management system manages its goods based on some "put-away" and "pick-up" processes. A put-away process is used to store the incoming goods in the warehouse. A pick-up process is used to select the goods from the warehouse to dispatch them to the customers of that warehouse. Suppose a specific warehouse (say Warehouse X) is using a different process known as cross-docking. A cross-docking process is used to receive the goods at the warehouse but these goods are not actually stored in the warehouse. These goods are planned to be shipped to the customers of Warehouse X the moment they arrive at the warehouse. Suppose the COTS product does not provide this cross-docking functionality. How will the COTS vendor then be able to fulfill the requirements of this customer (i.e., Warehouse X)? The vendor will need to develop/change the software product features to support the cross-docking functionality of Warehouse X. We can see that although cross-docking is one way of managing a warehouse among other warehouse management techniques (such as put away and pick up), it needs to be supported by the software functionality. The COTS vendor can achieve this cross-docking functionality by changing the existing put-away and pick-up functionality in the COTS product. This kind of work in which an existing functionality needs to be changed to meet the customer requirement is known as the customization of the COTS product.

COTS products are characterized by their extremely large size, focusing on a specialized business area, and capturing of all the business processes in that specialized business area.

The methodology for implementing COTS products is very similar to that for SaaS products mentioned earlier. However, there are some differences. The bulk of the COTS product is already developed and only a minor part needs to be developed or changed during an implementation. Most COTS products have been in the market for a long time; thus, they are very mature. Most business processes are already defined in the COTS products. Thus, during any implementation, only some minor changes may be needed in some of the already defined business processes of the COTS products. In contrast, most SaaS products have entered the market recently. They are still being developed and thus are not as mature as the COTS products. Thus, the effort required for making any changes or addition to a SaaS product (on an implementation project) is much larger when compared to that of a COTS product.

In COTS or SaaS, *implementation* means that the software product already exists and it needs to be customized and then installed for a customer to meet that customer's specific needs.

On any COTS or SaaS implementation project, many types of professionals are employed. However, the bulk of the work on these projects is done by the business analysts. Business analysts gather the customer requirements, create the blueprint, and also work in the realization and help in the go-live phases of the project. Hence, you can see that the business analysts work in all the phases of the implementation project. The work related to writing/changing the source code is done by the software developers (technical consultants). Source code writing/changing work is much smaller than other types of work on these projects.

## 2.12 Usage of Software Development Models

There are many factors that need to be considered while deciding which model is best suited for a given project. One such factor is time. Suppose a client wants to have a large software product developed within a short span of time; the incremental model will then not be suitable. Some software products cannot be built incrementally because all the features in them are essential or interdependent on each other; therefore, the incremental model is not suitable for them either. Similarly, for the projects sponsored by the government, the sponsors need to know the cost, time duration, complete features of the product, and so on before they can commit. For such projects, all the terms and conditions, including the size and duration of the whole project, are specified in the contract before the project starts. In such cases, the incremental model, again, is not suitable.

However, there are situations in which incremental approaches are best suited. The hypothesis in incremental approaches is that the clients initially do not know much about the kind of product they want and thus do not commit much about its features at the beginning of the project. This translates to the clients wanting to keep the commitment open ended so that they can continue giving more requirements to the project team as they keep a close eye on the demands from the users of the product.

For any software product development, there is no fixed model that can be adopted perfectly. There are many factors that affect your decision on which model to adopt.

## 2.13 Popular Process Standards

As per the online Merriam-Webster Dictionary, a process is a series of actions that produce something or lead to a particular result. Therefore, a software development methodology (such as Waterfall or Scrum) is a process because the methodology consists of a series of actions that lead to a software product. You already learned that a software development methodology consists of several phases. For example, design, construction, and testing are some phases in the Waterfall methodology. Therefore, all the phases in a software development methodology together represent a process (or a methodology). Since each phase consists of a series of actions and leads to an artifact as a result (e.g., the construction phase consists of a series of actions such as writing and refining the source code and finally produces the source code as the artifact), on the basis of the definition of *process*, you can treat each such phase as a process. Therefore, you can view a software development methodology either as a single process or as a chain of linked processes.

When all the phases in a software development methodology are described as a chain of linked activities, this is called a software development life cycle.

During the 1990s, some software practitioners realized that, although some popular models are used for developing software products, the processes vary from one organization to another because these models explain the process at an abstract level. The practitioners realized the need to define some standard models describing the processes and then used these models for software product development purposes. The Software Engineering Institute at Carnegie Mellon University was among the first to do so and produced a development model that is known as the Capability Maturity Model (CMM). CMM emphasizes the maturity of software development processes, defining five levels of maturity. If an organization wishes to know where it stands in terms of its software product development process, then it can employ a consultant who is an expert in assessing these processes. A detailed discussion on CMM and other standards is beyond the scope of this chapter.

## 2.14 Process Standards and Software Engineering Methodologies

Process standards such as CMM state that the processes evolve over time and it is possible to increase the quality of the software products as well as decrease the development cost and time of the products, by refining the software development processes further and further. These process standards explain how to refine the processes. However, before an organization starts its journey to improve its processes, it must first be assessed as to where it stands now: ahead of its peers or lagging behind. If it lags behind, then it is necessary to identify the areas where it lags behind and why.

Once the areas for improvement for an organization are identified, then a team of consultants can create a road map to make improvements in those areas. Once this road map is implemented, the organization will achieve a certain maturity level.

How the process standards for software projects are related to software engineering methodologies can be understood from the following example. Suppose an organization uses the Waterfall model for its software projects. The organization has also implemented the CMM model, and its requirement management processes have been rated at a very mature level. This implies that this organization will always execute the software projects where the requirement management is done perfectly with almost no defects escaping to the software design phase (from the requirement management phase). The level of quality of requirement specifications (which are the artifacts produced during the requirement management phase) could be, say, 99.99%. Thus, on any software project executed by this organization, requirement specifications will always have a 0.01% chance of defects.

This discussion underscores the importance of implementing the process standards. If this organization (having a high maturity level for creating the requirement specifications) contacts some customers for a new project, then these customers know that this organization can build the requirement specifications exceptionally well and will not have much of a problem giving their projects to it.

## 2.15 Software Process Activities

Some software development models were discussed in the previous sections. Each model has a different way of executing the software development processes. For example, in the Waterfall model, all the phases are carried out sequentially, in the order of requirement gathering, software design, construction, testing, and software release. In contrast, in agile models, these phases are not performed sequentially because, after software construction, software design can be taken up in the form of refactoring. In this section, we discuss various detailed activities that are carried out during each of the phases, that is, the process of software development.

The core (essential) software development processes include requirement gathering and specification, design, construction (coding), testing, and release. The peripheral

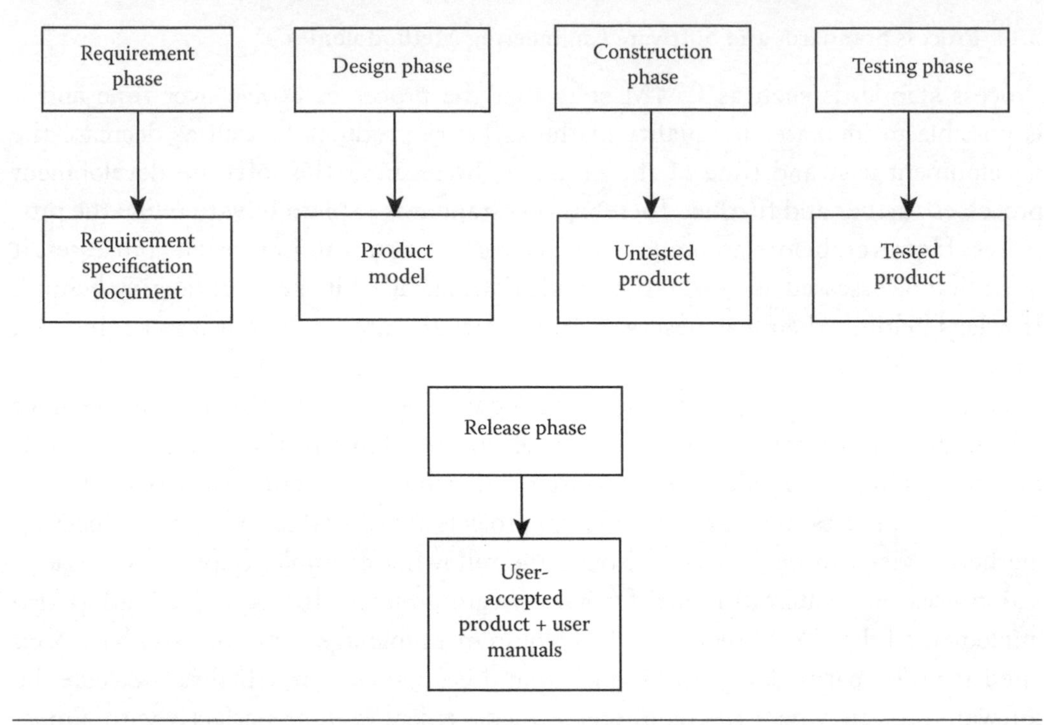


**Figure 2.7** Software development life cycle core processes and their work products (artifacts).

(noncore) processes include feasibility studies, configuration management, software maintenance, and project management. The output or result of a process is an artifact or work product.

During each core process, the project team works on building some work products (or artifacts or deliverables or outcomes). A work product is expected to be delivered after the completion of the corresponding core process. Figure 2.7 describes the core processes and their corresponding work products.

The project team works on requirement gathering and specifications during the requirement gathering phase and develops and delivers a requirement specification document at the end of this phase. Similarly, after the design phase, the project team delivers a software product model (design). After the construction phase, an untested software product is delivered. After the testing phase, a fully tested software product is delivered. Finally, after the release phase, the user-tested product is implemented at the customer site. During the release phase, user documents and technical documents are also created and delivered.

### *2.15.1 Feasibility Study Phase*

Feasibility studies are performed to ensure that the risks are identified and the remedial solutions are discussed. Feasibility studies are peripheral activities in software projects. In many cases, the feasibility studies are performed by a team that is different from the product development team.

A feasibility study can be carried out before a proposed project starts. Market analysis such as how well the product will be received in the market and the costs involved in developing it is conducted. On the basis of the results of the analysis, the company decides whether the proposed project is economically feasible.

Feasibility studies can also be carried out after the project starts. This kind of study is conducted by a company that is interested in developing an in-house product or when a client requests the company to develop a product. After getting the user requirements, they are evaluated to find out if it is feasible to implement the product. Such feasibility studies are technical in nature (i.e., the studies are not about market analysis) and so are known as technical feasibility studies. The outcome of a feasibility study is a report that outlines whether the project is feasible to implement. We will learn about feasibility studies in Chapter 3.

### *2.15.2 Requirement Phase*

A software product needs to be developed based on the user requirements. These requirements can be classified as functional requirements, such as the description of inputs and outputs, and nonfunctional requirements, such as performance and security. Some nonfunctional requirements such as product maintenance also need to be considered. A good software design, which can be easily changed or extended during software maintenance, definitely helps extend the life of the software product. During requirement gathering, all kinds of requirements mentioned above are gathered. Once all they are gathered, the requirement specifications are made. The software requirement specification (SRS) document, containing all the software requirement specifications, is prepared.

The SRS document is the artifact that comes out after the requirement phase. We will learn about requirement management in Chapter 4.

### *2.15.3 Design Phase*

The SRS document is used by the software designers to design and create the user interface layer (or front end), middle layer (or business logic layer), back end (or database), and any other layers that may be needed for the product. All these software layers are part of the complete software design for a software product. A large software product has a large number of requirement specifications. Therefore, the software design becomes notoriously complex for such a product because its large size brings complexity into the design. A complex software design will never be easy to construct (i.e., write the source code) and will lead to a lot of quality issues. A large number of software bugs will creep into the source code because of the complexity of such a design. To reduce the complexity, there are techniques such as creating separate designs for each part of the proposed product.

The product model created at the design phase is the artifact of this phase. We will learn about software design in Chapters 5 through 8. Chapter 5 deals with high-level design and modeling, Chapter 6 deals with the user interface, Chapter 7 deals with the middle layer, and Chapter 8 deals with the back end.

### *2.15.4 Construction Phase*

Software construction is the process of converting the software design into a software product by writing the source code. A user interface is developed based on the interface design, a middle layer is developed by writing the source code for the business logic, and a back end is created by creating the database entities such as tables and fields. When all three layers are properly linked, it results in a complete software product. This untested complete software product is the artifact of this phase. We will learn about software construction in Chapters 6 through 8.

### *2.15.5 Testing Phase*

The testing process can uncover software errors or defects or software bugs (some authors distinguish between these three terms but we use them interchangeably in this book). A software defect may result from a syntax error in the source code. A software defect may also result from many causes including wrong calculations in the source code (which may result in giving wrong results after the computation) or the environment in which the software product is operated. For example, if the software product takes a very long time to respond to a user action, then it is a performance defect. If a software product is not able to run on an operating system, then it is a compatibility defect.

The software developers perform debugging by compiling the source code. Even if the source code compiles properly, it may still contain a large number of functional and nonfunctional defects. Software testers, who are part of the project team, test the software product to ensure that it is free from these defects. If the defects are identified, then the source code is modified and tested, and this process continues until all the defects are removed.

In many projects, testing comprises more than just validation (in validation, the source code is executed to test it). In such projects, testing comprises verification and validation. Verification is done to verify if the artifacts available at different phases conform to the specifications. The SRS document is compared with the customer requirements to verify the following: (1) if all the requirements of the customer are covered in the SRS document and (2) if any part of the SRS document is incorrectly made or does not exactly correspond to a requirement of the client.

Similarly, the product model is compared with the SRS document to verify the following: (1) if all the specifications have been covered in the product model and (2) if any part of the product model is incorrectly made and does not exactly correspond to a requirement specification. In addition, the source code of the untested product is compared with the product model to see if the model has been covered entirely and correctly in the source code. The tested product is the artifact of the testing phase. We will learn about software testing in Chapter 9.

### *2.15.6 Release Phase*

Software testing, explained in the previous paragraph, is performed in-house by the project team. During software release, the software product is installed at the client's site. The end users of the product then test it. This process is called User Acceptance Testing (UAT). Once UAT is performed and no major quality issues are found, the end users start using the software product. During software release, the product development team creates user manuals and technical manuals for the product. If necessary, the product development team provides training to the end users on how to use the product. User-accepted products and manuals represent the artifact of the release phase. We will learn about software release management in Chapter 10.

### *2.15.7 Software Maintenance*

Software maintenance is not a core process of the software development life cycle. It comes after the core software development process is completed. When a software vendor signs a contract to build a software product for its customer, the vendor can also make a commitment toward providing maintenance support for the product after it is delivered to the customer. This commitment is not necessarily a part of any business contract.

During software maintenance, the reported software bugs are rectified. In some cases, additional support, such as adding new features or improving the performance of the product, may be provided. Similarly, the software product can be ported to a new version or to a new platform as part of the maintenance. We will learn about software maintenance in Chapter 11.

### *2.15.8 Software Configuration Management*

During the software development process, a large number of artifacts are generated. All of these artifacts need to be kept in a place where they can be easily accessed by the project people.

On most software development projects, many versions of the same artifact may need to be generated. For example, after receiving a change request from the customer, a change is incorporated into the SRS document. In this case, both the original version and the modified version of the SRS document have to be maintained. Since the requirement specifications have changed, the subsequent design documents and the source code may also need to be changed.

During software testing, there can be many cycles of software testing and defect fixing. This results in many versions of the source code because both the original versions and the new versions are kept. This means that many versions of artifacts are generated during the entire life cycle of software product development. To maintain all the artifacts and their various versions, a configuration management and version control system is used on software projects. We will learn about software configuration management in Chapter 12.

#### *2.15.9 Project Management Processes*

As we discussed earlier, there are several software development methodologies such as Waterfall, RUP, and Scrum available for software development. Each such methodology has different phases or processes that are known as software engineering processes.

Project management consists of different phases or processes, namely, project initiation, project planning, project monitoring and control, and project closure. If we accommodate the phases of a software development methodology within the framework of project management, then that project management represents software project management (or software engineering project management). Project management is a peripheral process or activity of the software development methodology.

We will learn about software project management in Chapter 13.

### 2.16 Chapter Summary

Software engineering methodologies are a set of well-established processes to help develop large and industry-strength software products. Software engineering methodologies help build software products at a competitive cost, with good quality, and within the budget limits. If software engineering methodologies are not employed, then it is difficult to achieve these goals.

Some popular software engineering methodologies include Waterfall, RUP, Scrum, and XP. These methodologies can be categorized as plan driven or risk driven. Plan-driven methodologies such as Waterfall are used for projects where risk is less and all the characteristics of the software product being developed are well understood. Therefore, it is possible to make a well-defined project plan in advance. On such projects, it is also possible to commit a fixed cost or fixed budget because the project risks are minimal.

Risk-driven software engineering methodologies are used for projects where there are risks such as change in the requirements, communication gaps between the developer and the customer, and the inability to design the software product as a whole as per the requirements. To cut the risks, agile methods as well as incremental approaches are used. Close communication between the developer and the customer can be achieved by having a representative of the customer as a core member of the project team.

Tasks on software projects include requirement gathering and specification, software design, software construction, software testing, and software release. These are the core processes. Peripheral processes include project management, configuration management, feasibility studies, and software maintenance. Depending on the software engineering methodology used, these processes can follow a sequential or an iterative order.

## QUESTIONS

1. Explain the benefits of plan-driven software engineering methodologies.
2. What kinds of risks are involved in plan-driven software engineering methodologies?
3. Why are risk-driven software engineering methodologies needed?
4. What is meant by "agile" in agile software engineering methodologies?
5. Why do stakeholders choose a plan-driven methodology over a risk-driven methodology?
6. What are the benefits of incremental software development?
7. Under what circumstances may an agile software engineering methodology not be used?
8. What are the processes involved in software product development?
9. What is the relationship between software engineering methodology and software development life cycle?
10. How are software engineering and project management related?

# Recommended Reading

Kent Beck (2002), *Test Driven Development: By Example*, Addison-Wesley Professional, Indianapolis, IN.

Federico Bergenti, Marie-Pierre Gleizes, Franco Zambonelli (Editors) (2004), *Methodology and Software Engineering for Agent Systems*, Springer, New York.

Joshua Dehlinger (2007), *Incorporating Product-Line Engineering Techniques into Agent-Oriented Software Engineering for Efficiently Building Safety-Critical, Multi-agent Systems*, Proquest Information and Learning Company, Ann Arbor, MI.

Martin Fowler, Kent Beck, John Brant, William Opdyke, Don Roberts (1999), *Refactoring: Improving the Design of Existing Code*, 1st Edition, Addison-Wesley Professional, Indianapolis, IN.

Wilhelm Hasselbring, Simon Giesecke (2006), *Trustworthy Software Systems, Research Methods in Software Engineering*, GITO MbH, Berlin, Germany.

Krzyszlof Sacha (Editor) (2007), *Software Engineering Techniques: Design for Quality*, Springer, New York.

Tomasz Szmuc, Marcin Szpyrka, Jaroslav Zendulka (Editors) (2012), *Advances in Software Engineering Techniques*, Springer, New York.

# 3
# Feasibility Study

**In Chapter 2, we learned**

- **What a software engineering methodology is**
- **Waterfall methodology, its benefits and drawbacks**
- **Rational Unified Process methodology, its benefits and drawbacks**
- **Scrum methodology, its benefits and drawbacks**
- **eXtreme Programming methodology, its benefits and drawbacks**
- **Boehm's Spiral methodology, its benefits and drawbacks**
- **What incremental software development methodology is**
- **What a process standard is**

**In Chapter 3, we will learn**

- **What a feasibility study is**
- **What a technical feasibility study is**
- **What an economic feasibility study is**
- **What a prototype is**
- **What a pilot project is**

## 3.1 Introduction

A feasibility study is conducted for all kinds of projects or new initiatives to assess whether the project or initiative is worth spending time and money on. Generally, when an idea is discussed, some goals and benefits are the targets of that discussion. To know if an idea is good enough and whether the targeted goals and benefits are achievable, some kind of feasibility study is conducted. Through feasibility studies, people try to find out possible and available alternatives. In our daily lives, we conduct feasibility studies for several activities. For example, if you wish to construct a new house, then you should conduct feasibility studies regarding technical matters such as suitability of land, availability of water and electricity, and distance from the nearby city. In addition, you should conduct feasibility studies regarding financial matters such as whether it is better to construct a house than to buy an existing one.

For any software project, a feasibility study is conducted to check whether the project is economically and technically feasible. An economic (i.e., commercial) feasibility study for a software project can be done either before a project team is slated to start working on that project or at any other time during the project. A technical feasibility study for a software project can be done generally after obtaining some information on user requirements, because only then can it be known whether such user requirements are technically feasible based on the available technology.

The technical feasibility of a project can be assessed by first building a prototype of the proposed software product. A prototype is built incrementally for the software product because, most of the time, the users of a software product are not able to define all their requirements at once. Only after seeing a prototype will they be able to provide their feedback as to whether the prototype correctly contains all their requirements correctly. The project team then refines the prototype on the basis of this feedback. This process goes on until the users are fully satisfied. Prototyping is further explained in Section 3.5.

Apart from the prototypes, pilot projects can also be used to assess the feasibility of a project.

Some of the business decisions related to the development of a software product, for which a feasibility study can be conducted, include the build/buy decisions or procurement.

The material presented in this chapter is focused on the feasibility study related to software projects alone.

### 3.2 Feasibility Study for Software Projects

Software projects are uncannily risky affairs. It makes absolute sense to have a feasibility study conducted for an upcoming project before the work on that project can start. At the same time, not all software projects need to go through a feasibility study. It is also a fact that the time and effort required to conduct a feasibility study will vary from project to project.

The feasibility study for a software project can be done either before or after the formal initiation of that project. This decision depends on various factors.

Some companies may conduct a preliminary feasibility study before gathering the user requirements. In such cases, the feasibility study is purely economic in nature. No technical considerations are involved. The company may conduct market research to find out whether its idea of creating a new software product is economically viable. If it finds that the idea is indeed viable, then it can start with requirement gathering. If not, then it will shelve the idea. An economic feasibility study is further explained in Section 3.3.

On the other hand, if a feasibility study is conducted after gathering the requirements, then it can be purely technical or both technical and economic. During requirement gathering and subsequently while creating the requirement specifications (explained in detail in Chapter 4), care should be taken not to include implementation details. Implementation details should be considered only during the technical feasibility study, where one can find out whether any requirement can actually be converted into a software design. A technical feasibility study is further explained in Section 3.4.

### 3.3 Finding the Economic Feasibility of Requirements

Economic (i.e., commercial) considerations are important while making a decision on building a software product or a feature of a product. If the project team realizes that the development cost of the product or a feature is turning out to be much more than any budget or commercial consideration can justify, then the development of that product or feature can be dropped. This decision can happen any time during the software development process. However, it is generally done at the early stages of the project when the software development expenses are still low.

Let us take into account some economic concerns that need to be considered when building a feature of a software product. Suppose there is a software product for the users (i.e., customers) of a bank for doing online banking transactions. The bank noticed that some users prefer not to use their credit cards while making purchases with online retailers because of the fear of credit card information theft. Hence, the bank decided to develop a new feature that will allow the users to create virtual credit cards through which they can make purchases online. A virtual credit card will have an expiry date as well as an upper limit on the dollar amount that can be spent using the card. Users can set both this upper limit and the expiry date as per their convenience. When a user wants to buy a product online on some website, he or she can create a virtual credit card and use it. For example, if the amount the user has to pay for a purchase is US$60 and the date of transaction is September 13, 2015, then the user can create a new US$60 virtual credit card with its expiry date as September 13, 2015. Once the user has used this virtual credit card then, even if that credit card information is stolen, it will not affect the user. Each time the user shops online, the user can create such a virtual credit card. This will provide added security for users who shop online.

The CEO of a bank is excited about this virtual credit card feature. This proposal finally reached the project management team of the bank, who were eager to find out how much effort and cost would be involved in developing this feature. They found out that the total effort required will be 20 man-months and the development cost will be approximately US$140,000. There will also be an annual maintenance cost (to maintain the software product once it is developed) at 20% of the development cost. This means that the maintenance cost will be US$28,000 per annum. The team also estimated that around 5000 users will use this product each month. The average transaction amount for each purchase is estimated at US$50. For each such transaction, the bank can charge the seller 1% of the purchase amount. Therefore, the revenues earned by the bank will amount to approximately (5000 × 50 × 1/100) or US$2500 per month.

Now, let us find out the return on investment (ROI) for this virtual credit card feature:

Development cost = US$140,000
Maintenance cost = US$28,000 per annum
Revenue = US$2500 per month

The formula used for ROI calculation is ((revenues/total cost) × 100)%.

ROI for the first year = ((2500 × 12/(140,000 + 28,000)) × 100)% = 17.8%.

Now, if you calculate the ROI for the first 5 years, then it is ((2500 × 12 × 5/(140,000 + 28,000 × 5)) × 100)% = 53%.

ROI for the first 10 years will be ((2500 × 12 × 10/(140,000 + 28,000 × 10)) × 100)% = 71%.

From these estimates, it is clear that building this software product (or feature) is not economically feasible because the returns are less than the investments even 10 years after this product is released. Thus, the idea of building this software product must be shelved.

As explained previously, some features look attractive technically, but because their development and maintenance costs do not justify them, they have to be shelved.

#### *3.3.1 Build/Buy Decision*

You need to decide on the build/buy (i.e., build/procure) decision while performing the economic feasibility analysis. If some software products available in the market can fulfill your requirements, then you need to decide whether it makes economic sense to buy one of them or develop a new product.

A downside of procuring an existing software product from the market is that it may not fulfill 100% of your requirements. This is simply because the existing product was not developed for your specific requirements. Generally, commercial software products (such as Microsoft Office) are made to suit the common requirements of a large number of users. Therefore, some specific requirements of some users cannot be fulfilled by those commercial software products. If this is the case, then you need to decide whether the unfulfilled requirements (i.e., the requirements you want but are not fulfilled by the product that is available in the market) are critical for your business. If they, then you need to look for other existing products. If none of the existing products in the market fulfill your crucial requirements, then you need to develop the product or contact a software vendor who can develop the product for you.

Generally, smaller software products that are available in the market do not have any customizable components, whereas larger software products generally have features to customize as per your requirements. Large Enterprise Resource Planning (ERP) software products are examples of customizable products. They can be customized to exactly meet your requirements.

Thus, your build/buy decision will depend on many factors as outlined above. You need to evaluate your requirements against what is available in the market, to make an appropriate decision.

### *3.3.2 Budget Instead of Cost Analysis*

In several countries, governments take up projects to facilitate their operations and provide better services to their citizens. Similarly, governments take up military projects because of necessity. The case is similar to government-funded initiatives for healthcare, education, and so on. In all such cases, cost–benefit analysis is not a criterion. The only criterion is social benefits or national interests. Budgets are allocated to government departments (i.e., government ministries) for the implementation of some plans. From this budget, the procurement cost for software development projects is derived. Bids are placed on a global or local basis. Suitable bids are selected by the departments. This selection is based on parameters such as cost, quality, and detailed implementation data furnished by the vendors. In such scenarios, the vendors conduct their own study on the costs and profits for developing and delivering such software projects.

In a nutshell, in most government projects, cost analysis is not the primary criterion for deciding the feasibility of the project; budget plays a more instrumental role.

## 3.4 Finding the Technical Feasibility of Requirements

In many cases, feasibility studies are conducted after a project is already started and its requirement gathering is completed. Once all your requirements are available, you can conduct all kinds of feasibility studies. You can collate the requirements and then assess if any of them are technically feasible. For example, one requirement states that a planning mechanism is to be provided for the production planning of all the plants that belong to the company. After some discussion, it is found that production planning has to take into consideration all the confirmed orders as well as forecasts. At the same time, production planning needs to be correct 95% of the time. From experience, it was found that no production planning can be accurate beyond 50% correctness. Naturally, this requirement cannot be implemented, and thus it needs to be dropped. Let us look at one more example. Suppose you need to develop a mobile phone application that can share the content with other mobile phones using a direct peer-to-peer sharing technology. The requirement is that the two mobile devices can be as far as 100 m away from each other and still share the content. However, the current peer-to-peer technology allows sharing only up to a distance of 1 m (e.g., Bluetooth technology). Thus, this requirement is technically not feasible and this kind of application cannot be developed.

In some other cases, the end users of the proposed system are not sure of their requirements. This can happen when the proposed system requires a lot of interaction from its users. In such cases, users may not be sure as to what their actions will be and how the system might behave. If they are asked how they perform their tasks, they are not able to describe all the activities involved in performing them simply because they do so mechanically. Now, if you want to know the user requirements, then the users will be at a loss to describe all the activities they perform on a daily basis. Thus, collecting their requirements becomes a challenge. In such cases, an evolutionary approach is helpful to find out the user requirements more specifically. A prototype can be made and shown to the users and then feedback from the users can be sought. Users provide their feedback and the prototype can be refined further on the basis of that feedback. When a prototype is shown to such end users, they are in a much better position to recall what activities they perform during their daily tasks.

The process of showing the prototypes and getting the end user feedback is repeated many times until the end users are completely satisfied with the prototype. Once this is done, the actual software product can be developed on the basis of that prototype. During this iterative prototype development process, the project can be evaluated for technical as well as commercial feasibility. At any stage of this prototyping, if the solution does not look feasible, then the project can be abandoned. For example, while developing the prototype, if it is noticed that the actual software product requires an exorbitant development cost that cannot be justified, then the project can be abandoned. Prototyping is further explained in Section 3.5.

### 3.4.1 *Difficult-to-Figure Scenarios*

There could be many occasions in which making a feasibility decision is not easy. For example, a company may be using a legacy system and realize that, although the legacy system is critical for the company's business, most of the hardware and software parts of that legacy system have become outdated. The company may realize that 3 years down the line, it will be difficult to maintain that legacy system. A solution to this issue is to scrap the legacy system and bring in a new one. The company may make financial calculations regarding the maintenance of the legacy system versus developing or buying a new one. However, a factor that worries company managers is the new system not working as expected. By all accounts, the existing legacy system is working fine, but there is no guarantee that the new system will work as well. This is a factor that accounts for a large number of legacy systems still being used despite the fact that they are quite outdated. The bottom line is that the company management never takes the risks associated with a new system easily.

### 3.4.2 *When Is Feasibility Study Not Required?*

Apart from government departments, some private companies also request a quotation mechanism to procure software products at competitive prices. The in-house technology teams of such companies are not equipped to develop the required software products on their own but can only run and maintain the implemented software products on their premises. Whenever these companies need new software products, they evaluate and contact some vendors (or service providers) of the software products. These vendors have worked with other customers in the past on implementing and delivering similar systems. Since these customers (i.e., these companies) in such cases know the reputation and track record of these vendors, they do not insist these vendors perform a feasibility study. At most, a conference room pilot demonstration is more than sufficient for the companies to make the decision about the procurement of the software product from such vendors. Generally, the conference room pilot program is a basic preconfigured software product that demonstrates the suitability of the requirements with matching software capability. The customer may also check the references cited by the vendor about the software product. These references describe the success stories of the software products that are developed by that vendor and are being used by other similar customers.

In most industries, if a particular vendor has a good reputation, then the customers are willing to have their software products implemented by that vendor. Since the customers know that a particular software product is already successful at similar businesses, the customers are assured, in advance, of the success of their proposed software products. Therefore, the customers may not need a feasibility study to be performed by the vendor while evaluating the feasibility of their proposed software products.

### 3.5 Prototyping

A prototype is a miniature product that has the same architecture as that of a full-blown product. A prototype is generally a functioning rough copy of a product and can be used for demonstration purposes. Prototypes are used to show what the product looks like and how it actually works (once it is developed). Showing a working prototype to the customers boosts their confidence in the project team's ability to deliver the product. As an example, in civil engineering, if someone asked you to construct a building, you may initially construct a toy version of that building using some cardboard pieces, glue, paint, and so on to show that person what the actual building will look like once it is constructed. That toy version is the prototype.

You can build a prototype of the product during the feasibility study and show it to the project stakeholders. On seeing it, the stakeholders will form a general impression of what they will get after the product is fully developed. They can also find out if the product looks and works like they expected and if they have problems or want to have some changes made.

There are many types of prototyping used in the software development processes. The prototyping types include throwaway prototyping, evolutionary prototyping, incremental prototyping, extreme prototyping, and so on. In this book, we discuss throwaway prototyping and evolutionary prototyping.

*3.5.1 Throwaway Prototyping*

Throwaway prototyping is one of the techniques used to develop prototypes for a software product. Throwaway prototyping is used to develop a prototype in several stages and in several iterations. These iterations are used to capture the user requirements bit by bit until all the user requirements are captured by the end of the process. When all the requirements are captured in the prototype, the prototype contains the complete software requirement specification. At this stage, the software requirement specifications are the only useful part in the prototype even though some software design and construction was also used in building it. The project team will start building a software product that matches the software requirement specifications contained in the prototype. Thus, the software design or construction used in building the prototype or any of its parts is not used in any design or construction of the actual software product. So literally, the prototype can be thrown away after developing the software product.

Figure 3.1 depicts the throwaway prototyping process. The process starts by getting the preliminary requirements from the end users. These may be crude but should be acquired quickly so that a quick design can be developed. Then, a prototype is built and shown to the users. The users give feedback as to what features look fine and what features require changes. They also state what changes may be needed. The users may also state what is missing in the prototype so that it can be included in the next iteration.

The process of refining the prototype will continue until the users stop giving their feedback to improve the prototype. Once the users are satisfied with the prototype, the prototyping process is considered to be complete.

In Figure 3.1, you will notice the following cyclical operation: quick design–build prototype–evaluate prototype–refine prototype–(back to) quick design. This cyclical operation represents the iterations for the prototyping process. Only when the prototype is considered complete will the process exit this cycle and proceed to build the final requirement for the actual product. You will also notice that the final requirement

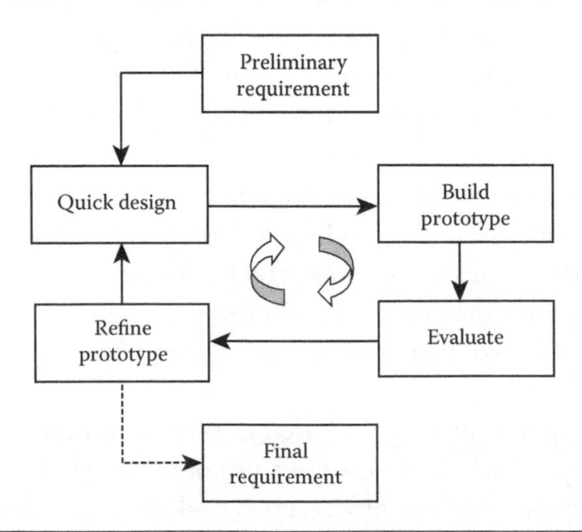


**Figure 3.1** Process for the throwaway prototype.

has a relationship with the iterations depicted. This relation is depicted as a dotted line. This dotted line signifies the fact that the relationship of the final requirements with the cyclical operation is not continuous. The final requirements are not considered part of the cyclical operation. Rather, the final requirements are outside of that cyclical operation. In other words, the final requirements are derived based on the final outcome of the cyclical operation. They are then used to create the actual software product. The throwaway prototype itself is not directly used in the software process and thus is a "throwaway," as its name indicates.

The fact that the prototype is never used directly in the software development signifies that the prototype does not contain any useful content that can be reused somewhere in the software development (of the actual product). This type of throwaway prototyping involves creating the user interfaces without writing any business logic. Navigation from one user interface to another is accomplished through some code that helps such navigation without using any business logic.

In the prototype, the output from the system for a given input (submitted by the user) is mimicked on the user interface through false output content. For example, suppose the initial interface shows a login page and the user is required to provide the username and password on this user screen. After the successful login, suppose the next screen is a user interface that contains the profile of that user. The prototype will show the login page, and when the user enters the username and password and hits the Submit button, the prototype will navigate to the user interface screen that contains the profile of that user, without doing anything with the business logic (in fact, there is no business logic developed).

Creating such a prototype will not involve much effort from the project team. However, this type of prototype will definitely help the developers find out what the users want in the proposed system that needs to be developed.

### *3.5.2 Evolutionary Prototyping*

Evolutionary prototyping is very different from throwaway prototyping. The main intention in using evolutionary prototyping is to reuse the source code, which is written for developing the prototype, in developing the actual software product. This implies that a lot of effort is put into developing such a kind of prototype. The prototype is in fact a working model of the software product. All the business logic that may need to be put in place in the actual software product also goes into developing the prototype. This is an expensive approach considering that if the project needs to be abandoned later, all this effort will go to waste. Nevertheless, this type of prototype can be developed for software products wherein there is a small chance that the project will be abandoned sometime during the prototype development stage.

Evolutionary prototyping can be used effectively on agile projects. This is because, on these projects, the software products are developed incrementally. For each increment of the software product development, the effort is just a fraction of the total effort required to develop the entire software product. At each increment, developing an evolutionary prototype thus will not require a large effort. Hence, even if the increment is completely abandoned, the effort that goes into the development of the prototype will not be much. For example, suppose a complete software product will be built over 10 increments (iterations). In one of the iterations, the evolutionary prototype is abandoned after putting in 70% of the effort for that iteration. This means a loss of only 7% of the total effort required for building the software product. These numbers indicate that the effort loss is not much in comparison to the total required effort. This example shows that evolutionary prototyping is very suitable for agile projects.

### *3.5.3 Incorporating Prototyping in Software Development*

In the software development life cycle, the prototyping stage can be considered the feasibility study phase. During this stage, if, for any commercial or technical reason, it is found that the software product cannot be developed, then the project can be abandoned. Abandoning the project at the prototyping stage will not cost much because the efforts required to develop a prototype are not much.

We have studied two types of prototypes (throwaway and evolutionary) in the previous sections. How can these prototypes be incorporated into a software development project? Let us find out.

In Figure 3.2, prototyping is shown integrated with a Waterfall model for software development. From this figure, you can see how a prototype is incorporated in the

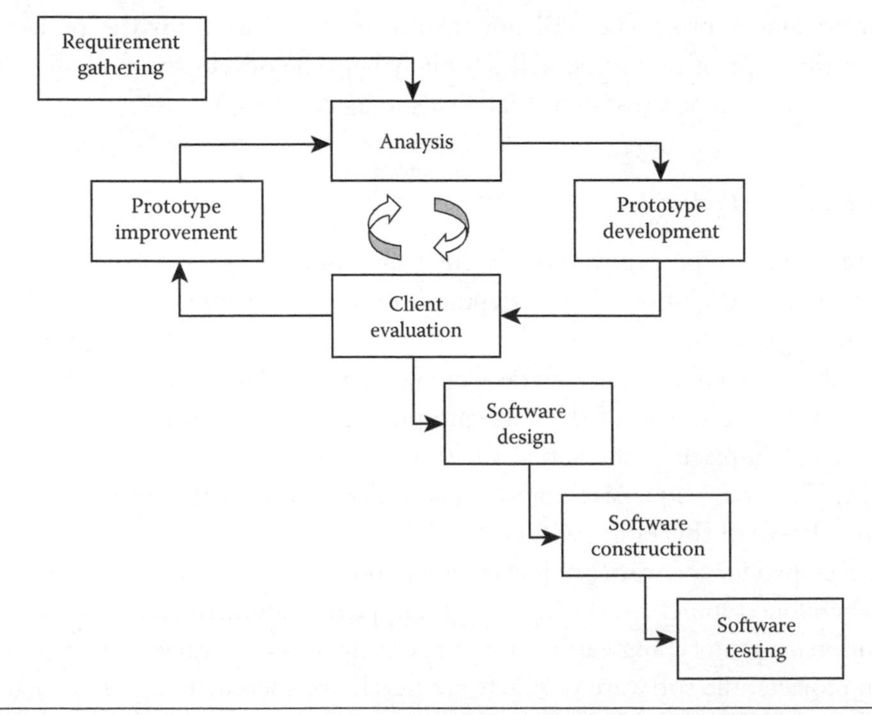


**Figure 3.2** Incorporating the throwaway prototype in the Waterfall software development life cycle.

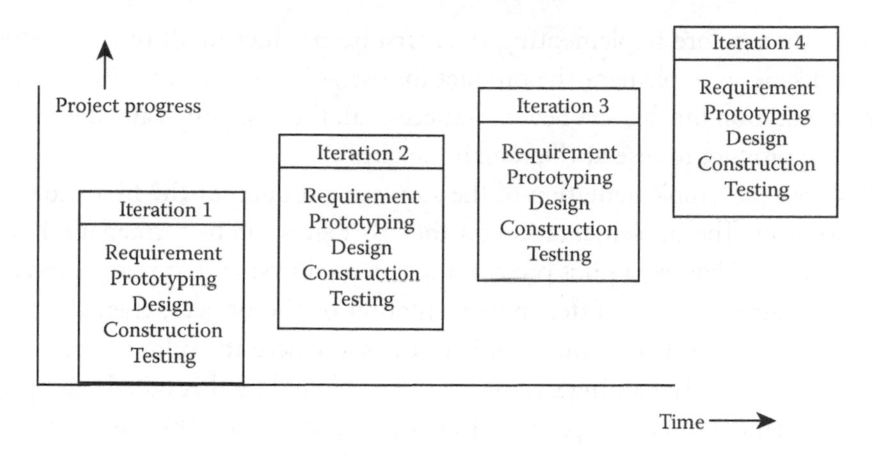


**Figure 3.3** Incorporating the evolutionary prototype in agile projects.

software development life cycle for Waterfall-based projects. Prototyping is done after obtaining the end user requirements and before developing the design of the proposed software product.

Waterfall-based projects use throwaway prototypes; that is, at the time of the prototype development, the project team is not prepared to write the source code for the business logic. As you can see from Figure 3.2, the prototype development phase is done at the beginning of the project. However, the software construction (coding) phase happens in the middle of the project. Thus, it is logical that source code writing is not possible during the prototyping stage.

Figure 3.3 shows the prototyping in agile projects. In agile projects, during each iteration, there can be a prototype. That prototype is meant exclusively for that specific increment (i.e., specific features) of the software product that will be developed during that iteration. Such prototypes are evolutionary in nature. The source code written for developing these prototypes will be reused in developing the software product. This is in contrast to the way in which throwaway prototypes are involved in the Waterfall projects.

Although the throwaway prototypes can also be used on agile projects, the usage of evolutionary prototypes on agile projects does not pose any real challenge. The source code written to develop the prototype can be readily used in developing the software product. This is because, in agile projects, writing the source code for developing the software product (in an iteration) can happen as the prototype development takes place.

### 3.6 Pilot Projects

Sometimes, a pilot project is done to assess the feasibility of a software product before it is completely implemented. A pilot project is developed with a full-blown software product but the scope of the pilot project is limited. For example, if a company has several sites (locations) and it wants to develop a software product to be used at all of these sites, then before implementing the software product to all of them, the company may decide to implement the product for use at just one of the sites first. When the implementation for this site becomes successful, the company can later implement the software product for use at the remaining sites.

In this example, implementation of the software product for the first site is known as a pilot project. The implementation for the other sites can be termed a rollout project. The benefit of having a pilot project implemented first is that if any problems are encountered during or even after implementation of the project, then the losses, in terms of a failed project, are limited only to the site where the pilot project was used. The customers in the remaining sites will not be affected by this failed pilot project.

If the implementation of a pilot project is successful, then the customer (i.e., the company) can start using that software product. During the usage of that product, some problems can also crop up. For example, suppose a SAP ERP software product is implemented, as a pilot, for one of the sites of a customer. During implementation, the project did not face any major risk and SAP was implemented successfully as a pilot project. Users started using the SAP ERP. During usage, it was discovered that the tax calculations were wrong. The project team was informed about this issue. The project team then studied all the implementation settings and found that the tax calculations were set up wrongly, and it was eventually corrected. However, in the meantime, the company already lost some revenue because of the higher taxes paid to the government.

Imagine if this problem was discovered after implementing the SAP ERP in all sites, in which case the customer could have lost sizable revenue because of the higher taxes paid in all those sites.

This hypothetical example shows that, even if things look good during project implementation, you never know if everything will run smoothly after the implementation. Generally, the pilot and rollout projects are used for Commercial Off-the-Shelf (COTS) or Software as a Service (SaaS) software products.

### 3.7 Chapter Summary

Feasibility studies are important to take care of the inherent risks involved in software projects. During the feasibility study, if it is found that the project is either economically or technically infeasible, then that project can be abandoned. If a feasibility study is not conducted, project stakeholders may be faced with a situation where they have spent a substantial amount of money and time only to realize that the project can no longer progress further and needs to be shelved. Feasibility studies ensure that this situation does not happen.

Economic (i.e., commercial) feasibility studies can be conducted either before the project even starts or after receiving the complete user requirements. Technical feasibility studies can be conducted only after the user requirements have been gathered. On the basis of these user requirements, a technical feasibility study can be conducted.

Both technical and economic feasibility studies can be done by building a prototype or using a pilot project.

A prototype is a miniature software product that may or may not have all the business logic implemented. A prototype can be developed by the project team by gathering the user requirements, after which it is then built and shown to the users. On the basis of the user feedback, the prototype can be refined and shown to the users again. When the users are fully satisfied with the prototype, the prototype building process is stopped and the project team can start building the actual software product.

A prototype can be a throwaway or an evolutionary type. A throwaway prototype is built in such a way that the project team will not spend much time or effort in building the prototype. The purpose of such a prototype is to just gather user requirements effectively. In the throwaway prototype, the prototype is not part of the final software product. An evolutionary prototype, on the other hand, is built in such a way that the source code, if any, written for building the prototype is reused when the software product is actually built.

## QUESTIONS

1. What is a prototype? What is the process of building a prototype?
2. What are the different types of prototypes?
3. What is a feasibility study?
4. What types of feasibility studies are done?
5. Provide an example of an economic feasibility study.
6. Provide an example of a technical feasibility study.
7. What is a throwaway prototype?
8. What is an evolutionary prototype?
9. What are the various methods of conducting feasibility studies?

# Recommended Reading

Department of Information Resources (1992), *How to Conduct a Feasibility Study for Information Technologies*, Austin, TX.

Ashraf El Sharkawi (2005), *Economic Feasibility Studies*, Cairo University, Cairo, Egypt.

# 4
# Software Requirements Specifications

**In Chapter 3, we learned**

- **What a feasibility study is**
- **What a technical feasibility study is**
- **What an economic feasibility study is**
- **What a prototype is**
- **What a pilot project is**

**In Chapter 4, we will learn**

- **What requirement management is**
- **What requirement gathering is**
- **What requirement specification is**
- **What a use case is**
- **How software design can be made from requirement specification**

## 4.1 Introduction

Creating a software product is a long process. It starts when the end users or customers provide their requirements (i.e., wish list) to the project team. Through these requirements, it becomes clear as to what features the proposed product should have and what kind of tasks it is supposed to perform. These requirements are the basis on which the final software product is created after going through the tasks of designing, writing the source code, and finally testing the product thoroughly.

First, the requirements are gathered from the end users who will use that product once it is developed. The requirements are provided to the development team by the customer who wants a product to be developed. The format of these requirements can be either formal (i.e., structured) or informal (i.e., unstructured or ad hoc). Requirements in an informal format need be converted into a formal format. Once the requirements are specified in a formal manner, they are known as requirement specifications. Generally, the requirement specifications are listed in what is known as the software requirement specification (SRS) document. In some cases, use cases can also be defined to represent the requirement specifications more formally to avoid any misunderstanding between the development team and the customer on interpreting the requirement specifications.

When the requirements are gathered, they can be in any form that is suitable to the person who gathered them. However, passing the unstructured data or unstructured information from one person to another can result in misunderstanding. Because of this, informal requirements are difficult to understand for anyone who is not the author of them. Requirement specifications need to be passed to the software designers for creating software designs. To make sure that the software designers have no misunderstanding about any of the requirements, the requirement specifications must be presented to them in a form that will ensure no misunderstandings. This is why requirement specifications should be made in a formal format.

**Usage note:** In this chapter, the terms *software*, *software product*, *product*, *software system*, and *system* are used interchangeably. Although a software product is different from a software system, at some points, it is better to use the term *software system* instead of *software product* to maintain the flow of the argument that is being discussed. When we use the term *software product*, it usually means a single software product. When we use the term *software system*, we generally mean many software products working with each other to help the user doing a task. A software system also denotes that a software product is in production and people are using it. It is a common practice to use these terms interchangeably. The same convention has been followed in this book.

## 4.2 Software Engineering Methodologies and Requirements Management

In Chapter 2, we studied some popular software engineering methodologies. Software engineering methodologies have a significant impact on the way we develop the software products. In software requirement management, we do things differently as per the chosen software engineering methodology. Let us discuss how software requirement management works differently for different software engineering methodologies.

### *4.2.1 Requirements Management in eXtreme Programming*

The people who are working on a project that is based on an eXtreme Programming (XP) environment have to work on iterations that are part of the software project. During each iteration, the product manager (sometimes known as customer or product owner) provides the requirement specifications that need to be implemented during that iteration. In XP, the product manager gathers and manages the requirements. The product manager is completely responsible for everything related to the requirement management life cycle. The requirement cycle includes requirement gathering, requirement analysis, requirement specifications, and sending them to the software designers (i.e., architects). In XP, there could be only three or four requirement specifications per iteration. Therefore, at any time, the project team is working only on those three or four requirement specifications. Once the current iteration is completed, the development team obtains the next set of requirements from the product manager. In XP, requirement specifications are known as user stories. Sometimes, they are also referred to as software product features.

### *4.2.2 Requirements Management in Scrum*

In Scrum projects, the person responsible for managing the requirements is the product owner. The product owner has two sets of requirement specifications. One is known as product backlog and the other as sprint backlog. The product backlog is not a complete set of requirement specifications for the entire software product but it includes all the requirement specifications that either have already been implemented in the software product or are currently being implemented in a sprint (iteration). The product backlog also contains future requirement specifications. Whenever the product owner gets a requirement from the marketing team, he or she converts it into a requirement specification and includes it in the product backlog.

The sprint backlog contains requirement specifications for the current sprint that is about to start. The product owner decides which requirement specifications are to be implemented in the current sprint and accordingly chooses the requirement specifications from the product backlog and inserts them in the sprint backlog. The product owner is responsible for everything related to the requirement management including requirement gathering, requirement analysis, creation of requirement specifications, and managing the requirement changes. More details on product backlog and sprint backlog are provided in Section 2.9.

### *4.2.3 Requirements Management in Waterfall*

On software projects that follow the Waterfall methodology, requirement management is very different from how it is performed in either Scrum or XP. There are two essential things about requirement management in the Waterfall model: first, the project team needs to gather all the requirements that will be implemented in the software product, and second, the project team always anticipates changes in the requirements because the customer may request such changes from time to time. Therefore, managing the changes in the requirements is an integral part of requirement management in the Waterfall model.

Once a request is received from the customer to change a requirement, the requirement and the specification for that requirement need to be changed. Once the requested requirement is changed, a verification of change in the remaining requirements needs to be performed because some of the remaining requirements may be dependent on the changed requirement. Any such dependent requirements and their requirement specifications also need to be changed accordingly. For example, if the

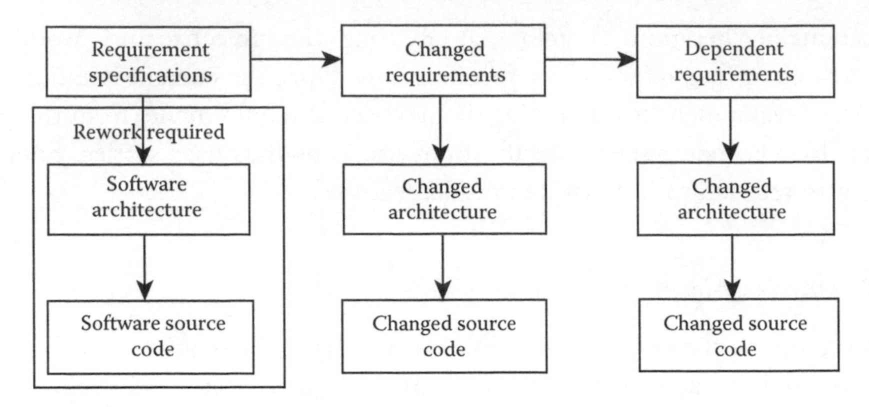


**Figure 4.1** A change in an already implemented requirement results in changes in the design and source code.

length of a window is changed on the basis of the customer's request, then the width of that window also needs to be changed if the width is dependent on the length. Only after such verification is performed can the complete requirement specifications be sent for making changes in software design.

If a large part of the software product is already implemented and the software design and source code already exist for a requirement that needs to be changed, then the source code and design also need to be changed (Figure 4.1). This is a tricky situation because the change not only results in a lot of rework but also may result in bugs (i.e., defects) entering the software product. As a comparison, in civil engineering, if the customer requests adding one more entrance to a building for which the design and most of the construction are complete, then it will need a lot of rework.

### 4.3 Implementation of the Requirements

While gathering the user requirements, the focus should not be on issues such as whether they are technically or financially feasible to implement. Just gathering the requirements, without considering other issues, is the best approach for many reasons. One reason is that requirement gathering should be done by business analysts. Business analysts are the people who have working experience of the industry for which the software product is proposed to be developed. They know the business processes well. Thus, they also understand the requirements well. For example, suppose the user states that the software should take care of the campaigns during their steel making. Someone who has never worked in the steel industry will not understand what a campaign is. Thus, an inexperienced person will not be able to understand and collect the requirements correctly.

The reason why the implementation details should not be considered during the requirement gathering stage is that the business analysts do not have proper understanding as to how a requirement can be converted into a good software design. It is a lot better for business analysts to concentrate only on collecting the requirements

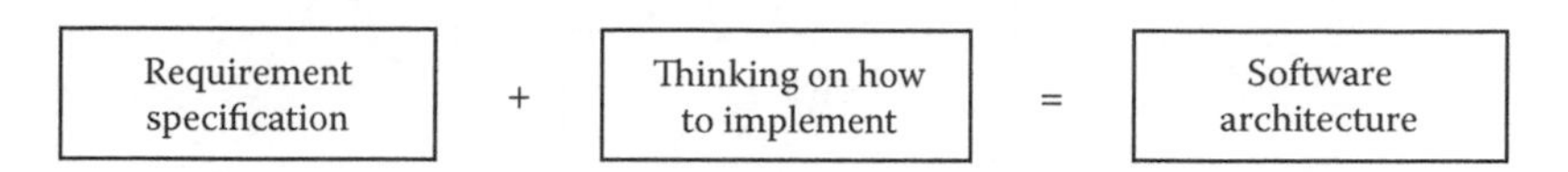


**Figure 4.2** Requirement specification together with implementation consideration is needed to make the software architecture of the product.

correctly and not on how these requirements will be implemented. For these reasons, it makes sense not to include implementation details during the requirement gathering stage.

Requirement specifications reflect what a software product should do. The question of how the software product should perform the specified tasks should be out of the scope for making the requirement specifications. Why? Well, there are two aspects for developing a software product. The foremost aspect is what the software product will do. The other consideration is how the software product will do those tasks. The "what" part is very important and it is the heart of the requirement specifications. The "how" part is an implementation aspect. The designers of the software product decide how the software product will do the tasks specified by the requirement specifications. Figure 4.2 illustrates the idea of separating the "what" and "how" parts of the software requirement specifications and their implementation.

When we completely separate the implementation from the requirement specifications, some questions arise. What if a requirement is technically infeasible to implement? What if it is economically infeasible to implement? The solution for these questions is that, during a feasibility study, all the requirement specifications will be evaluated to determine whether they are technically or economically feasible to implement. How a requirement can be implemented should be left to the system designers. They know very well how to convert a requirement into a software design (or part of a software design). For whatever reason (including technical or economical), if it is not feasible to implement a specific requirement, then the customer and developer will decide whether to drop that specific requirement, modify some requirements, or stop developing the product altogether.

## 4.4 Requirements Types

A software product is developed based on several requirements. They can be categorized as functional and nonfunctional requirements, as shown in Figure 4.3. Requirement categories (i.e., types) are explained in the succeeding sections.

### *4.4.1 Functional Requirements*

Functional requirements are the ones that are converted into features of the software product. Functional requirements help design the transactions and reports that the users may need while working with the product once it is developed.

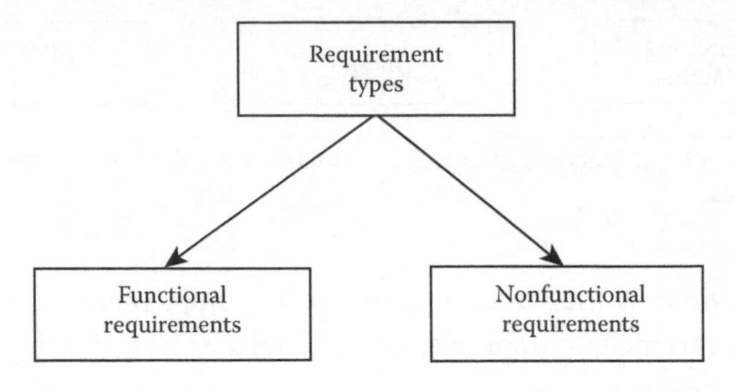


**Figure 4.3** Software requirement types.

A functional requirement generally translates into a software product feature that can accept some input from either the user or an entity. The software product feature will then provide some output (when the software product is running) to either the user or an entity. Here, an entity can be a program, a file, a database, or hardware. A functional requirement is the one that provides an answer to the question, what will a feature (of the software product) do? For example, assume that the proposed software product handles the academic activities of the students in a university. If a software requirement says "the system should be able to display the current GPA of a student after receiving the student ID," then it is a functional requirement because it provides an answer to what the student ID will do (the student ID feature will enable the system to display the GPA).

Some functional requirements are converted into user interface components that we can see while working with the product. Some functional requirements are also converted into business logic. Therefore, to design the software product easily, it is better to divide the functional requirements into user interface requirements and business logic requirements (or both). User interface requirements provide various elements on the screen, through which the user interacts with the software product. Sample elements on the screen include buttons, clickable links, menus, text fields (through which the users enter the inputs), text, graphic displays, sounds, and video (through which the users receive the output). The business logic requirements contain all the processing that will be carried out by the software product to provide the output(s) to the user. For example, if we consider the student GPA example given above, then the processing involves retrieving the grades based on the input (i.e., student ID), calculating the GPA from those grades, and finally displaying that GPA.

### *4.4.2 Nonfunctional Requirements*

Some of the most common nonfunctional requirements include user interface aesthetics, speed of computation, security, reliability, and scalability. Most of these are part of the product that is not visible to the users. In addition, most of these requirements do not form a feature of the product once the product is developed. However, they are essential for the software product to work. For example, the security requirements guarantee that the product is secure from hacking threats. Similarly, performance requirements ensure that the users do not have to wait for a long period to get the responses from the product.

Nonfunctional requirements are harder to measure. However, the project team and users should agree on some range of limits for them. For example, security is a big issue these days and it should never be compromised. Even with the best security mechanisms in place, it is possible that some data theft or identity theft takes place because of malicious hacking. If the software product needs to accept or transmit sensitive data, then security should be of utmost importance. The minimum agreeable arrangement for security requirements is that the software product should not allow unauthorized access. Although it is hard to measure a security requirement, if you state (in the requirement specifications) that unauthorized access should not be allowed in the software product, then it is a good enough security measure.

Nowadays, most software products are web based and perform tasks such as entertainment, conferencing, phone calls, social media, and business transactions. Therefore, nonfunctional requirements related to security issues become very important because any website can be prone to hacks and theft of data. Ensuring that a proposed Internet-based software product is secure and the data can be transmitted securely between the user's computer and the server (on which the proposed software product is hosted) is a critical requirement. The other requirement for a web-based software product is performance. If you state (in the requirement specifications) that the web pages, e-mails, and so on from the server should be loaded onto the users' computers within some acceptable time limit, then it is a performance requirement.

Software products generally need maintenance after they are deployed. Maintenance includes software bug fixing, porting the products to new versions, and enhancement of product features. To carry out any of these maintenance activities on a product, the people in charge of maintenance should have access to the technical documents of the product so that they know the architecture of the product and can make changes to the source code.

When a software product is built, provisions can be provided in the design so that the design or source code can be easily changed later on. For example, suppose some values appear on a menu through some hard-coded data placed in the source code. In this case, adding an additional menu in the menu structure will become difficult. However, if the menu structure is flexible or the values come through a database, then it will be easy to make additions to it at a later date. Providing a maintenance-friendly software design should be a goal during the requirement gathering stage. As another example, suppose a product is expected to be used by hundreds of customers presently. If it is designed (using an extendable design) to support thousands of simultaneous customers, then it will be easy to modify the product later if needed. In conclusion, the maintenance aspects of the proposed product are also an important part of nonfunctional requirements.

## 4.5 Sources of Requirements

Assume that you are part of a development team in an information technology company and your team is going to develop a new software product. The most frequently asked question in such cases is, where will the requirements come from to develop this new product? The reality is that the software requirements can come from many sources. Let us consider different scenarios as to where the requirements can originate.

- Your company was using some old software product. The product is no longer being supported for reasons such as obsolete technology, the vendor stopped providing support, or your company has grown and the old product is no longer able to support your business needs.
- Your company had no software systems in place. All the business work was performed by using paper or personal tools such as spreadsheets and word processors.
- Your company has acquired a business that has no software systems in place or the existing software systems are no longer viable.
- You are working in an information technology company and it has realized that there is a market need for a new software product.
- Another company has asked your company to develop a software product. This software product may be used by that other company or that other company may market this software product to its customers.

The requirements are gathered on the basis of the scenario that demanded the development of the new software product. If an existing system needs to be replaced, then you already have some requirements ready and need to find the additional requirements. If a completely new system has to be developed and no systems were being used previously, then all the requirements will be new. If a company asks another company to develop the product, then the company that asked should provide all the requirements to the other company. Figure 4.4 shows some sources of requirements.

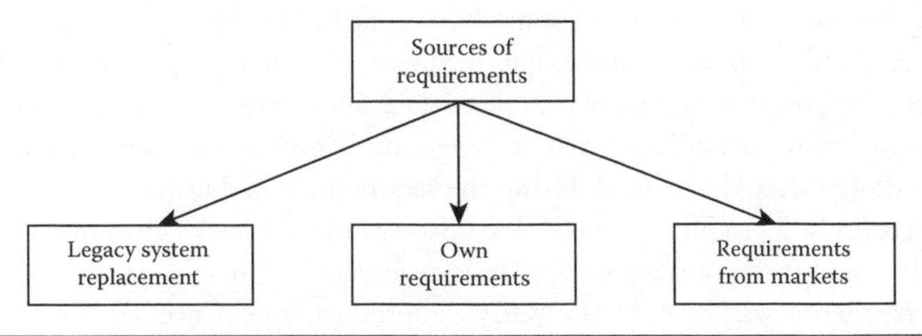


**Figure 4.4** Some sources of requirements.

### 4.6 Categories of Users

The users of a software product can come from several backgrounds. For example, a trader may use a website to receive the prices of some goods in real time, on his or her mobile phone. Some requirements for building such a website may be related to the availability of the goods' prices in real time, size of the mobile screen, connection speed of the mobile phone, maximum number of possible users of the website, and so on.

The users of a product can be categorized in many ways depending on the purpose for which the product is built:

- **Website:** Websites are built to be viewed by the general public from any location in the world. Generally, no user is blocked from viewing the websites. Some popular websites generate heavy traffic, amounting to millions of views per day. Requirements for the websites are generally gathered by using online polls, engaging marketing executives, or getting feedback directly from end users.
- **SaaS:** Software as a Service (SaaS) products are mostly used by the business houses (customers). The users of these products are the employees of these business houses. A SaaS vendor develops products for these customers, and the employees of the customers use the products not as a product but as a service. Once a SaaS product is developed, the customers buy the services of that SaaS product from the vendor who hosts the product on a website. The software vendor itself can host the SaaS product or a service provider can host it (by making an agreement with the vendor) and provide services to the customers of that vendor. A vendor who develops such a SaaS product may get the requirements from the potential customers by doing some market research. On the basis of the feedback on the customer requirements, the vendor then hires a project team that designs and develops the software product.
- **Custom built:** Sometimes a company wants to have a software product developed for its own use if the existing Commercial Off-the-Shelf (COTS) software products available in the market do not satisfy that company's requirements completely. In such a scenario, that company may initiate a project to build a custom software product in which case it is easy to obtain the user requirements from the people who will use the product for their daily work (i.e., the employees of that company). The people working for that company will use the product once it is developed. If that company was previously using a product, then some requirements for the proposed new product can come from the already implemented product. Additional requirements for the proposed new product need to be gathered for the areas in which the already implemented product was not used.

Nowadays, most custom-built software products are built for government projects. Government needs are very different from those of either the business houses or the general public. When a software product needs to be custom built for a government agency, there are many unique functional and nonfunctional requirements such as confidentiality, approval mechanism at many levels, and generation of documentation for later reference purposes, to name a few.

- **Mobile and handheld devices:** Because of the proliferation of mobile technology, a large number of software products are being built for mobile devices. The mobile devices have constraints such as small display area and less computing power (when compared to personal computers). Therefore, the requirements for the mobile devices should be prepared as per those constraints. The users of mobile devices are the general public. Requirements for developing the software products for such devices can be gathered by conducting user polls, engaging the marketing executives, or getting feedback directly from some end users.
- **Embedded systems:** Embedded systems are now part of many household gadgets. For example, they are embedded inside a smart television. These embedded systems enable the television to flick at the right time for the preferred channels automatically. They also enable the television to automatically adjust the video quality as per the viewer's preference. Embedded systems have their own set of requirements. Essentially, the software in an embedded device should be able to control the hardware to fulfill the maximum number of requirements. An embedded system is part of a hardware unit. The manufacturer of the hardware unit obtains the requirements (for the software that is installed in the embedded system) from the end users through polls, through user opinions, by engaging the marketing executives, and so on.
- **Customization:** Some vendors have developed large software products that mostly fit customer requirements. SAP, Oracle Applications, and so on are such COTS products. However, in many cases, these products do not exactly fit the customer requirements. To deal with such scenarios, the vendors provide interfaces in their software products through which customization can be done to make them exactly meet the customer requirements.

### 4.7 Software Requirements Life Cycle

Software requirements go through many changes and need to follow a life cycle pattern. When the requirements are gathered, they are mostly in raw form and are not of much use for the purpose of software development. They need to be refined further. Let us discuss the software requirement life cycle and the transformation that is needed to make the software requirements useful.

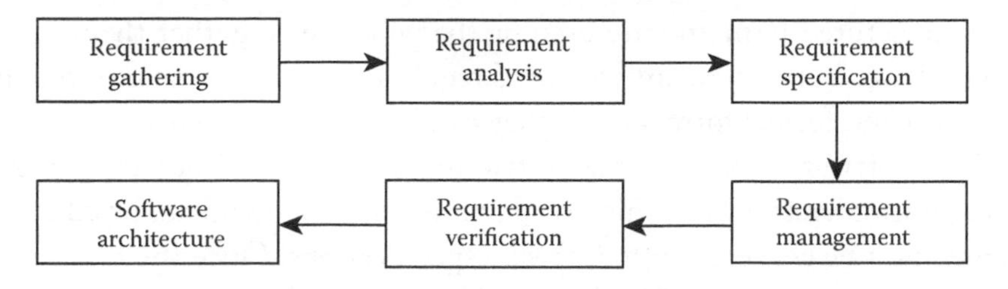


**Figure 4.5** Software requirements life cycle from requirement gathering to software architecture.

The requirements life cycle (Figure 4.5) is as follows:

1. Software requirements are gathered by the business analysts who are part of the project team. Business analysts meet the end users and elicit their requirements through meetings, questionnaires, e-mails, and so on. The end users describe their requirements to these business analysts and the business analysts record them. If the requirements are not clear in the first interaction, then they are elaborated further to make them clear. Once all the requirements are gathered, they can be analyzed in the next step. In Scrum, XP, and most agile environments, requirements are gathered by a person known as the product owner. The product owner is part of the project team, and at the same time, he or she is the representative of the customer. His or her word on the requirements is final and the project team (designers, developers, and testers) must clarify their issues regarding any requirements with the product owner.
2. The gathered requirements then go through some analysis. On Waterfall projects, the requirement analysis is done by the business analysts. On agile projects, the product owner is responsible for requirement analysis. The analysis is needed to assess whether the requirements have any inconsistencies or ambiguities. For example, if one of the requirements says that the background of the main web page is of blue color but another requirement insists on a red background, then there is an inconsistency. If a requirement states that the system should support a large number of customers, then it is ambiguous because there is no precise definition for "large."

   In agile environments, such as XP, the development company may not be gathering and keeping all the requirements before starting the initial design of the product. In some other environments, such as Scrum, the company can select a few requirements from a large number before starting the initial design of the product. However, for projects that are based on the Waterfall model, gathering all the requirements is a must before the design process can begin.
3. The analyzed requirements are converted into requirement specifications. Requirement specifications can be in the form of structured statements or use cases. Here, you must understand the difference between a software requirement and a software requirement specification. Software requirements are in

an unstructured form and are used by the people who gather them for their internal use. Later, these unstructured requirements must be converted into a standard and formal form so that they can be communicated to other people such as software architects and software designers, without the danger of being misunderstood. For this reason, all the gathered and analyzed requirements must be converted into software specifications. Once the requirements are converted into formal and standard forms, such standardized forms of the requirements are called requirement specifications. Requirement specifications can be communicated to other people without the fear of being misunderstood.

4. From time to time, some requests for changes in the requirements may come from the customer in the case of software projects that are based on the Waterfall methodology. In these situations, some requirements may need to be changed. If changes are made to the requirements, they need to be managed by the business analysts (the product owner in the case of XP or Scrum). This process is known as requirement management. In XP and Scrum, there is no room for changing the requirements that have already been developed (i.e., implemented) as product features.
5. Once the requirement specifications are complete, they can be verified by the software testers for correctness. When all the software requirement specifications are found to be correct, the stakeholders (project sponsors) can sign off the requirements to confirm what they want in the new product. The verified requirement specifications are sent to the software designers (i.e., software architects). The requirement specifications should be correct because if any of them are wrong, then the developed software that is based on these requirement specifications will also be incorrect. A requirement specification can be incorrect if what it specifies is incorrect (e.g., if a wrong formula is used in a computation).

In the next few sections, we discuss in more detail some stages of the requirement life cycle and the artifacts that are generated.

### 4.8 Requirements Gathering (Elicitation)

Requirements can be taken from the users in many ways. It could be through collecting the information via questionnaire, user interviews, meetings, and so on. Requirement gathering (elicitation) can be done using a piece of paper or through e-mail or just through any media suitable for that purpose.

Generally, a structured method is the best way to gather requirements. If you ask your questions to the users in a structured manner, then you will be able to organize the users' responses well. This way, it will also be possible for you to verify if you have received all the requirements or any of the requirements are missing.

*4.8.1 Requirements Meetings*

If you are gathering the requirements through meetings with the users of the proposed system, then you should be prepared beforehand to have your questions ready. Users generally use their own business language. If you are not familiar with the users' line of business, then you will face communication problems. Therefore, it is important to familiarize yourself with the line of business of users. Users are generally concerned about their own line of business and may not have knowledge about other lines of business of the same organization they work for. Therefore, your questions (i.e., requirement gathering) must be totally focused on the requirements of each specific user. For example, when you are meeting with a finance manager, your questions must be focused on the financial side of the requirements and not on the production matters of the organization.

*4.8.2 E-mail*

If you are gathering the requirements through e-mails, then you should be thoroughly prepared in your questions. You cannot expect users to provide information (which they would voluntarily provide during personal meetings) about things you have not asked for. Your e-mail should have all the questions related to the functionality required in the software product. The questionnaire should also include nonfunctional requirements such as performance expectations, security level required, and usability. It is advisable to divide a complex question into a set of smaller questions to avoid any misunderstanding.

### 4.9 Requirements Analysis

Requirement analysis is important because you need to find out if any of the requirements are ambiguous or incomplete. After you have checked all the requirements for correctness, you need to find out if any requirements are related to each other. The relationship could be at many levels. Generally, requirements should be separate from each other as much as possible. However, some requirements depend on other requirements. For instance, a financial report is always dependent on the transaction data. Therefore, if the transaction data are not available, then it is not possible to prepare the financial report, although there is no direct relationship between the requirement that generates the transaction data and the requirement that asks for the financial report.

In many cases, a requirement is dependent on other requirements. For example, a submenu is always dependent on the parent menu. Then, again, a small feature may be dependent on a larger feature. Requirements for these smaller features will always be dependent on the requirements of their parent feature. For example, assume that a part of the software product deals with procuring the goods from vendors. Here, the main feature of the product could be procurement management. If the procurement management policies change from local bidding to global bidding, then all the dependent requirements (such as currency and mode of delivery) will also need to be changed.

If the user interface is separated from the business logic, then any requirement changes to the user interface can be easily performed because these changes will not affect the business logic. However, if any changes to the requirements result in making changes to the business logic, then those changes will definitely affect the whole project.

### 4.10 Requirements Specification

Once the requirements are gathered from the customer, the next step is to create the requirement specifications. In many companies, the specifications are prepared in what is known as the SRS document. This document contains all the functional and nonfunctional requirements gathered from the users of the proposed system. Generally, the SRS document is written in a natural language, such as English or Russian. If the project team has been working together for a long time (on the same project or over many projects), then the SRS document is well understood by all the project team members because each team member has become familiar with the language used by the other team members. The SRS document is used for software designing and testing. First, the software testers analyze whether the requirement specifications in the SRS document are correct. If any of them are found to be incorrect, then they will be fixed by the business analysts. For example, if the requirement specification is made to calculate the taxes using an incorrect formula, then the business analysts would correct that tax formula.

Once the entire SRS document is correct, then it is used as follows: The software designers use the SRS document to create the design document for the product. Once the design is complete, the software testers use the SRS document for traceability testing of the design: the testers compare the design document with the SRS document to see whether all the requirement specifications in the SRS document have been converted into the software design. It is quite possible that some requirement specifications have not been converted. The SRS document is also used during User Acceptance Testing (explained in Chapter 9) once the product is developed. During the release phase, the end user (or customer) tests if the final product has all the features stated in the SRS document.

If the project team is working together for the first time, then the SRS document should be prepared carefully. Project team members are probably not yet familiar with the language used by the other members, and it will take time before they are. There should be no room for misunderstanding; otherwise, it will lead to the creation of a faulty design for the software product. If the project team is working together for the first time, it is better to use a formal language to create the SRS document.

A good-quality SRS document is useful to maintain the software once it is developed. To achieve a good-quality SRS document, rather than using a natural language, a formal and rigidly structured language is needed to specify the requirements. The use of a Unified Modeling Language (UML) ensures that the requirements are universally recognizable and unambiguous. The more formal and rigid the language used for creating the SRS document, the more precise the specifications will be. Depending on how formal you want to make your requirement specifications, there are several methods available. These include use cases, mockups, slideshows, and requirement specification document, to name a few. Mockups are fake screenshots mimicking what the screens will look like in the real product once it is developed. These mockup screenshots also include explanations in the form of external text (external text consists of headings and subheadings, bolded or italic words, etc.).

### *4.10.1 Use Cases*

Use cases are the depiction of an actor working with the software product. An actor could be a user (of that software product), a software product, or time. Generally, an actor and the product are depicted using UML to show a visual display of the requirements. How to use the use cases and how much to use them is a debatable topic. However, use cases definitely help us to understand the user requirements.

The approach in creating the use cases may differ based on how formal the use cases should be. Some authors such as Martin Fowler do not believe that there should be any standard way to depict the use cases. On the other extreme, Alistair Cockburn believes that the use cases should be very formal and should stick to the defined formats. Cockburn has suggested two levels at which the use cases can be defined using his templates, one of which he calls "fully dressed." This is a more formal and more rigidly structured way of defining the requirement specifications using the use cases. In the fully dressed type, the use case template includes details such as title, primary actor, stakeholder's interest, goal in context, scope, level, precondition, postcondition, success guarantee, minimal guarantee, trigger, extensions, and other related information. Cockburn also realized that making very formal and elaborate specifications may not be feasible for all projects. Therefore, he also defined another template for the use cases called "casual." In a casual template, there are only a limited number of fields such as title, primary goal, scope, level, and story.

Cockburn's method of using the templates for use cases is widely accepted and used in the software industry. Specifically, his casual method is popular in the software industry. Figure 4.6 presents a use case for a bank account balance check. Let us discuss some fields used in this casual method.

**Title:** This field, in any use case, specifies the title of the use case. The title should be short and should describe what the use case is all about. For example, we can have a title "account balance inquiry" for a use case of a customer of a

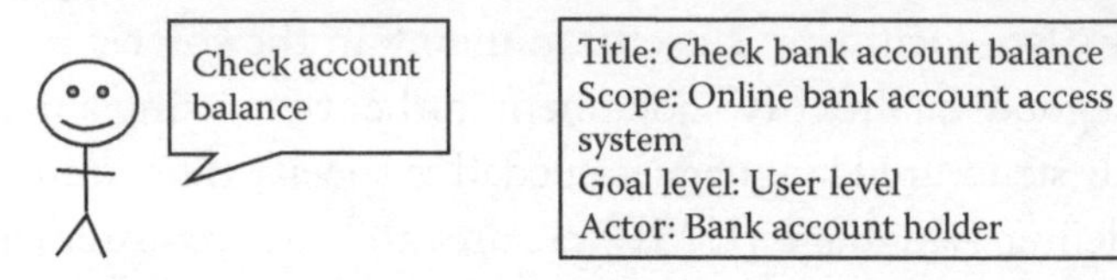


**Figure 4.6** Use case for checking the account balance.

bank who checks his or her bank account to see the balance in the account. In Figure 4.6, the title of the use case is "check bank account balance."

**Scope:** Each use case should be annotated with a symbol (or, alternatively, provided with a text description) to show the design scope of that use case. For example, a software component can be assigned a symbol of a screw. Similarly, a system can be assigned either a black box or a white box symbol depending on whether the system is integrated through some interface without or with the visibility of its source code. In Figure 4.6, we have provided a text description for the scope of a use case. The scope of a use case is important because it determines the size of the software component that will be designed based on that use case. For example, a web page for login is a software component and a corresponding use case may be there for this login web page. If the login web page is a component of a website, the scope for this login web page can be considered at the component level. Again, if the software designers will be creating a design for this component and the software developers will be writing the source code for this login web page, then this login web page becomes an integral part of the software system that is being developed. On the other hand, if this login web page is not part of the software system to be developed, then a suitable symbol or text description will describe this login web page as an external component (white box or black box). When the source code is available for the new component it will be represented as a white box; else, it will be represented as a black box.

**Goal level:** Goal level for a use case denotes the level at which the use case describes the requirement. Cockburn described different goal levels. Goal levels are depicted in Figure 4.7. Goal levels are extremely important because a use case can be at a summary level or at a lower level. Depending on the level of a use case, the corresponding software design can be a component, a class, a subsystem, or the complete software system itself.

**Summary level:** Summary level is the highest-level requirement and it represents the entire process involved. The time span to develop a system to fulfill this use case may take days, weeks, months, or years. For example, if the use case depicts that "a customer can access his or her bank account online," then this use case is at the summary level because this requirement involves many steps that need to be taken by the user as well as by the software system; thus, this requirement can actually be broken down into many subrequirements. In Figure 4.7, the "highest-level requirements" and "high-level requirement" are at the summary level. Summary-level requirements are important because

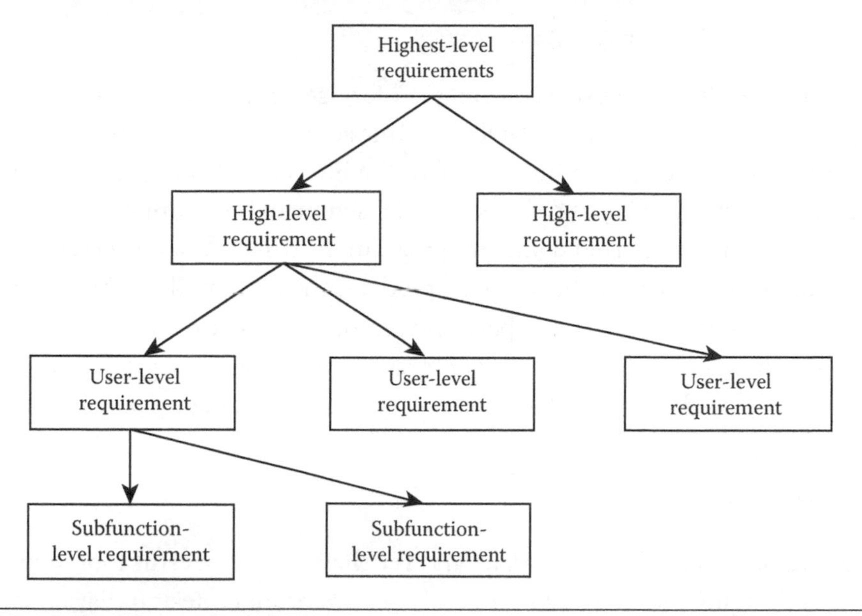


**Figure 4.7** Goal (requirement) levels for the requirements.

they depict the requirements at aggregate levels. In a software design, these summary-level requirements will translate into the software modules as well as the complete software system design itself.

**User level (also called user goal level or sea level):** This use case can result in just one action taken by the user with the system. In other words, the user will try to fulfill this use case in a single sitting. For example, if a use case describes "the customer can log into his or her account online," then it is a user-level use case. In Figure 4.7, the "user-level requirement" is depicted. User-level requirements are important because they depict the requirements at a level at which any user interacts with a software system. In a software design, these user-level requirements will translate into software components and software classes.

**Subgoal level:** Requirements at these levels are also known as subfunction-level requirements. In Figure 4.7, the "subfunction-level requirement" is at the subgoal level. Subgoal-level requirements are important because they depict the requirements at subuser levels. In a software design, these subgoal-level requirements will translate into software subcomponents and software classes.

**Actor:** As explained earlier, an actor in the Cockburn's use case template can be a human user or a system or time. In fact, that system can be another function in the same system or a function residing in some other external system. For example, a customer who wants to check his or her bank account balance online is the actor, whereas the online bank access system is the system with which the actor interacts. The use case here is account balance check. If the online bank access system has to find out the account balance from another program or function indirectly, then the account balance function itself will be another actor.

### *4.10.2 Relevance of Use Cases in Software Design*

Some people think that the use cases are good for user interface designs alone but not for designing the business logic or databases. This assumption comes from the fact that the use cases have an actor who interacts with the interface of the system. However, as explained earlier, this fact is not true because the actor can be a human user, an internal computer program, an external computer program, or time. When we create use cases for the business logic parts of the software product, the actor will not be a human user but a computer program. This computer program, in most cases, is part of the user interface that interacts with a computer program in the business logic. We can see that a use case can be used for designing business logic, a database, or a user interface.

### *4.10.3 Use Case Example*

Use cases are not very difficult to make and yet they are a powerful tool to be used as the input in designing a software product. Software product design diagrams such as a component diagram and a data flow diagram can be easily drawn by using the use cases as inputs. We will learn about software high-level design in Chapter 5.

Cockburn defines the use cases at many levels that can depict the requirement specifications at the aggregate level, the actor level, and even at a lower level than the actor level. These use case specification levels can easily correspond to the software architecture level, the software component level, and the software detailed design level. Hence, employing use cases at different levels will help you design your software product. Of all three levels, the actor (or user)-level use cases are the most useful. This is because, by using the actor-level use cases, you can build the component-level design, which is the foundation for building your high-level software design.

Let us see an example of a use case for a restaurant order management system. We are providing the actor-level use cases here. The corresponding component-level software design for this restaurant order management system is provided in Chapter 5.

This type of order management system is used by restaurants whose inside sales team takes the orders from the customers over the phone or at the sales counter. When a customer phones or arrives at a sales counter, the sales representative will take the customer order by using this order management system. The sales representative will log into the system and then search the menu and find the suitable menu as per the customer requirements. Then, the sales representative will confirm the order. The system will suggest the price the customer needs to pay. The sales representative will then take the money from the customer either in cash or swiping a credit card or entering the credit card details. Once the payment is done, the sales representative will close the order.

After closing an order, customer information, such as the customer's phone number, address, and name, is saved in the database. Next time, when the customer makes an order, the customer information is retrieved from the database.

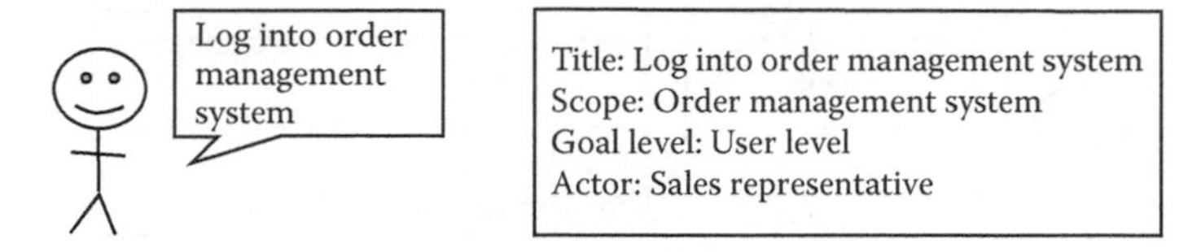


**Figure 4.8** Login to the restaurant order management system by the sales representative.

The first user (or actor)-level use case is provided in Figure 4.8. The user here is the sales representative who will log into the order management system for the restaurant.

Figure 4.9 depicts a use case for the sales representative taking the customer order either over the phone or at the sales counter. The sales representative will choose a food menu for the customer. There are two types of orders: take-out and dine in.

Figure 4.10 depicts the use case for confirming the customer order by the sales representative.

Upon confirmation, the order management system will display the amount to be paid by the customer for the order. The payment can be made as described above. This use case is depicted in Figure 4.11.

Once the payment for the order is completed, the sales representative will logout from the order management system. This is depicted in Figure 4.12.

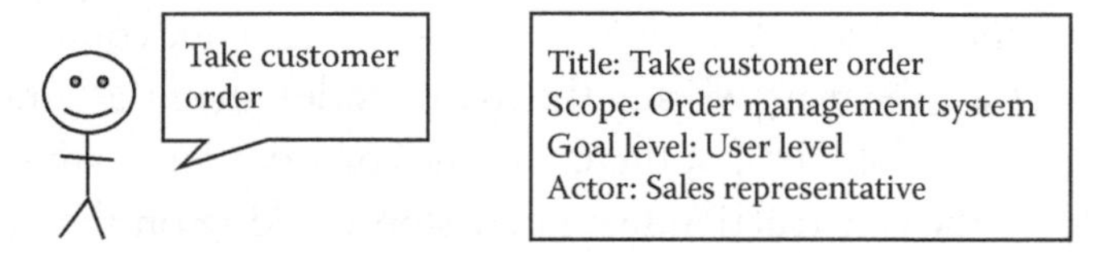


**Figure 4.9** Customer order being taken by the sales representative in the restaurant order management system.

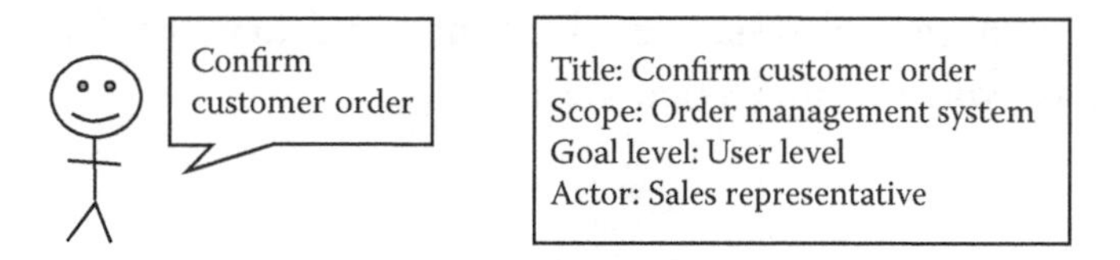


**Figure 4.10** Customer order being confirmed by the sales representative in the restaurant order management system.

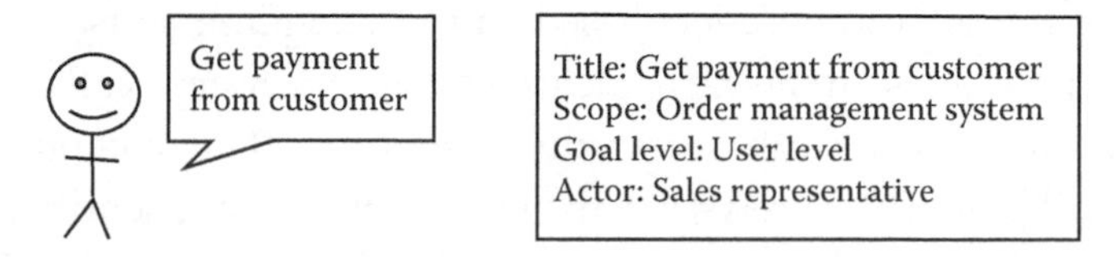


**Figure 4.11** Payment against the customer order being taken by the sales representative in the restaurant order management system.

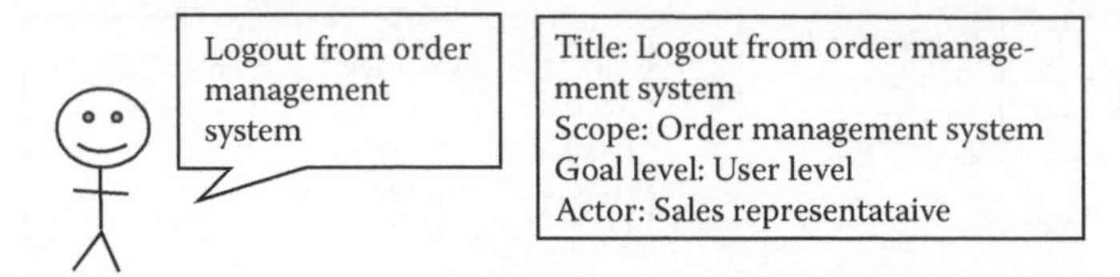


**Figure 4.12** Logout from the restaurant order management system by the sales representative.

As you can observe, a use case can depict vital information on how the proposed software product will function for each interaction of the user with the system. If all the use cases for the proposed software product are captured correctly, then it can become the complete information that is needed to build a design for the proposed software product.

### 4.11 Requirements Management

As explained earlier, if changes are made to the requirements, you need to manage such changes; this process is known as requirement management. Requirement management is very important throughout the software development life cycle. Requirements can be changed or additional requirements may arrive during the development life cycle. If the changed or newly arrived requirements are not managed properly together with the existing requirements, then the project team will have a hard time. For example, if a requirement is changed but is not communicated to the construction team (i.e., the team writing the source code), then the construction team will be working with the old design; therefore, the construction team will be creating the wrong product. The construction team may stop working on the old requirement and start working with the new design (which is based on the new requirement) upon discovery of this communication gap. This kind of rework is always costly and must be avoided.

Requirement management is handled differently in different software engineering methodologies, as explained in Section 4.2.

### 4.12 Case Study

A complete case study consisting of all the requirements for a medium-scale software product is provided. In the succeeding chapters, we will see the design, build, and testing phases of the software product based on these requirements.

The software product is an online bank account access system called OBAAS. The account holders of the bank can access this system on their computers by using a browser. They will point their browser to the Unique Resource Identifier (URI) of this system. Once they log into the system, they will be able to perform various functions provided by OBAAS.

We will be creating two versions of OBAAS for two markets. In one market, a request for new cheque books through OBAAS is free to the bank account holders. However, in the other market, there is a fee of US$5 for each new cheque book request. For this other market, there is also a requirement that allows a user to transfer money from his or her account to another user account if both accounts are with OBAAS.

To take care of this requirement of supporting the two markets with different software product requirements, we will create two separate versions of OBAAS. For the first market, the system will be known as OBAAS 1.1, and for the other market, OBAAS 1.2.

### *4.12.1 Assumptions*

Apart from the proposed system, the bank may have other information systems to perform tasks such as tax calculation, interest calculation, and administration. Most of these systems may be linked to OBAAS at the database level. Other than this possible common database, OBAAS is not linked to any of these other systems. In other words, OBAAS is not linked to any other information system. The system boundary of OBAAS is limited to its own functionality. Thus, each version of OBAAS is a stand-alone software product.

Most banks have sophisticated online systems that allow the customers to access their account and perform tasks such as checking account balance, paying bills, requesting cheque books, transferring money, and setting personal profiles. The online banking systems have strong security features; therefore, it is extremely difficult to hack the accounts of the customers. Unlike real banking systems, OBAAS has a very basic functionality and its goal is to demonstrate different aspects of software engineering. The complete requirement specifications for OBAAS are explained next.

### *4.12.2 Top-Level Requirements for OBAAS 1.1*

OBAAS should have functions for the following:

1. A welcome page that serves as the default page for the website of OBAAS
2. Creating new account
3. Checking account balance for existing users (customers) who have bank accounts
4. Requesting services (cheque book) from the bank
5. Paying bills from user accounts
6. Closing user accounts

All these functions should be available to the users of OBAAS 1.1. On the default page, and then on all other pages, all these functions should be available as a menu. If a user clicks any of the menu buttons, then he or she will be presented with the respective web page.

*4.12.3 Detailed Requirements for OBAAS 1.1*

OBAAS can be used by a bank's existing customers. These customers already have a bank account with the bank. When an existing customer wants to have online access to his or her bank account, then he or she contacts the bank directly by phone, e-mail, or letter and gets his or her username and password from the bank. Using this information (consisting of a username and a password), a customer can create his or her account in the online account access system (provided by OBAAS).

Note that most banks work in this manner. They issue a username and password to their customers when these customers want to have access to their bank accounts online. These customers already have their account numbers issued by the bank. In a real-world scenario, the username, password, and account number are checked when users want to create their online access account. In OBAAS, we have not made any such checks. Implementing such checks will make OBAAS too complicated. Our goal is to make OBAAS simple so that readers can understand the basics of software engineering to create some software products.

After the advent of online accessing of accounts, banks also introduced an article, which is known as customer number in most banks. This customer number can be given to the users (i.e., customers) by the banks or it can be generated by the online access system when the user creates an account. For simplicity, in OBAAS, we are using an account number that is generated when the user creates an online access account.

OBAAS will do the following:

1. When the user points his or her browser to the URL of OBAAS, the application should present a welcome web page containing different menu buttons representing different services of OBAAS.
2. When the user clicks the Create account menu, he or she should be taken to the Create account page. This page should contain a form where there are textboxes for Username, Password, Retype password, Phone number, and Address. There should be another textbox named Category for which the data should be noneditable and it should have a default value U. The textboxes for Password and Retype password should not display the characters and should show only an asterisk for each character entered by the user. The Create account page should have two buttons, Submit and Clear. If the user provides information in the textboxes for Username, Password, and Retype password and the information in the textboxes for Password and Retype password matches, and if the user did not leave any textboxes empty and clicks the Submit button, then the user information should be stored in the database of OBAAS. In this case, the user account should be created and the user is taken to a page containing a message confirming the successful account creation and the account number assigned to the user. If the Clear button is clicked, then the information provided by the user in the textboxes should be cleared and no

information from the Account creation form should be stored in the database. If the information provided in the Password and Retype password fields does not match and the user clicks the Submit button, then a dialog box should appear with the message "Entries in the password and retype password fields do not match. Please ensure password and retype password are the same" and an OK button. When the user clicks the OK button on the dialog box, the system should dismiss this dialog box and the user should be at the Create account page where the user can enter the information again. If a textbox was left empty and the user clicked the Submit button, then a dialog box should inform the user to fill that empty textbox. Since OBAAS is not connected to any real bank account, when the online account is created for a user, there will be no balance in the account. Testing such a system for functions such as bill payment or money transfer will not be possible. Therefore, we will deposit US$500 in the user's account when the account is created in OBAAS.

3. When the user clicks the Account balance menu, the Account balance page appears. On this screen, a form is displayed with textboxes for Account number, Username, and Password. The textbox for Password should not display the characters and should show only an asterisk for each character entered by the user. There are also two buttons, Submit and Clear, on this form. If the user fills these textboxes and clicks the Submit button, then the information submitted by the user should be checked for correctness by matching it with the information in the database, and if it is found to be correct, then the amount should be displayed in the next screen. If the user provides incorrect information and clicks the Submit button, then a dialog box should appear with the message "The username/password you entered is incorrect. Please enter correct entries" and an OK button. When the user clicks the OK button on the dialog box, the system should dismiss this dialog box and the user should be at the Account balance page where he or she can enter the information again. If a textbox was left empty and the user clicked the Submit button, then a dialog box should inform the user to fill that empty textbox. There should be a Clear button on the Account balance page to reset all the text fields at any time.
4. When the user clicks the Service request menu, the Service request page should appear. Currently, using the Service request page, a user can only request a cheque book from the bank. On receiving a cheque book request, the bank will send a new cheque book, through a courier service, to the address that was provided by the user while creating the account online. On this Service request page, a form should be displayed with textboxes for Account number, Username, and Password. There should be a Service request dropdown menu with only the cheque book request item in it. There should also be two buttons, Submit and Clear, on this form. The textbox for Password should not display the characters and should show only an asterisk for each

character entered by the user. If the user fills the textboxes and selects "request a cheque book" from the drop-down list for Service request and clicks the Submit button, then the information submitted by the user should be checked for correctness by matching it with the information in the database, and if it is found to be correct, then a message confirming the cheque book issue should be displayed. If the user provides incorrect information in the checkboxes and clicks the Submit button, then a dialog box should appear with the message "The username/password you entered is incorrect. Please enter correct entries" and an OK button. When the user clicks the OK button on the dialog box, the system should dismiss this dialog box and the user should be at the Service request page where he or she can enter the information again. If a textbox was left empty and the user clicked the Submit button, then a dialog box should inform the user to fill that empty textbox. There should be a Clear button on the Service request page to reset all the text fields at any time.

5. When the user clicks the Bill pay menu, the Bill pay page appears. Currently, using the Bill pay page, a user can only pay the bills for two different service providers (billers). On this Bill pay page, a form should be displayed with textboxes for Account number, Username, Password, and Amount. There should also be a drop-down menu for selecting the Biller. There should be two buttons, Submit and Clear, on this form. The textbox for Password should not display the characters and should show only an asterisk for each character entered by the user. If the user fills the textboxes and selects a biller from the drop-down list and clicks the Submit button, then the information submitted by the user should be checked for correctness by matching it with the information in the database, and if it is found to be correct, then that amount should be transferred to the account of the Biller and the same amount should be subtracted from the account of the user. A screen confirming the bill payment should be displayed to the user. The system will also check in the database if the account of the user has a high enough balance to pay the bill. For example, if the account balance of the user account is US$400 and the bill amount is more than US$400, then the system should not perform this transaction. The message "Not sufficient funds in the account. This transaction cannot be performed" should be displayed to the user. The user can dismiss this message box by clicking on its OK button. After dismissing the message box, the bill payment page will again be displayed. The user can pay another bill of any amount equal to or less than US$400. If the user provides incorrect information and clicks the Submit button, then a dialog box should appear with the message "The username/password you entered is incorrect. Please enter correct entries" and an OK button. When the user clicks the OK button on the dialog box, the system should dismiss this dialog box and the user should be at the Bill pay page where he or she can enter the information again. If a textbox was left empty and the user clicked the Submit button, then a dialog

box should inform the user to fill that empty textbox. There should be a Clear button on the Bill pay page to reset all the text fields at any time.

6. When the user clicks the Close account menu, the Close account page should appear. On this screen, a form should be displayed with textboxes for Account number, Username, and Password. There should be two buttons, Submit and Clear, on this form. The textbox for Password should not display the characters and should show only an asterisk for each character entered by the user. If the user fills these textboxes and clicks the Submit button, then the information submitted by the user should be checked for correctness by matching it with the information in the database, and if it is found to be correct, then the account closure page should be displayed to the user. If the user provides incorrect information and clicks the Submit button, then a dialog box should appear with the message "The username/password you entered is incorrect. Please enter correct entries" and an OK button. When the user clicks the OK button on the dialog box, the system should dismiss this dialog box and the user should be at the Close account page where he or she can enter the information again. If a textbox was left empty and the user clicked the Submit button, then a dialog box should inform the user to fill that empty textbox. There should be a Clear button on the Close account page to reset all the text fields at any time.

The use cases for OBAAS 1.1 are provided next.

#### *4.12.4 Use Cases for OBAAS 1.1*

Here are the use cases for OBAAS 1.1 described in Section 4.12.3.

The use case in Figure 4.13 sets the boundary of the system.
The use case in Figure 4.14 is used to represent Item 2 of Section 4.12.3.
The use case in Figure 4.15 is used to represent Item 3 of Section 4.12.3.
The use case in Figure 4.16 is used to represent Item 4 of Section 4.12.3.
The use case in Figure 4.17 is used to represent Item 5 of Section 4.12.3.
The use case in Figure 4.18 is used to represent Item 6 of Section 4.12.3.

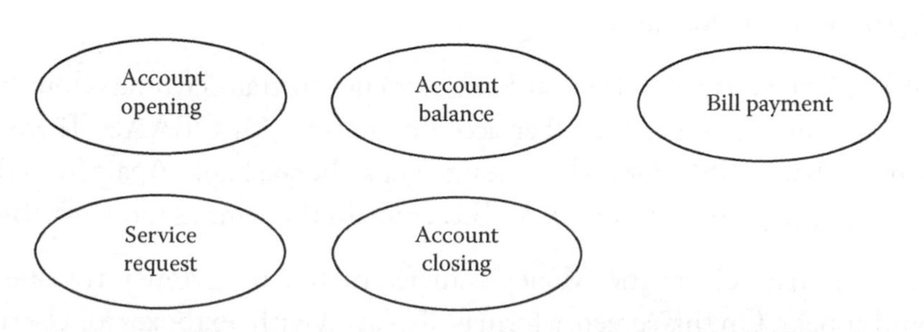


**Figure 4.13** System boundary for OBAAS 1.1.

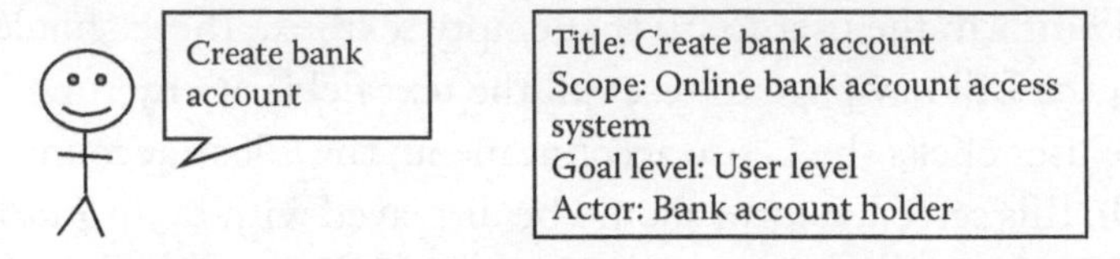


**Figure 4.14** Use case for creating a bank account by a user.

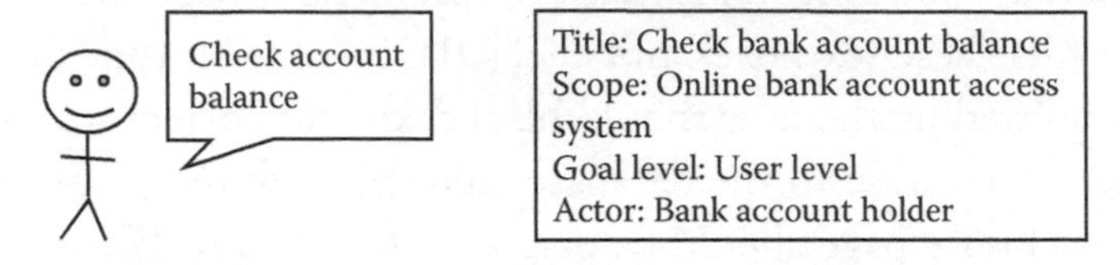


**Figure 4.15** Use case for checking the account balance by a user.

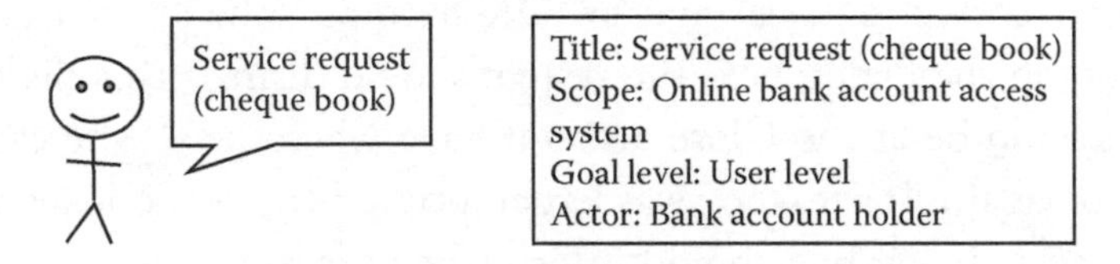


**Figure 4.16** Use case for service request (cheque book) by a user.

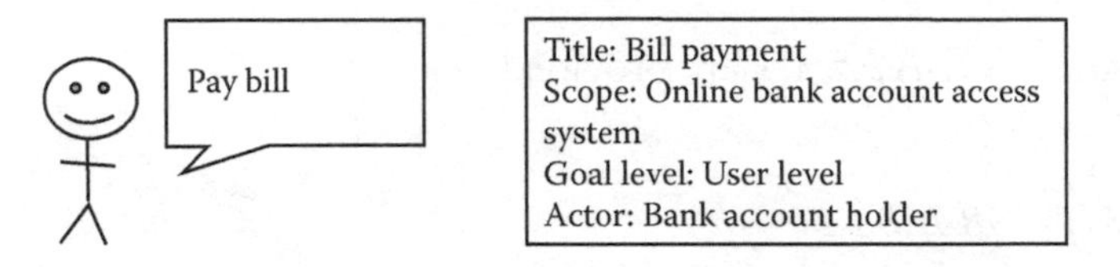


**Figure 4.17** Use case for bill payment by a user.

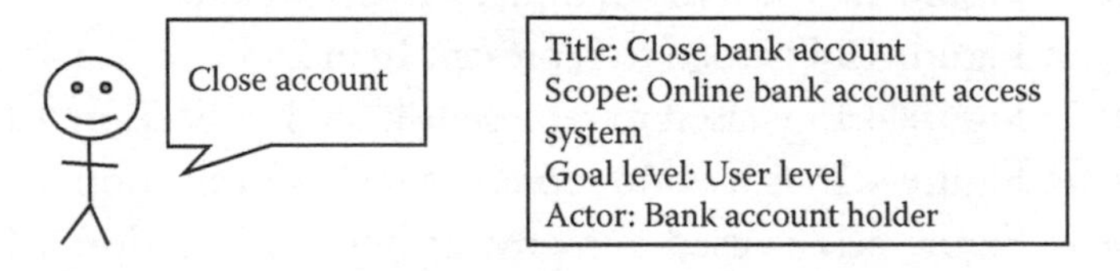


**Figure 4.18** Use case for closing the account by a user.

*4.12.5 Requirements for OBAAS 1.2*

In OBAAS 1.2, there is a requirement for a customer to transfer funds from his or her account to another account if the other account is also with OBAAS. There is also a requirement to charge US$5 for each request to get a cheque book. Apart from these two changes, all other requirements for OBAAS 1.2 remain the same as those of OBAAS 1.1.

1. When the user clicks the Money transfer menu, the Money transfer page should appear. On this screen, a form is displayed with textboxes for Username, Password, From account (the user account number), Amount, and To account

(the account number of the party to which the money will be transferred) fields. There are also two buttons, Submit and Clear, on this form. The textbox for Password should not display the characters and should show only an asterisk for each character entered by the user. If the user fills the textboxes and clicks the Submit button, then the information submitted by the user should be checked for correctness by matching it with the information in the database. If the account balance of the user, before this transaction, is higher than or equal to the amount that needs to be transferred, then the system should proceed with the transaction. When the transaction is successful, the system should display a message containing the amount that was transferred and the account number to which the money was transferred. If the account balance is less than the amount to be transferred, then an error message should be displayed on a dialog box informing the user of the insufficient funds. If the user provides incorrect information and clicks the Submit button, then a dialog box should appear with the message "The username/password you entered is incorrect. Please enter correct entries" and an OK button. When the user clicks the OK button on the dialog box, the system should dismiss this dialog box and the user should be at the Money transfer page where he or she can enter the information again. If a textbox was left empty and the user clicked the Submit button, then a dialog box should inform the user to fill that empty textbox. There should be a Clear button on the Money transfer page to reset all the text fields at any time.

2. When the user clicks the Service request menu, the Service request page should appear. Currently, using the Service request page, a user can only request a cheque book from the bank. On receiving a cheque book request, the bank will send a new cheque book, through a courier service, to the address that was provided by the user while creating the account online. On this screen, a form should be displayed with textboxes for Account number, Username, and Password. There is also a Service request drop-down menu. There are also two buttons, Submit and Clear, on this form. The textbox for Password should not display the characters and should show only an asterisk for each character entered by the user. If the user fills the textboxes and selects "request a cheque book" from the drop-down list for Service request and clicks the Submit button, then the information submitted by the user should be checked for correctness by matching it with the information in the database, and if it is found to be correct, then a message confirming the cheque book issue should be displayed. If the user provides incorrect information and clicks the Submit button, then a dialog box should appear with the message "The username/password you entered is incorrect. Please enter correct entries" and an OK button. When the user clicks the OK button on the dialog box, the system should dismiss this dialog box and the user should be at the Service request page where he or she can enter

the information again. If a textbox was left empty and the user clicked the Submit button, then a dialog box should inform the user to fill that empty textbox. There should be a Clear button on the Money transfer page to reset all the text fields at any time. In OBAAS 1.2, there is a service fee of US$5 for each cheque book request. This fee should be deducted from the bank account of the user. If the user's bank account balance is below US$5, then the service request for a cheque book should not be entertained. A dialog box with a message "insufficient funds" should be displayed to the user in such case.

### *4.12.6 Use Cases for OBAAS 1.2*

Here are the use cases for OBAAS 1.2 described in Section 4.12.5.

Figure 4.19 represents the system boundary for OBAAS 1.2.

The use case in Figure 4.20 is used to represent Item 1 of Section 4.12.5.

The use case in Figure 4.21 is used to represent Item 2 of Section 4.12.5.

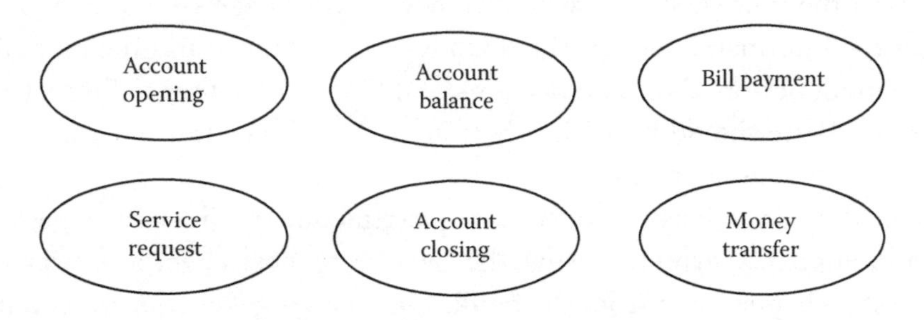


**Figure 4.19** System boundary for OBAAS 1.2.

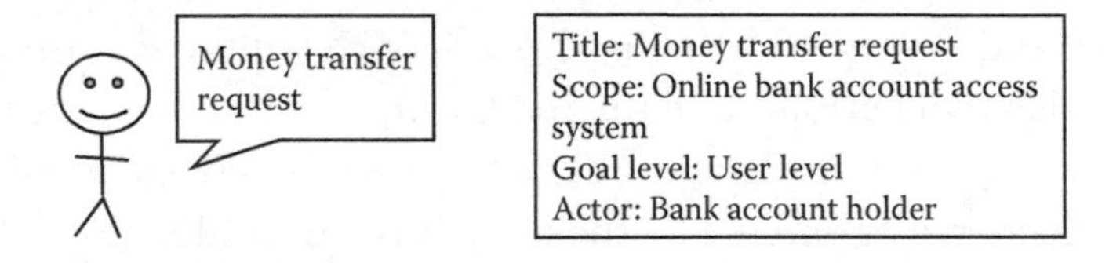


**Figure 4.20** Use case for money transfer by a user.

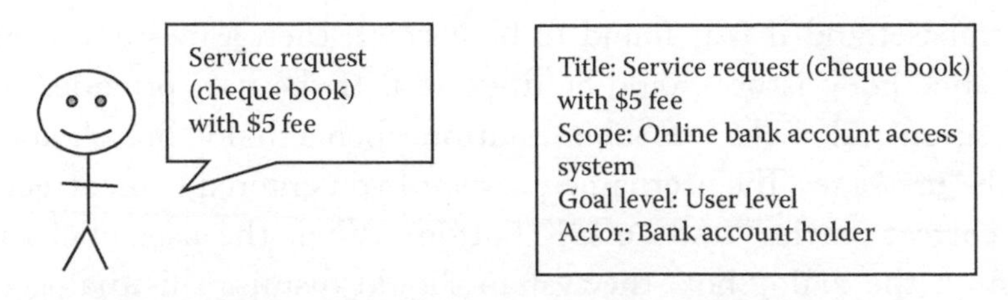


**Figure 4.21** Use case for service request (cheque book) by a user.

#### *4.12.7 Future Enhancements for OBAAS*

The product features of OBAAS can be extended to meet many customer requirements as well as to provide product features for the internal management of the bank through OBAAS. Some of these features include the following:

1. Provide a facility for customers to change their password. Generally, it is a good idea to keep changing passwords to help secure access to the account. In fact, some banks require their customers to change their passwords after a fixed period.
2. Provide a facility to transfer funds to other banks. This kind of functionality requires the use of web services.
3. Introduction of services: Currently, OBAAS offers a new cheque book issue service to the customers. A bank generally offers many services such as opening or closing fixed deposits (also known as certificate of deposits in North America) using online banking, mobile banking, linking of bank accounts, and so on. All these kinds of bank services can be implemented by developing new product features in OBAAS.
4. Remittances from other sources of funds: Many banks offer remittance services where it is possible to transfer funds using sources such as Western Union and money transfer agencies into a bank account.
5. Currency conversion for fund transfer: If a fund transfer is required from a foreign location, then the currency of the funds needs to be converted into the local currency before fund transfer can be done. In some countries, it is also possible to keep the funds in a bank account in multiple currencies. In such cases, the bank account itself needs to manage all the funds in multiple currencies.
6. Security: Presently, OBAAS offers minimal security features in the form of username/password authentication. OBAAS displays asterisks when the user enters some characters in the password field. Security features can be enhanced by utilizing the security socket layer for the entire website so that the communications and data exchange between the web server and any web browser is totally secure.

## 4.13 Chapter Summary

Software requirements are the basis on which a software product is built. The project team elicits (gathers) the requirements from the users by using e-mails, meetings, and so on. Users can specify the functional and nonfunctional requirements. When implemented, functional requirements provide features (to the software product) the users can use to do their tasks. Nonfunctional requirements are those that specify the performance levels, security levels, and so on of the software product. The requirements (either functional or nonfunctional) can also come from the existing legacy systems that need to be replaced with the new (i.e., proposed) system.

Software requirements are analyzed for inconsistencies, internal relationships, ambiguities, and so on, after they are gathered. Once the software requirements are analyzed and the corrections are made, then the requirement specifications are made. Software requirement specifications can be in any form depending on the needs of the project. If the project needs very formal requirement specifications, then techniques such as use cases are used. If the requirement specifications need not be very formal, then they can be specified in a document using a natural language that everyone in the project can easily understand.

Software requirements can change if the customer requires any changes to the existing requirements at a later date. Changes to the requests can be accommodated if the software design is either not yet started or completed (or being completed) but the changes to the requirements will affect the design in a minimal way. Otherwise, a lot of effort may be needed to redo the work based on the changed requirements.

Different software engineering methodologies deal with requirement management in different ways. In the case of XP, there is no provision for change requests (for the features that are already implemented). This is because only three or four requirements are taken to build a partial software product during each iteration in XP. This partial product is integrated with the main build of the software product; thus, incremental building of the software product takes place. Once these software features are built in an iteration, the succeeding requirements will be considered.

In the case of Scrum, the project team keeps all the requirements in a system that is known as product backlog. The Scrum team, in fact, uses two systems for keeping the requirements. The main system, where all the requirements are kept, is known as product backlog. The other system is known as sprint backlog, where only those requirements that are being used to build the software product in the current iteration (sprint) are kept. If any requirement is changed, then the product backlog or the design needs to be changed accordingly.

In Scrum, like in XP, only three or four requirements are taken and a partial software product is built according to those requirements. This partial product is integrated with the main build of the software product; thus, incremental building of the software product takes place.

In the Waterfall model, all the requirements are gathered and then the requirement specifications are made. The software design phase starts when all the requirements are specified and then handed over to the design team. If any of the requirements are changed after the project team has received them, then the project team may have to do rework if the design and source code are developed (or being developed).

UML is a useful tool to depict the software requirement specifications in a concise and standard manner. Use cases are a type of UML and they are used to describe the software requirement specifications. Use cases consist of actors that interact with the software system. The requirement specification itself can be set at a summary level, user level, or subuser level based on whether it corresponds to a software component, a complete software system, or something in between.

## QUESTIONS

1. What is a software requirement?
2. What are the different types of software requirements?
3. What is software requirement management?
4. How are software requirements managed in various software engineering methodologies?
5. What is a use case?
6. How are software requirement specifications made?
7. At what level can a use case be set?

## Recommended Reading

Alistair Cockburn (2000), *Writing Effective Use Cases*, 1st Edition, Addison-Wesley, Professional, Indianapolis, IN.

Alan Mark Davis (2005), *Just Enough Requirement Management: Where Software Development Meets Marketing*, Dorset House Publishing, New York.

Elizabeth Larson, Richard Larson (2013), *Practitioner's Guide to Requirements Management*, 2nd Edition, Watermark Learning, Minneapolis, MN.

Dean Leffingwell, Don Widrig (2003), *Managing Software Requirements: A Use Case Approach*, 2nd Edition, Addison-Wesley Professional, Indianapolis, IN.

Dean Leffingwell (2011), *Agile Software Requirements: Lean Requirements Practices for Teams, Programs and the Enterprise*, 1st Edition, Addison-Wesley Professional, Indianapolis, IN.

# 5

# SOFTWARE HIGH-LEVEL DESIGN AND MODELING

**In Chapter 4, we learned**

- **What requirement management is**
- **What requirement gathering is**
- **What requirement specification is**
- **What a use case is**
- **How software design can be made from requirement specification**

**In Chapter 5, we will learn**

- **What a software high-level design is**
- **What a software architecture pattern is**
- **What software architecture is**
- **What a software design pattern is**
- **What a software component design is**
- **What high-level software design technological considerations are**

## 5.1 Introduction

You have gone through a feasibility study for the proposed product. You gathered the requirements and prepared the requirement specifications for that product. It is now time to think about the implementation details of that product. The implementation details should provide the procedure on creating a software product design that will fulfill the functional and nonfunctional requirements of the product. Apart from keeping an eye on the implementation of these requirements, you also have to think about the technology that will allow you to implement the requirements in a cost-effective and quick manner.

The implementation should first consider the system model (or software model) to be developed. Here, you should think about the system structure of your system model. The system structure will include designing the components in such a way that they incorporate the product features, which in turn will fulfill the requirement specifications. Next, your system model should include the data flow in the system when a user interacts with it. When a component of the system interacts with another component, how the data will flow also comes under the purview of data flow in the system.

Another consideration during the design is the use of templates that will help build a large number of similar components from a single template. This area is known as design pattern. The idea of using design patterns is to reuse already existing expertise in terms of well-defined templates instead of designing the components from scratch. Design pattern is an important consideration for building software products. We will use design patterns extensively in this book.

The software model can be thought of at three levels. At the top, you can think of breaking down the model horizontally. For example, you can think of the software model as consisting of a user interface tier, a middle tier, a back-end tier, and so on. The tiered structure is a kind of pattern known as architecture pattern.

Each horizontal tier can be further divided into many parts vertically. At this level, you will design the components inside each of the layers (tiers) you created at the higher level earlier. This level of software design is known as software component design. Division of the software design vertically can be based on some pattern. This pattern is known as component design pattern.

At the bottom level, the software design is concerned with creating the classes and their class members including methods and class variables. This level is known as software detailed design. We will learn about software detailed design in Chapter 7.

If you look at the technologies available for implementing the requirements, you will also find that many do not work well with each other. You will need to choose the right technologies for your databases, business logic, and user interfaces. You need to make sure that the technologies you choose for these parts work well with each other.

You also know that software designs tend to be complex. It is not an exaggeration to state that software designs always tend to be complex. Designing a software product is always a challenge for even experienced software designers. If you are designing a software product, you need to address the following two fundamental challenges:

- Even when you have very detailed software requirement specifications, you need a lot of imagination and creativity to think of a corresponding software design. For instance, although the requirement may state that the software product will be a web-based application and the requirement has provided detailed information on all the functions and features of the product, a lot of thinking is still needed on your part on how all those functions and features can be designed.
- Since the design will always be complex, you need to provide mechanisms to make it less complex. This is because when software developers start writing the source code, they find it difficult to implement a complex design. If they get a simpler design, writing the source code becomes a lot easier and better. A simpler design also results in fewer software bugs (or defects) creeping into the software source code.

For these reasons, it is important to plan how to implement the requirements into a good design and then make the design as simple as possible.

In software projects involving agile methodologies, a complete software design is not required at the beginning of the software project, but even with agile methodologies, where the software product will be built incrementally, you need to provide a high-level software design. This is because you need to provide a framework and a template for the entire product. Without this template, it is difficult to build large products.

Because of the need to discuss software design at many levels, this chapter is divided into three parts:

- Software architecture: Software design at the highest level, for example, a tiered architecture that includes a user interface, a middle tier, and a back end. We will cover software components in Sections 5.4 through 5.8.
- Software component design: Software design at the component level, for example, the components that make the middle tier. We will cover software architecture in Sections 5.9 through 5.12.
- Technologies and programming languages: Available technologies and selection of programming language affect the overall software design. It is important to learn about the considerations related to the available technologies and programming languages so that we can create a good design for the software product. We will cover software architecture in Sections 5.13 through 5.16.

**Usage notes:** In this chapter and many other places in this book, we have used the terms *software tiers* and *software layers* interchangeably. Some people like to differentiate between these two but, for our purpose, we will use them interchangeably. A software tier or a software layer is a separation (or division) of software architecture into logical parts so that the entire software product can be divided into manageable parts.

We have used the term *software architecture* to denote a software design at the highest level. For example, a software design may consist of a user interface, a middle layer, and a database layer, when we present the software architecture. When we use the term *software component design*, we basically mean the software design at a level lower than the software architecture. As explained earlier, a software product can be designed using software components. If the architecture of a software product consists of different layers (user interface, business logic, etc.), then each of these layers can be broken down into software components. A software design at the software components level will thus provide a lower-level detail (about the software product) than the software design at the software architecture level.

There are essentially three levels at which a software product is designed. We introduced software architecture design and software component design in an earlier paragraph. The third level is below the software component design and it is known as software detailed design. Software detailed design is done to build the classes and objects from the software components. As said earlier, we will learn about software detailed design in Chapter 7.

#### *5.1.1 Modeling Languages*

When we model a software product, we use a modeling language. Many researchers have developed modeling languages for different kinds of software products. Some of these modeling languages include Business Process Modeling Notation, Express, and Jackson Structured Programming. Unified Modeling Language (UML) is the most commonly used. We will use this language to model the software products.

### 5.2 Methodology Used

Recall that we discussed software development methodologies in Chapter 2. The software engineering methodology that is used to develop a software product will influence the decision about the technologies to be used to develop the product. Some software products need to be developed incrementally. In those cases, only some technologies are well suited.

#### *5.2.1 Rational Unified Process*

In methodologies such as the Rational Unified Process (RUP), during the software development process, the requirements go through evolution and thus the design may need to be changed several times. With some technologies, it is not easy to change the design, but it is easy with other technologies. For example, object-oriented design (OOD) allows making changes more easily than any other technology currently available. Therefore, object-oriented technology is a good fit when a project is based on RUP.

#### *5.2.2 Incremental Iterative Methodologies*

OOD is good for incremental product development as well. In OOD, it is possible not to have a complete product design initially. A partial product design can be prepared initially for projects where incremental methodologies such as XP or Scrum are used. When the size of the product increases through the incremental addition of new features, some of the classes defined initially may not support the new features. However, after refactoring, it is possible to integrate the new features. Refactoring was briefly explained in Chapter 2. We will learn more about refactoring and OOD in Chapter 7.

### *5.2.3 Waterfall Methodology*

In the Waterfall model, the design should have provisions for incorporating the changes in the future. At the same time, a complete design of the proposed product should be created at the very beginning of the project. Almost all types of programming and designing methods work well with projects that are based on the Waterfall methodology. A primary requirement is that the time required to design the software product should be known in advance. This is because the projects that are based on the Waterfall model require a clear visibility (complete project plan including the timelines for doing the entire project work) and the development company has to give the cost and effort estimates to the customer at the beginning of the project. A complete software design helps determine how much time it will take to implement (construct) it. This up-front information is vital to provide visibility of the entire project to be done. Thus, the entire project can be planned with much accuracy.

## 5.3 How to Reduce Complexity in Software Design

Complexity in designing software makes the software development process difficult. Reducing the complexity is the biggest challenge in software design. How is this done? The best way to reduce complexity is by dividing the entire design into manageable parts. This can be done at two different levels:

- At the top level, the design can be broken down into layers such as a user interface layer, a business logic layer, and a database layer using any of the software architecture patterns.
- At each of these layers, again, the design can be broken down into smaller manageable modular parts using any of the software design patterns. Modularity is a significant tool to use here. One of the best ways to design a software product is to design the software parts into modular components and hide most of the implementation details (of the components) and expose only the details that are needed to integrate the components with each other or with some other external systems.

Figure 5.1 depicts the user interface layer. The user interface, again, can be broken down into many parts such as user interface components (windows, textboxes, dropdown boxes, etc.), client-side scripts, and client-side validations. We will learn about the user interface layer in Chapter 6.

Figure 5.2 depicts a middle layer. The middle layer itself can be broken down into many components such as Component 1, Component 2, and so on. These components in turn are broken down into many parts. In object-oriented programming, these parts correspond to classes. Server-side scripts are also part of the middle layer. We will learn about the middle layer in Chapter 7.

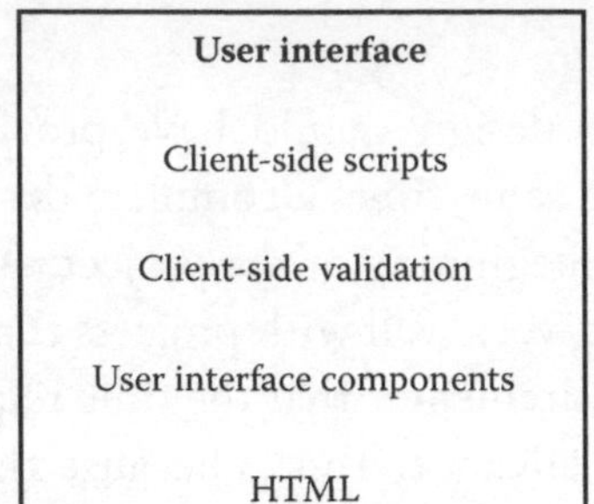


**Figure 5.1** User interface layer.

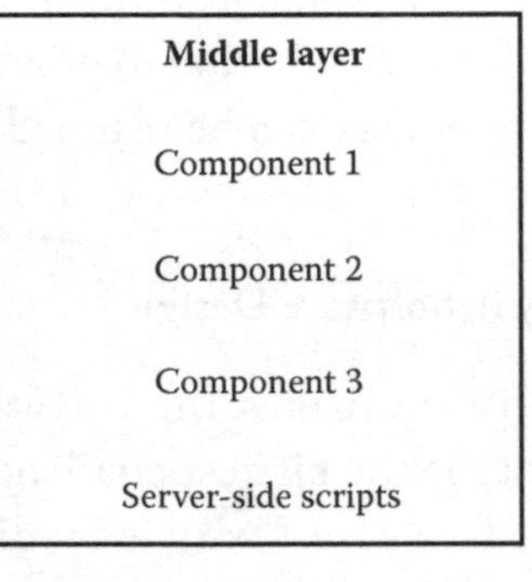


**Figure 5.2** Middle layer.

Figure 5.3 depicts a database layer. A database layer itself can be divided into schemas, tables, columns, and many other database entities. We will learn about the database layer in Chapter 8.

The first move in reducing complexity is the division of the software product into layers. This division also increases the productivity of the product development process. The software industry has achieved maturity in designing software products by dividing the product into layers. The most common layers are in the form of user interface, business logic, and database. In some cases, additional layers such as the presentation layer

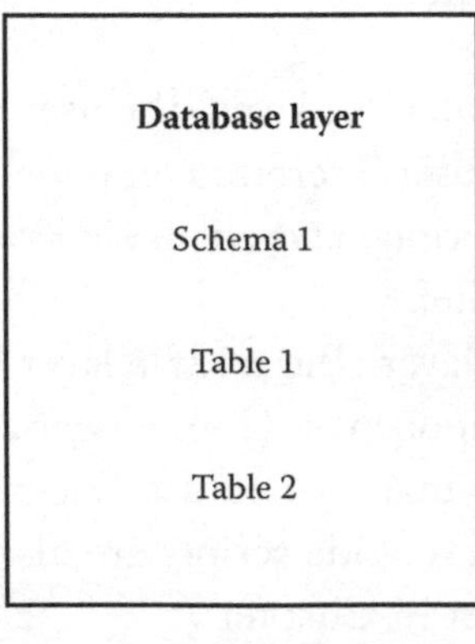


**Figure 5.3** Database layer.

can be included. Layering also promotes expertise in each specific area. For instance, user interface designers are experts in designing the interfaces. The sophistication in user interface design has been increasing because of this factor. Similarly, we have programmers to write the source code for implementing the business logic. Creating the databases and managing the data are also very specialized areas. All of these areas gained immensely because of partitioning of the software design into layers.

A user interface can be based on a simple model–view architecture. The inputs provided by the user are processed by the middle layer and the output can then be displayed to the user on the user interface. In case the output on the user interface needs to be controlled on the basis of some criteria, you can implement a controller by using technologies such as the Model–View–Controller. We discuss these technologies in Chapter 6.

The middle layer or the business logic layer is an area where you need to think about how well to incorporate the modular design. For example, if you are using object-oriented programming, then each piece of business logic can be implemented using one or more classes. Depending on the granularity level you want, you can decide how many classes are needed and how large each class will be. Generally, small classes are better because they lead to better software design. Small classes are also easy to maintain. We will learn about the middle tier in Chapter 7.

A lot of thinking about how to design the database layer is required. If you are working with relational databases, then you know that you need to come up with a good entity-relationship diagram to design the core of the database. Another area that needs a lot of attention is the normalization of the database tables so that the data stored in the tables are atomic, nonredundant, and easily searchable. We will learn about the database tier in Chapter 8.

## 5.4 Logical Design for Software Architecture

When we speak of layers and high-level designs, we also need to know how the software layers or components are physically separated. You can build a software component that may be installed on one computer and you can plug it (e.g., by using web services) with other software components that may be installed on some other computers. It is perfectly possible to have the user interface, business logic, and database on the same computer or they may reside on different computers. Depending on needs such as load balancing, performance, and user preferences, you can have any number of physically separated layers or components. If different components of a software system are not on the same computer, then we can present such a software system differently in the logical design.

Figure 5.4 depicts a typical three-tier software architecture. You can see the logical separation of the layers into user interface, business logic, and database. We will learn about multitier software architecture (three-tier is a type of multitier software architecture pattern) in Section 5.5.

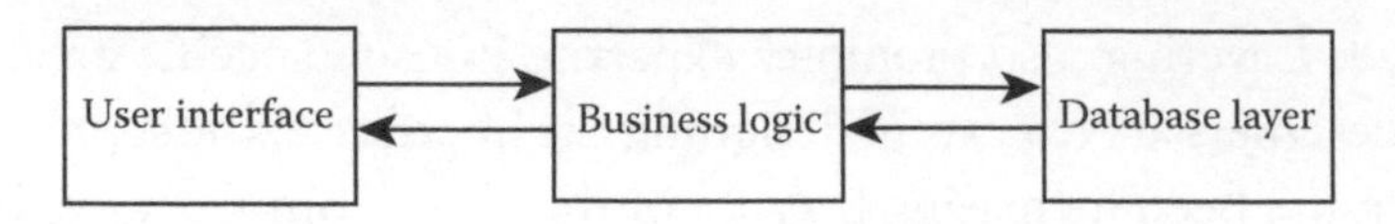


**Figure 5.4** Typical three-tier software architecture presented with logical separation of user interface, business logic, and database.

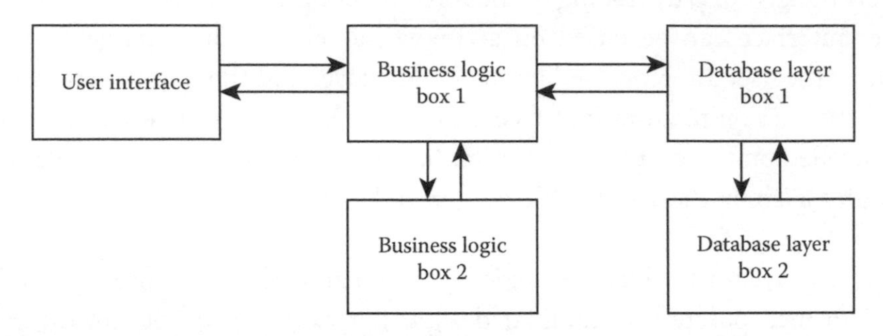


**Figure 5.5** Software product with actual physical layer separation.

Figure 5.5 depicts a three-tier software architecture with a physical depiction of computer servers and the software architecture layers on top of these servers. It is the same software architecture depicted in Figure 5.4 but with physical systems shown as well.

There are many benefits of depicting the design of a software product in a logical manner. The logical design simplifies the software design if the software product needs to be deployed over many physically separated computers. A logical design depicts the entire design in a simplified way by hiding the physical locations of different components but completely captures the essential ingredients of the design.

## 5.5 Architecture Design Patterns

Some software products are monolithic and have no layers. For example, some simulation programs or gaming programs may not use a database. The business logic may also be minimal; therefore, there is no need to have a separation between the user interface and the business logic. All the source code for the user interfaces and business logic will be written in the same space. Thus, this kind of software product is monolithic in nature. Many embedded software products belong to this category. However, monolithic software design has many limitations. The most obvious one is that this kind of software architecture is useful for very small software products. The monolithic architecture is not suitable for mid-sized or large software products.

The reality is that many software products are very large. A monolithic software design cannot be used for such software products. These kinds of software products need a software design that is scalable. A scalable software product design must have separation of concerns so that each concern can be scaled independent of other concerns.

Different software products need different kinds of software architecture.

We have many software products that are large in nature and thus have a software architecture consisting of many layers. Each layer is completely separate from the other layers. The layers in most commercial systems are the user interface layer, the business logic layer, and the database layer. This three-layered architecture is thus very common. Sometimes, a software product needs to be integrated with other software products. In such cases, one more layer is added on top of these three layers.

When we divide a software architecture into layers, we are essentially following one of the software architecture patterns. A software architecture pattern is a well-defined template to break the design of a software product into many parts. For example, when we have to build a web-based software product, there is an architecture pattern that we can follow. This architecture pattern states that there should be a web-based user interface in the form of a web browser, the middle layer is put inside an application server, and the back end could be any database.

Similarly, when we need to build a software product that can be divided into three parts, we design it using a three-tier architecture template. We also have architecture patterns for two-tier, *n*-tier, and so on.

Some other architecture patterns include event-driven architecture, rule-based architecture, and pipes and filters architecture patterns, to name a few. In this book, we concentrate only on the multitier architecture pattern because it is the most widely used architecture pattern for building software products. Service-oriented architecture is also described in this chapter, but in the succeeding chapters, we will build examples and case studies based on the multitier architecture pattern.

### *5.5.1 Two-Tier Architecture Pattern*

Even if your software product needs to have a user interface, business logic, and a database, you can still have a two-tier software architecture pattern. In this case, your business logic and the user interface will be on the same logical layer and the database will be on a separate layer. Small software products can be built using two-tier architecture.

As you can see from Figures 5.6 and 5.7, there is a big difference between a two-tier client–server architecture and a two-tier web-based architecture. In web-based architecture, the user accesses the application on his or her browser. The application itself runs on a server that is a combination of a web server and an application server. The application may consist of static and dynamic content. The static content is a Hyper Text Markup Language (HTML) code and the dynamic content comes from programs written in any programming language. The application server runs the dynamic business logic and converts it into HTML content. While

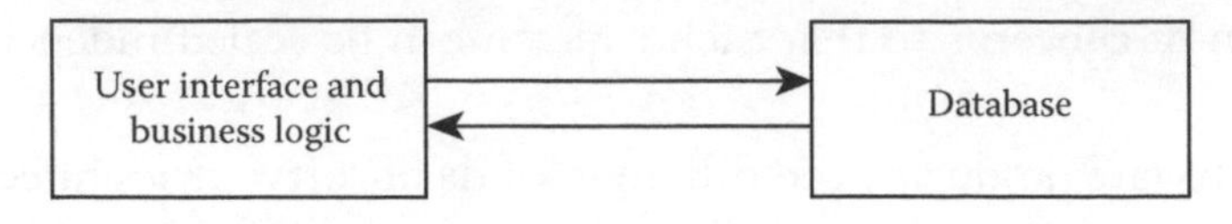


**Figure 5.6** Two-tier architecture pattern for client–server software products.

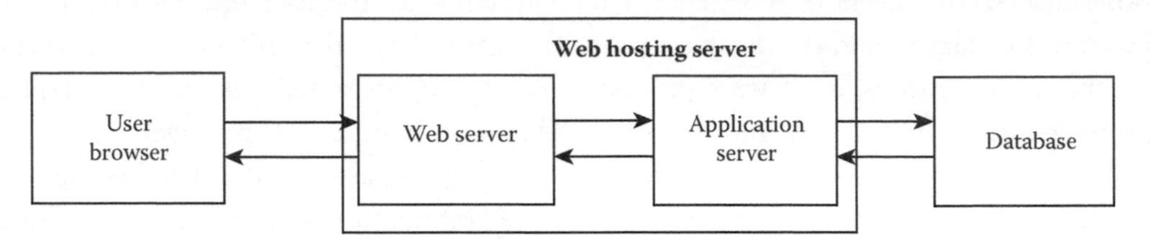


**Figure 5.7** Two-tier architecture pattern for web-based software products.

running, the business logic code may also connect with the database to manipulate some data. This converted static content is then passed to the web server. The web server then combines all the static content (residing directly on the web server and the content coming from the application server) and this static content can then be displayed on a web browser. We will learn more about web-based architecture in a later section.

### *5.5.2 Three-Tier Architecture Pattern*

If your software product needs to have a user interface, business logic, and a database as separate layers, then you need a three-tier software architecture pattern. A three-tier architecture pattern can be both a web-based and a non-web-based (client–server) architecture pattern. A three-tier client–server architecture pattern is shown in Figure 5.8 and a three-tier web-based architecture pattern is shown in Figure 5.9.

The three-tier architecture pattern is the most common type of software product architecture. If you want your product to be deployed as a website, then you need a

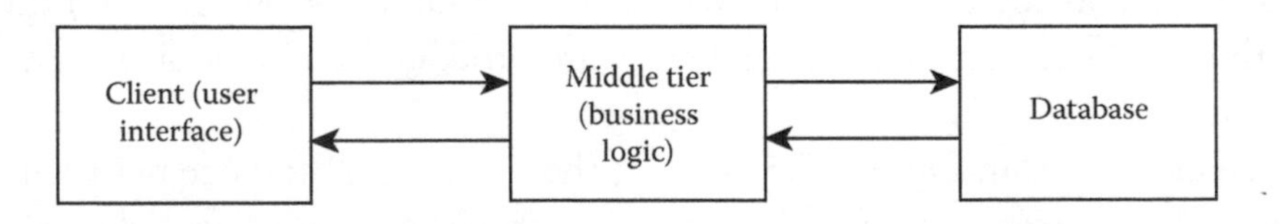


**Figure 5.8** Three-tier client–server architecture pattern.

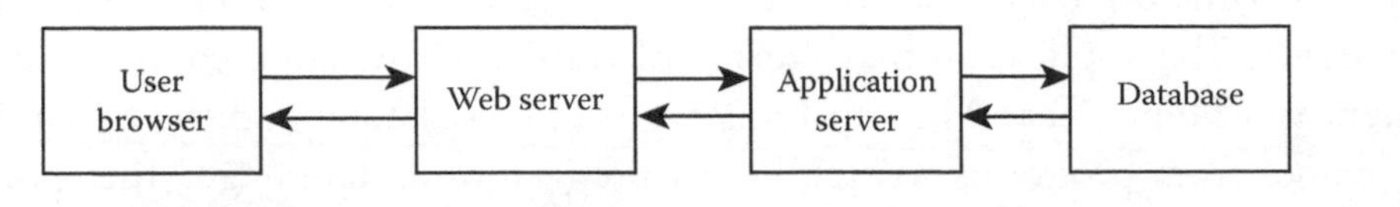


**Figure 5.9** Three-tier web-based architecture pattern.

web server, an application server, and a database in it (in addition to your local computer and a web browser installed on it). The web server hosts the web pages (of your product) in HTML format and contains static information. Static information does not change unless the web developer updates the HTML content of the web pages hosted on the web server. The application server hosts the business logic (dynamic content). The database will host your permanent data.

As explained in Section 5.4, the physical arrangement of various layers can be dependent on the hardware configuration of various computer servers. For example, it is possible that the web server, the application server, and the database can all be deployed on the same computer server or they can be deployed on different computer servers.

When a user points a Universal Resource Locator (URL) to the website (of your product) on his or her web browser, it will access the web server. If the user clicks any link or button on the web pages and if this clicking activity triggers some business logic to run, then the web server passes the command (which represents the clicking activity) to the application server. The business logic, which is in the form of source code, processes the command. While processing the command, if the database needs to be accessed (to retrieve some data from the database or to modify some data in the database), then the command will connect to the database. After getting or manipulating the data in the database, the command will pass the data to the business logic on the application server. The application server will again process the data and then pass the processed data to the web server. The web server in turn will send these data to the web browser. The web browser will display those data on the screen of the user computer.

There are many benefits with this kind of architecture. The presentation layer or user interface (in the form of web pages) is completely separated from the business logic layer. Similarly, the business logic (the program code that runs on the application server and provides the dynamic content) is completely separated from the database. The software development can be done in parallel for all three layers and it results in faster development of the software product. This separation also helps specialized people to develop each of these layers.

### *5.5.3* n-*Tier Architecture Pattern*

Sometimes, the software application you are developing is not a full product and it needs to be integrated with another software product. In such cases, the resultant architecture will have many layers depending on how many software applications are integrated with that software product. The integration can be done using any suitable technologies such as CORBA, .NET, RPC, and web services. The *n*-tier software architecture pattern can be a client–server or a web-based architecture pattern. An *n*-tier web-based architecture pattern is shown in Figure 5.10.

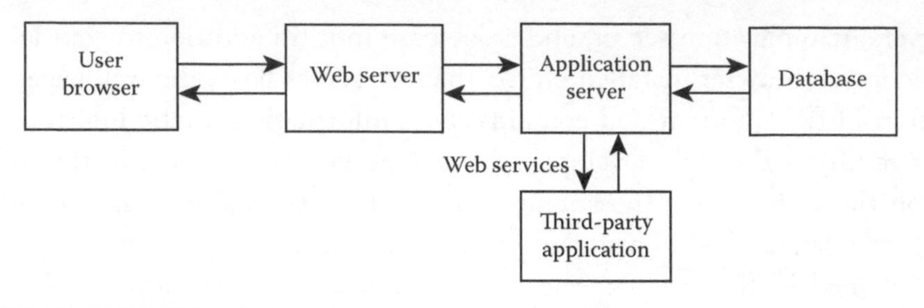


**Figure 5.10** *n*-Tier web-based architecture.

### 5.6 Client–Server Architecture

One of the oldest computers is the mainframe type. In mainframe computers, there used to be a server and many consoles. The consoles used to have no processing power on their own; they used to work as a display unit for the output generated from computer processing done on the server.

Later, engineers started connecting a computer over a network to another computer. In this kind of arrangement, one of these computers works as a server and the other as a client. If a software product needs to be developed for this kind of arrangement, where there is a client separate from the server, it can be divided into a client part and a server part. This arrangement is also termed *front end* and *back end*. A front-end design usually consists of developing the programs that are installed on client computers. The front-end design mostly consists of a user interface design, but in some cases, it can also include a part of the business logic code that performs processing on client computers. However, most processing is done on the servers. Therefore, in client–server architecture, the main software product that is developed is installed on a computer, which will work as the server or the back end while the user interface part of the software application is installed on client computers.

Client–server architecture uses a client that is basically an operating system–dependent program. For example, we can use a rich Windows-based graphical user interface for software products that use the Windows operating system on client computers.

Client–server architecture is an implementation of an *n*-tier architecture pattern. Depending on whether a database is a part of the architecture, the client–server architecture can be a two-tier, three-tier, or *n*-tier architecture. It is possible that another software product is integrated with the software product that is being developed. In such cases, the client–server architecture can be an *n*-tier architecture.

After the advent of the Internet and the World Wide Web (WWW), the usage of client–server products is dwindling.

### 5.7 Web-Based Architecture

After the advent of the Internet, websites have become very popular. Once you create a website and host it on the Internet, anybody, from anywhere, can visit your website and buy the products or services that you offer. Most websites have some static content and some dynamic content. The dynamic content is generally produced using the business logic and a connection to the database.

The architecture of a software product, which is generally hosted as a website, is very different from a client–server architecture. The user accesses a website using his or her browser. The content that should come to the user screen is hosted on a web server. The web server keeps all the static content in its file system. The web server can also generate some dynamic content. This dynamic content, usually written in a scripting language, runs on the user browser. This script is known as client-side script as it can run only on the web browser of the user's computer. The application server can also generate dynamic content, which can run when a user request arrives. This dynamic content can be written in a scripting language or can be inside some compiled files. The content in the compiled files is written using a programming language such as Java (and saved either in byte code format or in machine code format). The scripts that run on the application server are known as server-side scripts. The application server can run both server-side scripts and compiled code. At the end of this architecture is a database that keeps the permanent data. Figure 5.11 outlines this architecture.

The web architecture is not easy to understand for a novice. Let us discuss this technology more. As can be seen in Figure 5.11, the web browser displays the content coming from the web server as well as the content that is generated after running the client-side scripts at the web browser. The web server does not run any program (scripts or compiled codes) on its own. A web server only hosts HTML content in the form of HTML files. Thus, it is only capable of displaying the static content that is in the form of HTML. If any client-side scripts (also known as client-side dynamic content) are stored on a web server, then this script is actually downloaded on and runs on a web browser. The application server runs both server-side scripts and programs (in the form of a compiled code) stored on the application server. Together, the server-side scripts and programs can also be called "server-side dynamic content." The application server is connected to the database whenever any server-side script or program needs to access the database. The content generated from running the scripts and programs on the application server is sent to the web server. This content is in the form of HTML. Whenever the web server needs to send some content to a web browser, the web server combines the static content that is stored on the web server itself as well as the content coming from the application server.

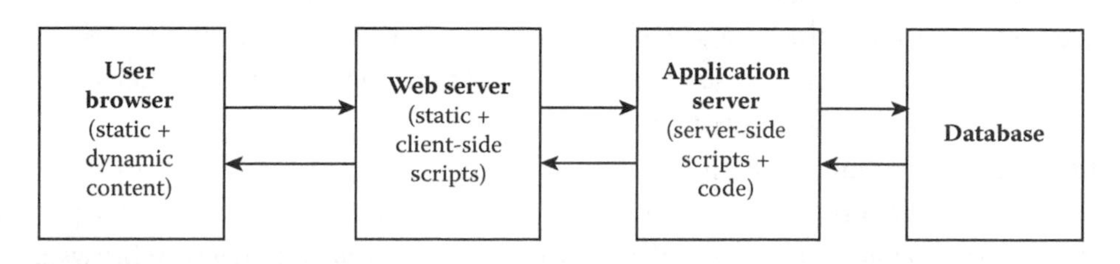


**Figure 5.11** Web application architecture with substructures.

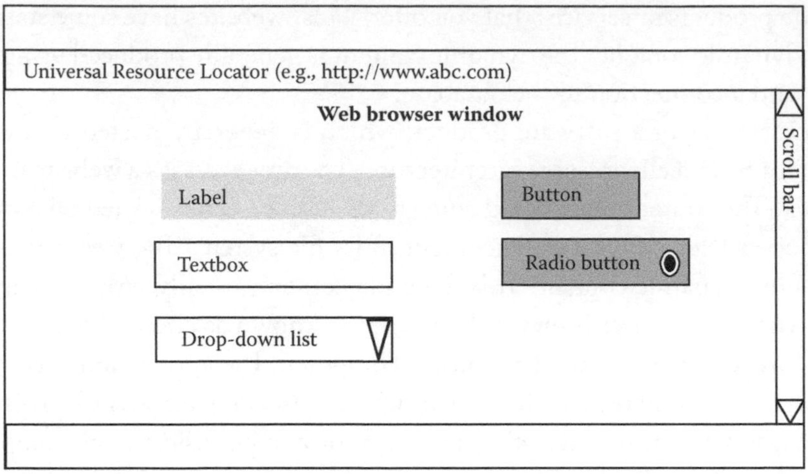

**Figure 5.12** Web browser.

*5.7.1 Web Browser*

A web browser (as shown in Figure 5.12), such as Microsoft Internet Explorer or Mozilla Firefox, is installed on user machines. Web browsers are also simply known as browsers. A browser can display any content that is in HTML format. A browser displays images, text, and videos and can play music if all the plug-ins are available in the browser. A browser has a URL text field. URL is also known as Universal Resource Locator.

The content on a web browser comes from any pages hosted on a website (which are themselves located on a web server). If the user types the URL of a web page, then the content available on that web page is downloaded onto the user's web browser.

A web browser is an important component of any web-based software product. However, a web browser itself is not developed by a software project team that builds a website. The browsers are built by companies such as Microsoft (Internet Explorer) and Mozilla (Firefox).

*5.7.2 Web Server*

A web server is the computer that hosts the websites and their web pages. A web server can host many websites at the same time. Each website has its own URL. When a user points his or her browser to a URL such as http://www.microsoft.com, the browser downloads the default web page located at this website. Each website may have many web pages and the user can navigate from one page to another using the web links. Each web link points to a web page.

The static content on a web page is displayed, in its formatted HTML form, on the web browser of any user. To make a web page more user-friendly, some scripts may also be included on any web page. When a web page is displayed on a web browser, the script on the web page runs on this browser. It is known as client-side script. Client-side scripts are used to perform many utilities. One such utility is to maintain the user information throughout a session on a website so that the user activities on the website can be tracked back whenever needed. One of the techniques to keep track of the user activities on a website is to use cookies. A cookie is a small piece of information stored on a user's computer when a web page loads on this computer for the first time. This cookie may store personal information such as username, password, and personal preferences of the user. When the user visits the website next time, the script on the web page reads the information stored in the cookie and loads the page accordingly. For example, if a user had consented to save her username and password after visiting a website, then the next time she visits that website, she may not have to type her username and password again. This information will be read and passed to the web page through running the script. This script in turn reads the data already saved inside the cookie to get the username and password of the user. JavaScript and Python are commonly used for writing client-side scripts.

### *5.7.3 Application Server*

An application server can store the server-side scripts and compiled codes. The server-side scripts can be called directly in the URL at the user browser. For example, consider http://www.mycompany.com/abc.asp. In this URL, abc.asp is a web page that contains both HTML content and server-side scripts written in Active Server Pages (ASP) script. Although this web page has been called on the user browser, the scripts run on the application server. The output of the script is shown on the user browser inside the abc.asp page. The output of any server-side script is always compatible with the HTML syntax.

Application servers can also host a compiled code in the form of computer programs. These programs can be called inside any application server-side script. For example, we can have JavaServer Pages (JSP) scripts and some source code written in pure Java language on the same application server. These JSP pages may make calls to some Java executable code (in the same application server) and the Java code will run whenever such a call is encountered. In the case of Java, the piece of source code that runs on such a call is known as a Servlet.

There are many technologies available in the market for server-side scripting. Some popular technologies include PHP, ASP, .Net, and JSP. To support each of them, you need to install the appropriate application server. For example, a web page that is written in ASP needs an ASP application server to interpret this type of scripting language. Application servers are not compatible with each other or with any scripting language that the application server does not support. Thus, when designing a web-based software architecture, you may need to choose the scripting language, and accordingly you need to choose the right application server.

*5.7.4 Benefits of Web-Based Architecture*

The client–server architecture has some problems. The first problem is that each user machine needs to be installed with the client software. This requires a lot of effort. The second problem is that, while building a client–server product, the project team needs to build the client part as well, as part of the software product. This, again, requires extra effort. The third problem is that whenever there is an update in the server-side software product, the client parts also need to be updated.

Web-based architecture has overcome all the problems associated with client–server architecture. The client is always there with user computers in the form of a web browser. Web browsers are sophisticated software products and can do all the stuff required by the client in any software product. Therefore, the product development team does not need to develop the client part for any software product. When any updates are made to the server, the project team does not need to worry about updating the client parts. Browsers can connect with any website because the content that goes to the browser from any web server is always in HTML format. This makes them truly platform independent.

Another benefit with web-based architecture is that you can deploy a web-based product on a company's intranet if it is not required to be deployed over the Internet. Here, again, the users can access the software product running on a server using the web browser on their computers.

Because of these benefits, web-based architecture has replaced the legacy client–server architecture in almost all software products.

## 5.8 Service-Oriented Architecture

Many vendors provide the components or products specifically to be used by other vendors. These components or products are developed using a technology known as Service-Oriented Architecture (SOA). A schematic presentation of an SOA is depicted in Figure 5.13. Here, one software product acts as a service provider while another software product acts as a service consumer. For example, some vendors have developed software products that provide current airfares for different airlines. If you are developing a software product to provide online air ticket booking facility for the end users, then it is not an easy task to keep updating the latest airfares for various airlines. Your job will be easier if you plug in your product with an available software product that provides the current airfares for different airlines. By doing this, you do

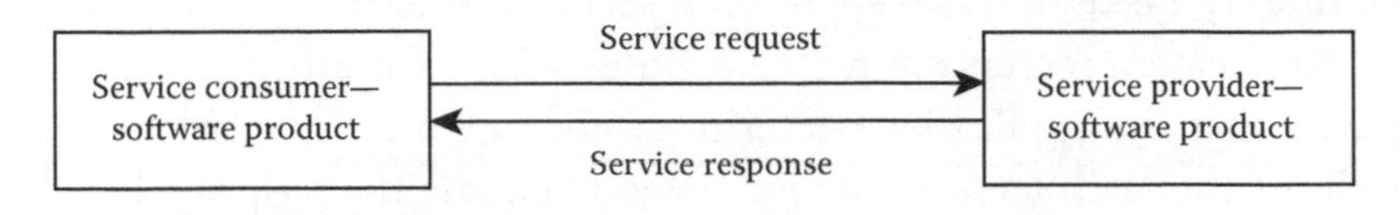


**Figure 5.13** SOA in which a software product acts as a service provider and another software product acts as a service consumer.

not need to keep updating the latest airfares because your product will be constantly receiving the updated airfares from that available software with which you plugged in your product.

SOA products (i.e., products or components that are developed using SOA) are deployed over the Internet, which is why SOA products are also known as web services. Any consumer who wants to avail the services of these SOA products can easily integrate his or her own product by plugging in to these SOA products. The open interface of an SOA product, where any product of the consumer can plug in with that SOA product, makes it platform or vendor independent. Therefore, anybody can plug in their software products to these SOA products.

An SOA product has its software services available through its open interface. A software product that consumes the service provided by an SOA product should also have the same open interface as the SOA product. Only then will communication be possible between these two software products. Therefore, if you are building a software product that will be availing the services of an SOA product, then you need to provide an interface (to the software product that you are building) so that this interface will consume the software service of the SOA product (when you plug in your software product with that SOA product).

A software product that depends on SOA architecture (whether a consumer of SOA services or a provider of SOA services) has a distinctly different architecture when compared to the software products that do not depend on SOA services. If your software product is a consumer of SOA services, then you will have to consider an *n*-tier architecture because one tier of your software product will be exclusively used by an SOA service provider. You will have to create your software components so that they are able to consume an SOA service. On the other hand, if you are building your software product as an SOA service provider, then you will have to build your software components in such a way that the consumers of your SOA service can be easily plugged into your software product.

Specific details of SOA are out of the scope of this book. For a more in-depth discussion on and a clearer understanding of SOA, we refer readers to a book on SOA.

## 5.9 Software Component Design Fundamentals

Thus far, we have seen how the architecture of a software product is designed. Now, we will see how the components (i.e., modules) of the software layers of the software architecture are designed.

One of the approaches in modeling a software system is the use of components. This is very similar to building the electronic products where you create the integrated circuit (IC) boards and then assemble them to build the complete electronic product (e.g., a television). When you design an IC board, you have to ascertain what kind of processing it will do (e.g., voltage regulation or image brightness control). Later, when you have built all your IC boards, you can then wire those IC boards together to assemble your product (in this case, the television). Similar to how you build the electronic products, you can build the software products using components. In the case of electronic products, IC boards are the components. You will need to think on similar lines to build software components to be used in building the software products. All the attributes and behavior belonging to each software component will be contained within its boundaries. They will be accessible only through the integration considerations in the form of well-defined component integration specifications.

Component design for user interfaces is out of the scope of this book. Components for user interfaces such as textboxes, labels, and windows are available as prebuilt components with most programming languages. We will design and build user interfaces using these prebuilt components. User interface design and construction is discussed in Chapter 6. Component design for databases is carried out using entity-relationship diagrams. We will discuss database design and construction in Chapter 8. Here, we discuss component design for the middle layer (i.e., the business logic layer) only.

When we model a software system, we need to consider many things.

- Compatibility: If a system is composed of components, then it is possible to replace a component from a system with an equivalent component and the system will still work. This is because the interfaces of equivalent components will be the same regardless of the internal composition differences (if any) among the equivalent components. Thus, if a new component needs to be added to an existing software product, you can either build it or purchase it from the market if its specifications are well defined and exactly match the requirements. Generally, if a component has the same specifications as is required to be added to a software product, then this component will be compatible and can be easily integrated with the software product.
- Abstraction: If you have a collection of things and want to know what information is common among them and what information is different, then you will need to find and classify all these pieces of information first. On the basis of the classification information, you can group things (whose information is similar although not exactly the same) together. Later, you can put a label for each group of things so that each group can be identified. This concept is known as abstraction. For software components, you can make a group of similar but not exactly the same components. You can first create a master component and later create all similar components from this master component. The information that is common with the master component can be copied to these other components. The information that is different and unique for each component can be included with each of these specific components. Abstraction is in fact a very powerful tool. It allows a software product to be built to cater to the needs of a wide range of users by building a software design that allows the development of similar but different software features using a minimum amount

of source code but is able to create many features in the software product. Abstraction is further explained in Chapter 7.

- Refinement: Refinement is the opposite of abstraction. While abstraction helps in aggregating similar things, refinement helps in segregating dissimilar things. If you have already performed the abstraction to build a master component, then, through refinement, you will find out how all the components that will implement the actual functionality can be developed from the master component. The relationship among all the components also needs to be defined. Generally, this relationship will eventually result in a hierarchy structure for all the components. Abstraction and refinement are complementary to each other. They are further explained in Chapter 7.
- Modularity: Software architecture is divided into components called modules. Modular design generally means dividing the software product design into many components that can be managed individually without affecting the other components. For example, a modular component can be extended in the future without affecting its relationship with the other components of the product. In a modular software design, a component can easily have loose coupling with other components, and at the same time, that component will have high cohesion among all the parts inside it. Tight coupling between two components means that there are many connections between those two components and that there is a lot of interaction between them. Tight coupling will lead to problems because too much information exchange leads to the failure of many components at the same time if one component fails. Tight coupling is a bad design. Self-contained components with only required (minimal) information exchange with other components are a better software design. This is known as loose coupling. At the same time, if a component contains many parts that are dissimilar to each other, then it will create maintenance problems. For example, if a class contains methods for managing the database connections and performing the business logic computations, then it will be difficult for a person to maintain that class because of the difficulty in finding this class and its methods and fixing the source code. It will be a lot better if a separate class takes care of the database connection management functions, and when any changes related to the database connection management are needed, the person will know which class takes care of this (database) functionality. Hence, that person can easily find the class and change its code. This technique of keeping related methods in one class and moving unrelated methods to other classes is known as keeping high cohesion inside each class.
- Data structure: Data structure is a representation of the logical relationship between the individual elements of data. Data flow happens in a software product when some event or a user action leads to the running (execution) of the components of that product. For example, if the user has provided some

input data on the user screen and pressed the Submit button, then the data provided by the user pass to business logic components. The business logic components will do some computation and send the data back to the user screen. How the data need to be presented to the user and what kind of processing needs to be performed on some input data are governed by the data structures defined in the business logic.

- Information hiding: Components should be designed in such a way that the information contained within one is not available to other components that have nothing to do with that information. This is very important because it prevents unauthorized access of data by other components. Data belonging to a component should not be exposed at all or else other components can view or modify the data without authorization. If another component wants to view or modify these data, then it should be given access only after some strict authorization.

## 5.10 Component Diagrams

Component (i.e., structure) diagrams deal with the items that must be present in the system that you are modeling. Since component diagrams represent the structure of the system, they are widely used in documenting the architecture of software systems.

The structure of a software system can be represented at many levels. At the top level, the complete structure of the software system can be drawn. This diagram can be called the system diagram. A system diagram itself consists of many smaller parts that are joined to each other. The system diagram, showing its parts and the connections between them, can be best described by thinking in terms of how an electrical system and its components are wired to provide a complete functional electrical gadget. Similarly, a software system can be shown as consisting of software components and their connections with each other. A drawing showing a software system composed of the software components and how these components are connected to each other is known as a software component diagram.

The best way to draw the component diagram is to use UML. We covered UML in Chapter 4. UML can again be used to draw the component diagrams.

Component-based development is based on the idea that the previously developed components can be reused and, if necessary, those components could be replaced by other equivalent components. OOD heavily uses component diagrams. Components in OOD are a more abstract depiction of objects. Using abstraction, you can create a template of a component. This template can be later used to create actual classes and objects by your implementing the template in various ways as per your need. This results in reuse and thus will lead to the best software design. A software design with the least amount of duplicate source code (to build the components) results in a defect-free, maintainable, and easily extensible software product. This also results in a very productive project team because software developers will need to write less source

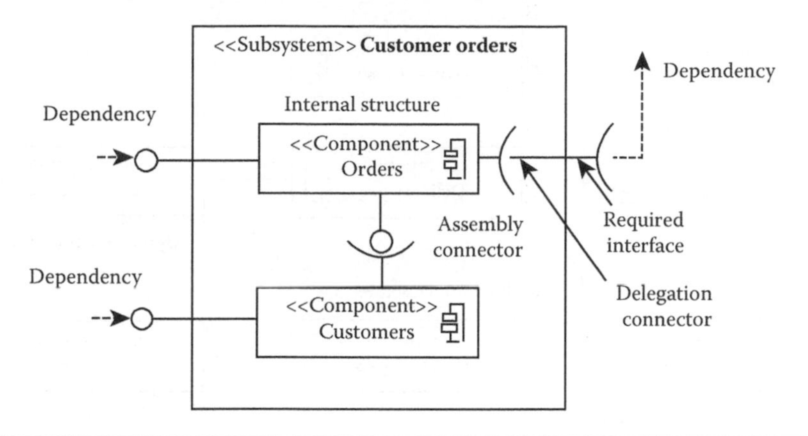


**Figure 5.14** Subsystem with components and connections.

code. Thus, if you are developing a software product using OOD, then component diagrams can be the best way to show the structure of your software product.

In Figure 5.14, we have depicted a subsystem. This subsystem is part of a larger system. You can see the dependencies at three places. This means that this subsystem shares information (both ways) with other subsystems that are part of the same larger system. You can also see joints (couplings) between the components. Two types of couplings are shown here: delegation couplings and ball joint couplings. Ball joint couplings (the coupling shown between the Orders and Customers components) are used to denote tight integration between the components. Delegation couplings, on the other hand, are used to denote loose couplings between the components. In this figure, delegation coupling is shown as a delegation connector with the Orders component to be integrated with an external component. The external component is not shown because this kind of coupling can be used to integrate with just any external or internal component even at runtime. Tight coupling between the components can be used when the components are parts of the same system. Loose couplings, on the other hand, can be used to integrate the components with other components that may be part of another system.

### *5.10.1 Component Example*

Let us take a look at a component diagram given in Figure 5.15. This figure depicts the component diagram for an order management system for a restaurant. We saw the use cases for this system in Chapter 4.

To translate the use cases into component design, we need to recall what the use cases were depicting about the interaction between the user and the system. Then, we have to think of how the use cases will help in building the software components. Hence, let us have a look.

- The user (sales representative) would log into the restaurant order management system. For this to happen, a login page is needed in the order management system.

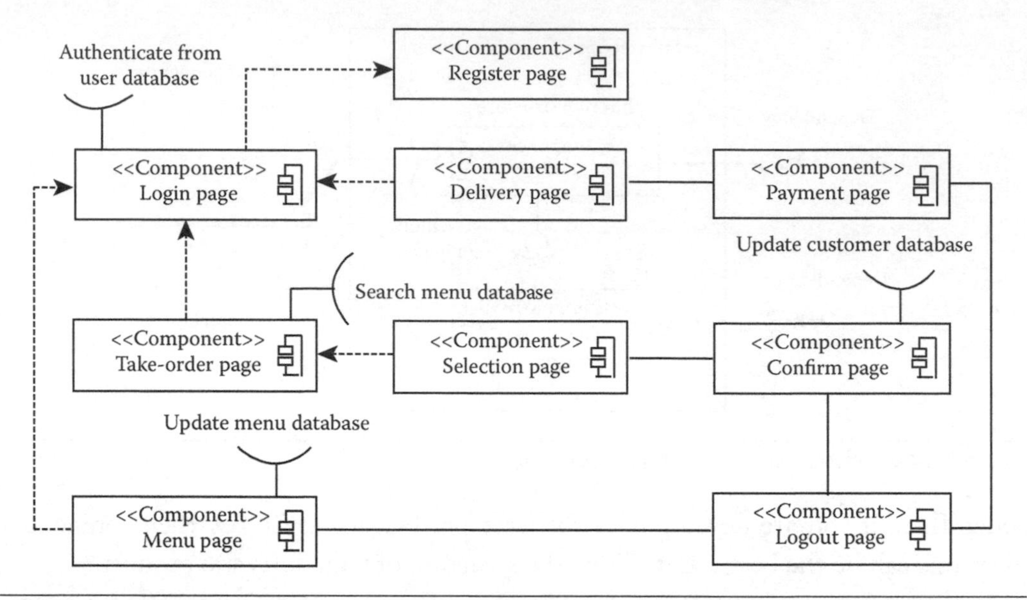


**Figure 5.15** Component diagram for a restaurant order management system.

- The user will take the customer order. For this to happen, we need to provide a food menu page from where the sales representative will pick the appropriate food menu for the customer. There should also be a menu search and selection page so that the sales representative can find and select the appropriate menu for the customer.
- The user will confirm the order. Once the customer agrees with the menu choices, the order can be confirmed. For this purpose, we need to have a confirmation page in the system.
- The user will collect payment from the customer. After confirmation, the system will suggest the price for the order. The user will inform the customer and the customer will pay for the order. The user will then update the order with the payment information. For this functionality, we need to have a payment page in the system.
- The user will logout from the order management system once the payment for the order is received from the customer. Hence, we need to have a logout page in the system.

There is another requirement for this system. When an order is saved, the customer name, phone number, and order information are saved in the system. This is done after the order is confirmed. This functionality is needed because if a customer makes an order the next time, the customer information saved from the previous order can be used to serve the customer better. For example, the customer is asked about his or her phone number first. After entering the phone number, the system automatically detects whether the customer made any earlier orders. Customer name and address are also pulled up from the database based on the phone number.

On the basis of the use cases and their corresponding design requirements, we need to build the component design for the system.

We have depicted the component diagram for the restaurant order management system in Figure 5.15. The website consists of nine web pages. Each web page can be treated as a component. All these web pages are connected to each other through a link. On some web pages, there is also a connection to a database. In fact, all the web pages will be connected to some database to display or modify some data through the connection to that database. For example, the login web page will authenticate a user when the user tries to log into the web page.

The relations between different components are depicted in Figure 5.15. When a user accesses the restaurant management system, he or she is always presented with a login page. Only after successful authentication is the user able to view and perform the functionality available in the system. Once the user authentication is successful, then the user can navigate to another web page. A dotted line connecting a web page and the login page, with an arrow pointing toward the login page, represents the fact that there is a connection between the login page and that web page. In addition, the arrow represents the fact that if the user tries to access that web page directly (without login at the login page first), then the system will navigate the user back to the login page. There is again a dotted line with an arrow connecting the take-order page and selection page. This means that if the selection of a menu is not correct or a user wants to change the menu selection, then the system will navigate the user back to the take-order page from the selection page. There is also a dotted line with an arrow connecting the register page and the login page. This arrow is pointing away from the login page. This is depicted to convey the fact that the register page will take the user to a system (or part of a system) that is not part of the restaurant order management system. The solid arrows signify that the user can navigate from one web page to another without the need to login again. That means a session object will track and authenticate when a user navigates from one web page to another web page if both web pages are connected through a solid line (session objects are explained in Chapter 7). Notice that even when the user is on the logout page (i.e., logout request state), it does not mean that the user has logged out. The user can still go to other pages that are connected to the logout page.

When you model your software products, you need to create the components and their relationships carefully. In OOD, components are used as the template to create the classes. We will learn about classes in Chapter 7.

## 5.11 Data Flow Diagram

A data flow diagram depicts how the data flow from one component to another (in a software system) during data processing when a user or another software component interacts with a component. These interactions between components also show the behavior of a component; thus, a data flow diagram is also known as a behavior diagram.

A data flow diagram deals with how the data flow in a system. It specifies things such as where the data are coming from, where will they go, and where will they be stored. A data flow starts from a user input and then travels through the business logic where some computation occurs and finally to the data store where some data manipulation happens. A data flow diagram captures all the details about the data transformation that happens through this process.

A data flow diagram consists of three elements: process, external entity, and data store. A process is a business activity through which the data are modified or transformed. For example, a debit transaction is a process through which a customer's bank account is debited. After the debit transaction, the customer's bank account balance changes. A process can have many details that can be further elaborated later. However, when you draw the data flow diagrams, you should at least capture the major changes and the transformation of data during a transaction. In Figure 5.16, a process is depicted.

An external entity is external to the system that we are dealing with. It can represent a human or another system and it is the place where the data come from or where the data will go. Data generation by an external entity can happen through some event (e.g., a mouse click) or time (e.g., an automated batch input) or some command (through another software product) seeking data from the system. Data are received by an external entity through the process explained in the previous paragraph. An external entity is depicted in Figure 5.17.

A data store is the place where the created or manipulated data (through the business process) will be stored permanently. A data store can be a database or a file. In Figure 5.18, a data store is depicted.

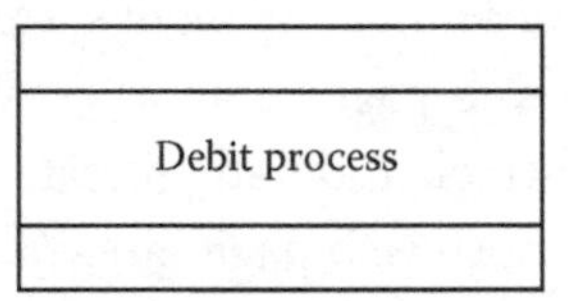


**Figure 5.16** Process in a data flow diagram.

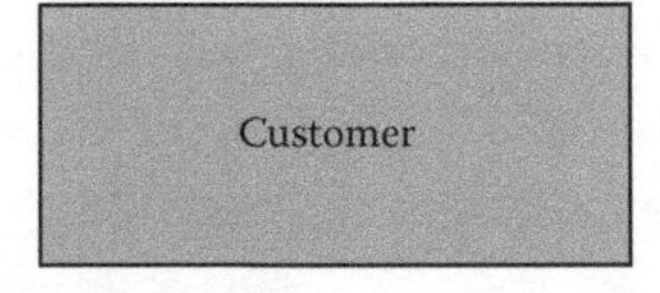


**Figure 5.17** External entity in a data flow diagram.

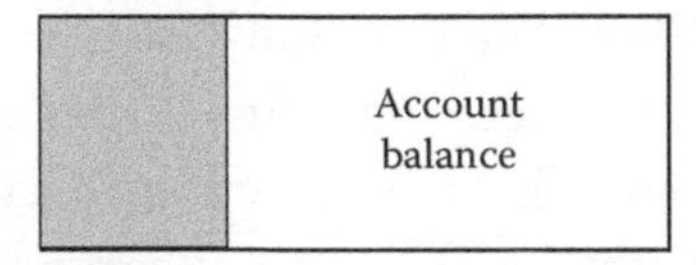


**Figure 5.18** Data store in a data flow diagram.

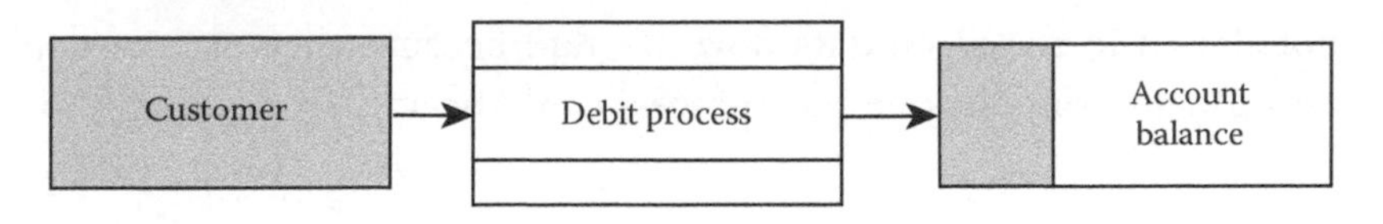


**Figure 5.19** Data flow initiated by an external entity to a data store.

In most cases, a data flow is initiated by an external entity. The software system processes and transforms the data. Finally, the data are saved in a data store. This relationship among the external entity, process, and data store is depicted in Figure 5.19.

### *5.11.1 Data Flow Example*

Figure 5.20 depicts a data flow diagram for the restaurant order management system. We saw the component diagram for this system earlier. The sales representative logs into the user data store using the login process and takes the customer order. A customer makes a purchase and the system processes this order. The purchase transaction is recorded in the transaction data store. The required food items on the order menu will be delivered from the restaurant and will be handed or shipped to the customer. The shipping activity will be handled by the food delivery process. The transaction data store will be updated as the customer pays for the transaction. Finally, the sales representative will logout from the system using the login/logout process.

Please note that we have shown data flows for only the processes that are described in the use cases (provided in Chapter 4). There could be some other processes also, but

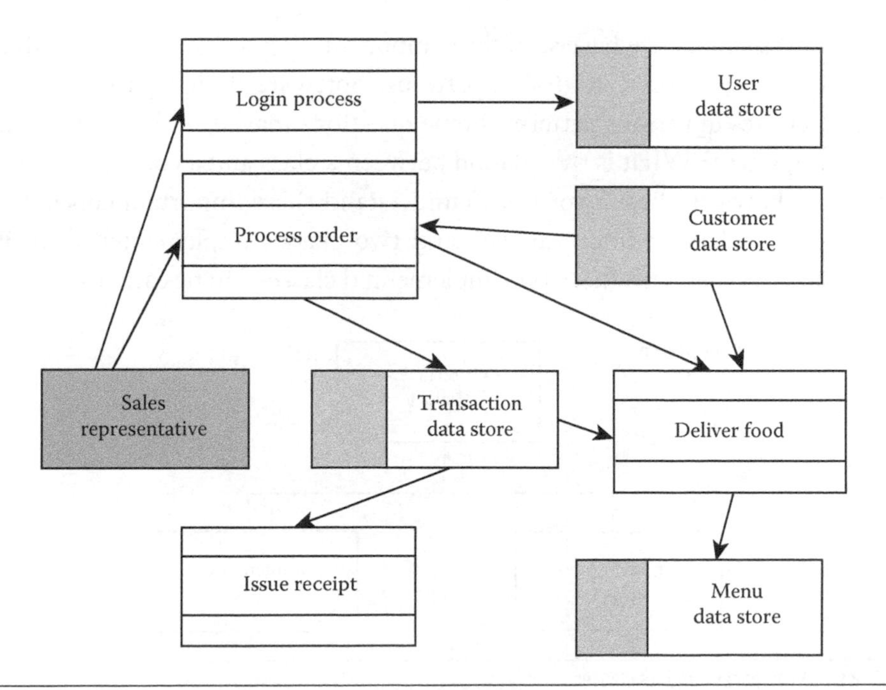


**Figure 5.20** Data flow diagram for restaurant order management system.

they are not shown in the given data flow diagram because they are not depicted in the use cases. As an example, one such process could be adding new users to the user data store.

### 5.12 Software Design Patterns

We learned about software architecture patterns in the previous sections. Software design patterns are similar concepts with the purpose of solving problems at a more minute level. Software design patterns are applied at class and object levels in OOD. We will learn about classes and objects in Chapter 7. Here, we will learn about design patterns that can be used to create classes and objects as per our requirements. Software design patterns are a great way to design the required classes and objects.

A software design pattern is a template for designing the classes. People use various kinds of templates to design various kinds of things. For example, a website template is used to design a website. Similarly, a software design pattern is used to create software classes. Software design patterns have been created based on the experience of the people who have worked in creating classes for various purposes. For example, we have a software design pattern for configuring the orders in some industries. In these industries, an order can consist of many items but the pattern of these items is fixed. Using a design pattern, you can create this order pattern using a hierarchy of classes and configure the orders.

#### *5.12.1 Difference between a Class and a Component*

In the previous sections, we have seen the component diagram and data flow diagram. Now, we will discuss software design patterns. Software design patterns are about grouping of classes in various patterns. Some questions may arise: What is a class and what is a component? What is the relation between a class and a component? What is the difference between these two? Let us understand these important concepts.

Figure 5.21 depicts an interface class and two classes implemented from it. The interface class together with those two implemented classes can be called a component.

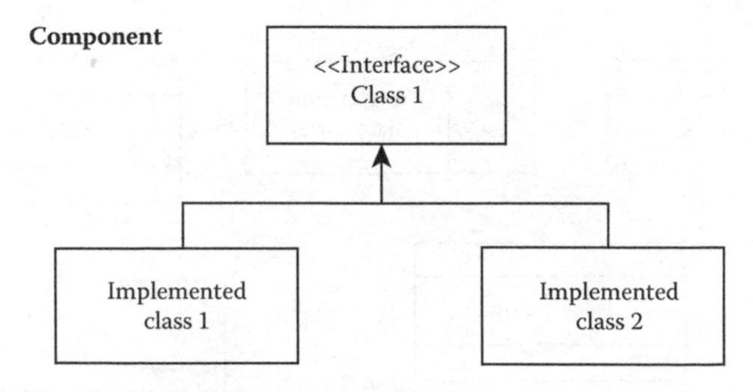


**Figure 5.21** A component and its classes.

The interface class does not have any implementation details (i.e., a programming code that does some processing). The implementation details are programmed in class 1 and class 2 only.

Why do you need an interface class if it is not doing anything? The interface class has a very important function. It allows you to create many classes that are similar (with only some differences). The interface class allows you to reuse the source code in many instances where you need to perform almost the same kind of computation as an existing class but it still has some differences. If you are not using an interface class, then you need to create another class. Thus, you may end up having many classes that are essentially doing almost the same thing. Maintaining so many classes becomes an issue in the long run. If you have an interface class, then creating a new class is easy. You just need to implement the interface; thus, you can reuse the code.

Software design patterns use the concept of reuse, and at the same time, there are many types of design patterns that can be used for different purposes.

### *5.12.2 Software Design Pattern Types*

Software design patterns are primarily divided into three types: creational design patterns, behavioral design patterns, and structural design patterns. As we know, in object-oriented programming, we create objects (during runtime) from the classes. These classes interact with each other to perform some computations. We can use the design patterns to create different types of classes from a single parent class (or interface). We can also use the design patterns to control the behavior of objects during runtime. Similarly, we can use the design patterns to control the interaction among objects during runtime.

We will see several design patterns in Sections 5.12.3 through 5.12.5.

### *5.12.3 Creational Design Patterns*

Creational design patterns are used to control how the child classes are created from the parent classes. During runtime, it is also possible to control how and when the objects can be created. These controls are important because, this way, many types of classes and objects can be created from a small pool of classes. This feature enhances code reuse by using the techniques such as inheritance and polymorphism. Some of the design patterns related to creational design pattern types include abstract factory, builder method, factory method, object pool, prototypes, and singleton, to name a few.

*5.12.3.1 Abstract Factory* The abstract factory design pattern is used to create concrete classes from abstract classes that are totally encapsulated from each other. To implement an abstract factory design pattern, you can create a single abstract class. Later, you can create interfaces or abstract classes as the children for this class. Finally, you can create the concrete classes that will implement these abstract classes.

For example, you can create several user control elements such as buttons and textboxes from a single class. Since the user control elements are very different from each other, their implementation will require the creation of classes that are very different from each other. At the same time, some user controls are used for the same purpose; thus, there should be some similarities among the classes that will implement these user control elements. The best scenario in this case is to create one abstract or interface class and then implement similar user control elements from this abstract class or interface. For example, a radio button and a checkbox work in a very similar way. Hence, we can create one abstract class for these user control elements and then implement this class by creating the concrete classes for the radio buttons and the check boxes. The same is true for the list box and the drop-down box, which work in a similar way. Figure 5.22 depicts an abstract factory design pattern. In this figure, you can see that we have created a main class, Class 1, which can be used to create many user control elements. At the next level, there are two interfaces: SelectionControl and ListControl. The SelectionControl interface can be used to create two concrete classes, RadioButton and CheckBox. Similarly, the ListControl interface can be used to create two concrete classes, ListBox and DropDownList. All the implementation details will be contained inside these concrete classes. During runtime, all the objects created from these concrete classes will behave differently from each other. Even during the design time, these objects will look differently from each other (e.g., a list box looks different than a drop-down box). This is how abstract factory design pattern works.

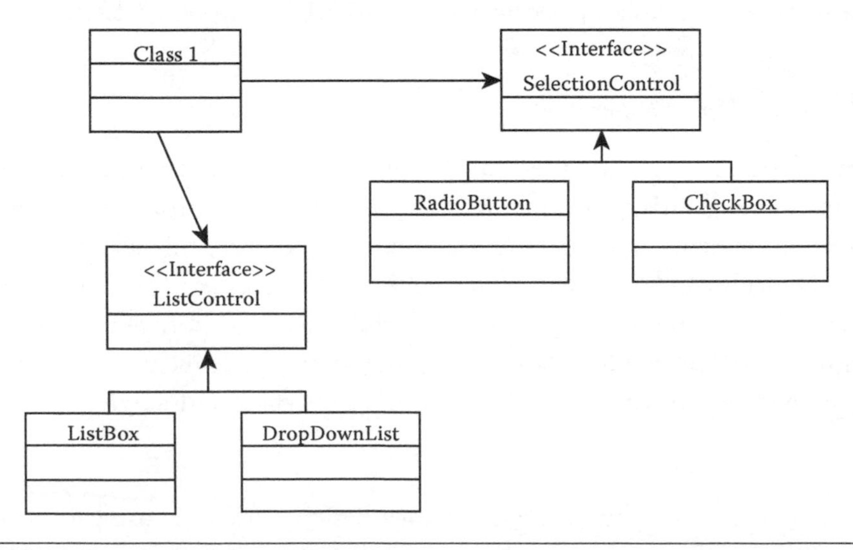


**Figure 5.22** Abstract factory design pattern for a class.

*5.12.3.2 Builder* The builder design pattern is used when you need to create many objects (during runtime) that are related to a single parent object. During runtime, it is possible to create dynamic relationships among these objects. For example, a builder design pattern can be used to create a representation of a configurable computer system. The top class can be implemented to represent a computer model number. Many classes (which are not the child classes of that top class) can be created to represent the parts of a computer such as the hard drive, memory card, and motherboard. The relationship among these classes is not hard-coded with the top class. The relationship among these classes will actually be established during runtime. Suppose a customer makes an order for a particular computer model. The customer specifies the configuration he or she needs for this computer model. Now, the relationships between the objects during runtime can be established. Thus, it will be possible to create an order (i.e., sales order) management system where configurable items (such as computers) can be ordered.

*5.12.3.3 Factory Method* The factory method is used when a new operation needs to be taken care of. For this new operation, there are no previously available classes. A super class is created to implement it. This super class will take care of the common operations. To take care of specific operations, we can create child classes that will be derived from the super class. For example, let us discuss the order management system for a restaurant. Suppose that currently they have two order types: take-out orders and dine-in orders. Hence, we have two classes to take care of these order types. Now, if the restaurant introduces a delivery-type order as well, then you will have to create a new class to take care of this new operation. In Figure 7.13, there is a super class Order and it is linked to two subclasses.

### *5.12.4 Structural Design Patterns*

Structural design patterns are used for managing the interfaces of the objects to control their integration with other objects during runtime. Some of the design patterns related to the structural design patterns include adapter, bridge, composite, decorator, façade, flyweight, private class data, and proxy.

*5.12.4.1 Adapter* Generally, software products witness several releases in their lifetime. Software vendors may change the underlying architecture of a software product each time a new release of that software product happens. To ensure that the older releases still work with the new releases, the software vendor introduces what is known as backward compatibility. To implement backward compatibility, the software vendor keeps the interface of the new release open so that the older release of that software product still works with the new release. The adapter design pattern is used to create this backward compatibility. For example, an adapter class can be created so that two different Word processing applications (an older release and a newer release of the same application) can become interoperable so that a user can open a document in one application although it was created in the other.

*5.12.4.2 Proxy* As the name implies, a proxy design pattern is used where an object needs to be replaced by another object during runtime. For this to happen, both objects should have the same interfaces although their internal structure can be different. For example, when we need to do some calculation, the precision of the outcome of that calculation can be different for different purposes. Suppose we need to draw a graph based on some calculations. One of the graphs shows columnar results while the other graph shows the same results in pie charts. The precision of calculation for these two graphs may be different. To resolve these differences, two classes can be created to perform the calculations. On the basis of the format of the results' display, the appropriate class needs to be called. During runtime, the suitable object will be loaded to take care of the right precision for the calculation.

*5.12.5 Behavioral Design Patterns*

Given an input, different classes behave differently because the outputs are computed differently by the methods contained inside those classes. At the same time, the difference in the outputs can be managed by controlling the methods contained inside those classes. Behavioral design patterns are used to control the behavior of the classes so that the objects created from those classes can interact with each other properly. Some of the design patterns belonging to behavioral design patterns include chain of responsibility, command, interpreter, iterator, mediator, and memento.

*5.12.5.1 Command* The command design pattern is used to pass a command from one object to another so that some computation or another command can be generated. The command design pattern is used to control the data flow that happens inside a software system after some event occurs. To understand this concept, a comparison is provided. An office manager asks his or her employee to prepare a report. The employee, in turn, asks one of his or her assistants to do it. This assistant then prepares the report and hands it over to the employee. This employee, in turn, sends the report to his or her manager.

*5.12.5.2 Iterator* In some software applications, the user needs to evaluate some options before selecting one of them. Each option should be associated with some preview so that the user can have some idea as to what that option does. In such cases, the iterator design pattern can be used. For example, in Microsoft Windows, there is a preview pane in Windows Explorer. When a user selects a file that has some preview associated with it, the preview of that file is shown in the preview pane. The preview helps the user find out what a file is all about, even without opening it.

## 5.13 Programming Language Considerations

Once you are given the task of designing and building a software product, the first thing you need to do is think about the technologies that will be used to design and build the product. There are many technologies available in the market and choosing the right one will affect the project.

In the previous sections, we have seen how a software product is modeled based on the given requirements. We also need to consider how the model of a software product can be implemented when the technology considerations are given. In the succeeding sections, we will discuss the technical aspects of implementing the models we have created.

From a purely business point of view, the technology that provides the best productivity and thus saves time and money is the best one to choose for product development. However, practical constraints may hamper decision making and thus choosing the right technology may become difficult. Not all technologies work well with each other. Therefore, there is a genuine need to choose a technology that allows the written source code to run on any platform. Otherwise, you will end up repeatedly writing the source code for various platforms on which the software product will run. This is definitely not a good option. Some other factors that may influence your decision on choosing the technology include the following:

- Size of the software product to be developed
- Type of the software product to be developed
- Technologies available in the market
- Code reuse
- Considerations for nonfunctional requirements

As explained in Chapter 4, requirement management should not include implementation details. Now that the requirements have been gathered and the requirement specifications made, it is time to think about implementation details. Let us see how the factors mentioned above influence implementation details.

### *5.13.1 Size of Software Product*

The size of the proposed software product is an important factor while deciding the technology for designing and building the product. Not all technologies scale well. Similarly, if you choose a technology that is well suited for building large products but you use it to build a small product, then it will be overkill and such technology is definitely not required. For example, if your software product will be used by a few people, then making an elaborate multitier architecture for that product will be overkill. Therefore, if your product is small, then never go for the technologies that are well suited for building large software products only. Similarly, if a technology is suited to build small products, then it is not necessary that it can scale well to build a large software product.

Some well-known technologies need people with specialized skills. These technologies are well suited to build large software products. Using them to build a small product will not only create a problem in arranging the right people but also lead to a waste of time and money. For example, Microsoft Access is a database well suited for building small software products. People with low skills can be used to create and use databases in Microsoft Access. If your software product is small, then using Microsoft Access is the best fit for this scenario as it will allow the software product to be developed cost-effectively as well as in less time. However, if you are developing a large software product, then Microsoft Access is not suitable. You will need to choose a database such as Oracle or Microsoft SQL, which are well suited for large software products.

Sizing the software product (i.e., determining how large the proposed software product will be) is important for estimating the cost and time factors as well as for finding the right technologies to build that product. If a proposed software product will be used by just a few users, then choose the technologies that will allow the rapid development of the product using the least amount of skill and money. However, if the proposed software product will be used by thousands of people, then careful planning is definitely needed to choose the appropriate technologies to implement the project.

### *5.13.2 Type of Software Product*

Software products are of many types and they are used in many industries such as defense, medical, business, communication, education, entertainment, and journalism, to name a few. Each industry has a different kind of need for software products. These needs translate into specialized software products for specific industries. For example, if you are building a software product for children's games, then you do not need to build the security components for that software product. On the other hand, security components are the most crucial components for building software products for the defense industry.

Software architecture will depend on several factors such as whether a database will be attached to the system or whether the user interface needs to be controlled using a button or a scroll bar.

### *5.13.3 Technology Availability*

If you want to develop a software product, there is a plethora of technologies available to choose for this purpose. You may want to choose the best technology that fits your needs. For example, if your project requirements indicate that you need to store a huge amount of data, then you need to find a database large enough to store it. If your requirement is that you need to provide rich content to users on their browsers, then you need to find out if there is any browser that can support rich content. If no such browser exists in the market, then you may need to create a plug-in for the browsers to get the desired functionality. The creation of a plug-in may increase the budget and time for your project considerably.

Sometimes, there is no suitable technology available to support the development of your software product. For example, assume that your customer wants to have compression software that can compress any file to 1% of its original size without reducing the quality of the file content. It is impossible to fulfill this requirement because not all files can be compressed to 1% of their original size without sacrificing quality.

Let us discuss the importance of technology in software design and development.

*5.13.3.1 Integrated Development Environment Infrastructure* If you are involved in software development work, then you need to have a good software infrastructure so that software development work will be productive. The infrastructure includes compilers, database connectors, application servers, debuggers, configuration management servers, and other components that are part of an Integrated Development Environment (IDE). If any part of the IDE is missing or not performing well, it will hamper your work. Thus, the selection of IDE is critical to the success of the project.

*5.13.3.2 Platform Independence* During the time when platform-independent software development technology was not yet available, most software development projects relied heavily on the infrastructure available in a specific platform. For example, if you decided to use the Microsoft platform to develop a client–server software product, then you would use the libraries provided by the Windows platform itself. You could write the code that would run on any Microsoft operating system. However, the same code will not run on other platforms such as an IBM operating system. After the advent of programming languages that are platform independent, the platform constraint was resolved. Once the code is developed, the same code can run on any platform. For example, the Java language used to build software code will run on any platform.

*5.13.3.3 Library Infrastructure* Libraries are collections of source code files that are provided by the inventors of a programming language. These libraries help provide an infrastructure that programmers can use while writing their source code.

The inventors of the operating system also provide libraries of source codes that, again, can be used by the programmers while writing their source code.

There is a difference between the above two types of libraries, though. If a programmer uses the libraries provided by an operating system, then his or her code will run only on that operating system. This is a big drawback. In this scenario, the programmer has to write as many versions of his or her source code as the number of operating systems on which the source code may need to run. However, if the programmer uses the libraries provided by the inventors of the programming language, then this code will run on all operating systems. The problem is that not all inventors of programming languages provide libraries. Over the years, this constraint has also been solved. Companies (or people) that developed programming languages, which are platform independent, have developed libraries specific to those programming languages. These libraries provide the infrastructure on which the code can run on any platform without problems. For example, the java.awt.* and java.sql.* libraries of Java language allow the software developers to build products in Java language on any operating system. If you are part of a software development team and these libraries are not available, then you will end up writing the code to communicate with the operating systems. This creates extra effort, which is undesirable.

#### *5.13.4 Code Reuse*

Code reuse, also known as source code reuse or software reuse, is a major consideration nowadays. Code reuse speeds up the development of a software product. There is already an existing code available in the form of a large number of software products in the market. Some of these products can be used, in part or in full, in the development of your project, if any software features in such products meet the requirements of the software product you need to build. If the software product you are developing involves a large number of software components that have similar functionality, then it is possible to reuse those components.

Code reuse helps increase the productivity of software engineers tremendously. If there are existing libraries of software code available, then they can be used in developing your software product. The usage of libraries will reduce the chances of errors (software defects) in the source code. Libraries are generally free from any software defects, because most of the time, they have been used in software development for earlier projects. If they had any defects, then they have already been removed. Thus, they are mostly defect-free.

There are two situations in which code reuse can take place. One is when the code is an intrinsic part of your own product, in which case the code is already available to you. Code reuse takes place by building new code libraries or using the existing libraries. The other situation is when you are going to use an existing product (or components) from some vendor and the required source code is not available to you. In this situation, you integrate the product or components with your own product through the open interfaces available in the said product or components. Most software vendors make the information about the interfaces of their products or components available. These interfaces are known as Application Programming Interface (API). The API documentation of the products or components specifies how you can integrate those products or components with your own product. Thus, the existing code of this other software product becomes available to you and you do not need to write any source code for the features that are available through this software product. Thus, you are essentially reusing the source code of the said product.

OOD especially helps in reusing the source code because it is possible to abstract the design consideration of all similar components and then use a common interface or abstract class to derive all the classes in your product. This way, the source code is reused in the most effective way.

### 5.14 Types of Programming Languages: A Brief Review

Another high-level design decision in developing a software product is the selection of programming language. There are many programming languages, which have their own benefits and shortcomings. For example, if your requirement is to perform heavy engineering and mathematical calculations in your software product, then a programming language such as MATLAB® will help you build your software product very quickly. Similarly, if your requirement is to perform user validation before the user submits a form on his or her web browser, then client-side scripts should be used. Some programming languages are good for efficient calculation of specialized routines while some other languages keep the software applications separate from the operating systems and thus become truly platform independent.

Programming languages can be divided into two different types, imperative and declarative, based on the way in which we can write the code using the programming languages. Imperative languages tell the computer how to perform a task, whereas declarative languages tell the computer what to do. Imperative languages can be further divided into procedural languages and object-oriented languages. Programs written in procedural languages consist of procedures and structured actions (e.g., Fortran, Pascal, and C). Programs written in object-oriented languages consist of objects and classes rather than actions (e.g., C++, C#, Java, and Python). Declarative languages can be further divided into functional languages and logic-based languages. In functional languages (e.g., Common Lisp, Erlang, and Haskell), programs consist of functions. A function, as we know in mathematics, accepts a set of arguments and returns a value. Logic-based languages (e.g., Datalog, Logtalk, and Prolog) consist of statements expressed in a logical form by using some facts and rules that are true in the given problem domain. Figure 5.23 presents this categorization of programming languages. Some programming languages belong to more than one (sub)category. For example, Logtalk is a logic-based language as well as an object-oriented language.

There are other ways of categorizing programming languages, based on the characteristics we use for such categorizations. For example, programming languages can be categorized as static versus dynamic, based on when they execute several common programming behaviors. However, these types of discussions are beyond the scope of this book.

Object-oriented programming languages are used extensively nowadays because of their inherent advantages over procedural programming languages. More details on object-oriented languages are provided in Chapter 7.

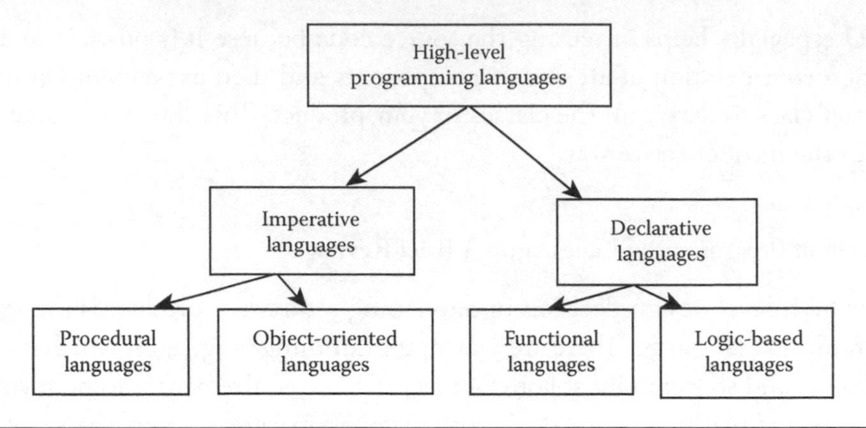


**Figure 5.23** Categorization of programming languages.

### *5.14.1 Special Languages for Web-Based Systems*

Some programming languages are developed to do special tasks. For example, there are programming languages developed for scripting on web pages. Some of these scripting programming languages include JavaScript, VBScript, and Python. Markup languages such as HTML were developed to create a format for the web pages. There are also languages such as JSP that embed the programming code inside the web pages that run on the server but not on the user browser (contrary to scripting languages [such as JavaScript] that run on the user browser). We will discuss scripting languages in Chapter 6.

We will discuss some nonfunctional design issues, namely, security and performance, in Sections 5.15 and 5.16, respectively.

## 5.15 Security

A top-priority nonfunctional requirement for web-based software products is security. Ensuring that users can use a software product without worrying about data theft is a real concern for software engineers. Another concern is that a person authorized to view or execute some specific transactions should be allowed to view or execute those transactions only. Security can be enforced in many ways. Let us discuss some of them here.

### *5.15.1 Authorization*

Any system can be secured by providing access to authorized people only. Authorization is implemented by presenting a login screen to users. The user enters a username and a password. The system checks the values for the username and password in the database. If the values match with the data stored in the database, then the system is unlocked for the user. Now, the user can successfully enter the system and access it.

Nowadays, a highly secured login access can be implemented for a software product that needs foolproof security. If you need to implement a software product that has a strong security mechanism, then you can use software components such as security sockets that help provide a strong security mechanism.

### *5.15.2 Role-Based Security*

Resource management software products (e.g., Enterprise Resource Planning [ERP] software products) are used by people whose roles are well defined by their companies. We can define the roles and attach them with the specific functionalities of the software product. When a new user (i.e., user ID) is created in the software product, a role is attached to that user ID. Whenever the user logs into the software product, the user will see only the functions that are enabled for the role attached to his or her user ID. For example, suppose we have a procurement management system. In this system, a buyer role can create, view, and edit the purchase orders. If a user ID is attached to this buyer role, then the user can do these transactions related to the purchase orders. There is also a role defined as inventory assistant. This role can receive, dispatch, and view the inventory in the warehouse. If a user is assigned this role, then he or she can perform these transactions. However, the user assigned this inventory assistant role will not be able to do anything related to purchase orders.

### *5.15.3 Security Based on the User Data*

In secure software systems, the security is tied to each user ID. A user can only view and perform the transactions that are applicable to the data of that user ID. The user will never be able to do any transactions on the data that are linked to some other user ID. For example, the online banking system allows a user to do transactions on the data related to that user ID only.

### *5.15.4 Data Encryption*

Encryption is the process of converting information from its original form to another form so that it cannot be easily interpreted by anyone except the designated user(s) of that information. There are basically two types of encryptions: symmetric key encryption and public key encryption. In the first type, the sender and receiver(s) of a message will have the same key to encrypt and decrypt the message, respectively. In the second type, the encryption key is publicly available, but the decryption key is available only to the receiver(s) of the message.

Data encryption may take place during an entire session when a user interacts with the software system. This type of encryption is implemented through Secure Socket Layer or Transport Security Layer. All the data sent from the system (to the user) and received by the system (from the user) are encrypted and decrypted using any available

64-bit or 128-bit encryption technology. If a hacker hacks the system, he or she will get the encrypted data only. You can see an example of encryption when you visit a website. If the website has implemented encryption, then you will be directed to a URL such as https://www.abc.com instead of http://www.abc.com. The user interaction sessions with "https" URLs are always encrypted. A detailed discussion on data encryption is beyond the scope of this book.

### 5.16 Performance

Another important nonfunctional requirement for web-based software products is performance. Web-based software products are accessed by the users through the Internet. Connection speed greatly determines the performance of any website. If the connection speed is low, then the web pages take a long time to load. This can create problems for the users. Sometimes, when the website is being accessed by a large number of users simultaneously, even if the connection speed is high, the web pages load slowly. These issues should be considered while designing the software product.

From a software design aspect, some of the factors that affect the performance of a system include the following:

- Large software components take a long time to read or load (depending on whether they are read on the client side or the server side). Therefore, always keep the component size of your software products as small as possible.
- Connecting to the database is a time-consuming process. If your source code connects too often to the database, then the response time from the system will be long and the user has to simply wait for the response. Use data caching when the data are fetched from the database of the system so that you do not need to connect to the database often. For example, if you need to get only a few records in one operation from the database, but the system finds that you will be using nearby records from the same database table frequently, then fetch as many records as possible (from that database) in one connection. The system can cache the unused data or records, but when required by the user, the system can pass these data to the user screen. A group of data can be fetched from a database by means of a data set. A software design can be built where the data set is larger than the required data. The unused data can then be cached. When a succeeding request from the user comes for some data, the required data can be fetched from this cache. This way, the response from the computer will be faster and the user will have a better experience.
- Do not use images or blobs (objects) that are large in size. Loading them on a user browser will take more time. Instead, if possible, use partial images that may be smaller in size. They will load faster and the response time will be better.

## 5.17 Case Study

After considering all the factors regarding the high-level design, we have decided to build OBAAS with the following features.

OBAAS will be a web-based software product. Users can access the software product using a browser (any web browser will work). OBAAS will be deployed on a web server having an application server as well. There will be a database (any relational database will be fine) at the back end to store the permanent transaction data. The web server will host the website and all its web pages. The web pages will have some client-side scripts for formatting. The web pages will also have server-side scripts that will run on the application server. There will also be some executable source codes on the application server. These codes will be called through web pages that have server-side scripts.

Let us see the high-level specifications for our solution.

### *5.17.1 Three-Tier Architecture*

The three-tier architecture will give us an opportunity to use the client-side scripts, the server-side scripts, an executable server code, and a database connection. This architecture will be flexible and we can choose any database or any application server. Maintaining this infrastructure will not be a problem.

### *5.17.2 User Interface*

A user interface will be built for web browsers and the details are provided in Chapter 6. The web pages are built using HTML. Client-side scripts are used for user input validation. We will use static content + client-side scripts + server-side scripts + executable server code in OBAAS. Once all the client-side and server-side scripts as well as executable server codes are run, a web page is composed and the user will be able to see it.

### *5.17.3 Middle Layer*

The middle layer will be composed of server-side scripts written on each web page. There will be some executable compiled binary code on the application server that will be called from the server-side scripts written on each web page. The database connection information, database connection management, and servlet control will be done through the executable server code.

### *5.17.4 Database*

The Oracle database has been used for OBAAS 1.1 and OBAAS 1.2.

*5.17.5 Component Diagram*

In Chapter 4, we have seen that there is a requirement for two versions of OBAAS. These versions are named OBAAS 1.1 and OBAAS 1.2. Accordingly, Figure 5.24 shows the component diagram for OBAAS 1.1 while Figure 5.25 shows the component diagram for OBAAS 1.2.

For web-based software products, a web page can be considered a component. This is because each web page serves one purpose and can be safely represented as a component.

In Chapter 4, we have seen the requirement specifications for OBAAS (Versions 1.1 and 1.2). We have also seen the use cases for it. On the basis of the use cases, the component diagram for OBAAS is made.

- Account balance: The use case states that the user can find the account balance of his or her bank account. In the component design, the account balance page searches the account database to find the account balance for the user.
- Service request: The use case states that the user can request a new cheque book from the bank. The user can go to the service request page. The user selects the cheque issue service request from a drop-down list at the service request page. When a cheque book is issued, the service database is updated.
- Bill payment: The use case states that the user can pay the bills using OBAAS. In the component design, when the user wants to pay a bill, he or she goes to the Bill payment page. The user selects a service provider from the list and then can pay the bill amount from the Bill payment page. After bill payment,

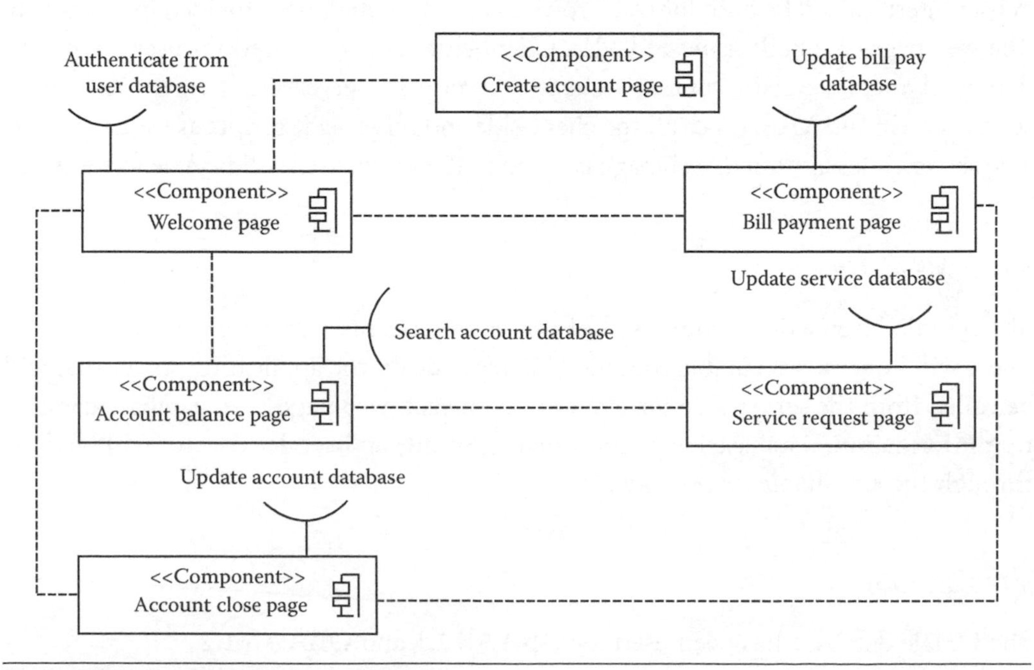


**Figure 5.24** Component diagram for OBAAS 1.1.

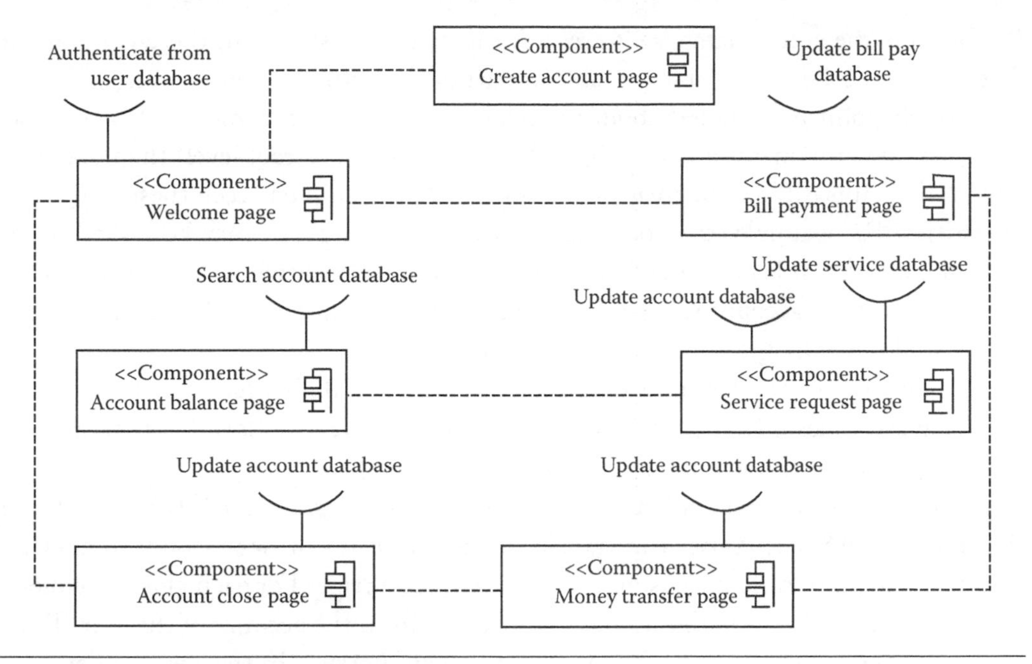


**Figure 5.25** Component diagram for OBAAS 1.2.

the user account is updated because the bill payment is done from the user account.

- New account: The use case states that a user can create a new account in OBAAS. In the component design, the account creation page allows the user to create his or her account in the system.
- Account closure: The use case states that the user can close his or her online bank access account. In the component diagram, an account closure page is provided so that the user can close the account.

In Figures 5.24 and 5.25, you can see that each web page is connected through dotted lines with some other web pages. However, unlike in Figure 5.15, there are no arrows on these dotted lines. This is because all the web pages in OBAAS are independent of each other. A user can go to any of these web pages directly. When a user selects any web page, then he or she needs to provide his or her username and password (along with other required information) on that web page itself. This kind of functionality for a web-based software product can be achieved without using the session objects. Notice that in Figures 5.24 and 5.25, due to simplicity reasons, only some sample connections are shown between the web pages. In reality, there are connections (i.e., dotted lines without arrows) between any two different web pages.

You will notice that we have used the terms *service database* and *user database* in the component diagrams. When the component diagrams are made, the actual database is not yet finalized. The component design activity is carried out before the detailed design activities. Actual databases are created during the detailed design process.

By the time the actual database is created, the databases shown in these component diagrams could be just a database table or part of the actual database table(s).

Another point to be noted about the relationship between components and databases is that components are at a higher level (i.e., more abstract level) than database tables. Thus, it is difficult to create a one-to-one relationship between these two types of entities. For example, the component "Account close page" may be linked to an account table, but it is not sure whether the account table may also contain some other entities or just the account-related data. This information becomes clear only after the classes and the entity-relationship diagrams are made.

The same is true for data flow diagrams. They are at a higher level than the classes and database tables. Hence, the "data stores" shown in the data flow diagrams are just approximations of the actual database entities.

There are two differences between the component diagram for OBAAS 1.2 and that for OBAAS 1.1. A user can transfer money from his or her account to another account. The other difference is on the cheque request page. For each cheque request, the user is charged US$5. This amount is deducted from the account of the user. Thus, on the cheque issue page for OBAAS 1.2, there are updates in the account database and in the service database.

### *5.17.6 Data Flow Diagram*

A data flow diagram for OBAAS 1.1 is shown in Figure 5.26. The user (customer) initiates all the transactions (processes). The bill payment process updates the account data store. The account balance process accesses the account data store to display the account balance of the customer.

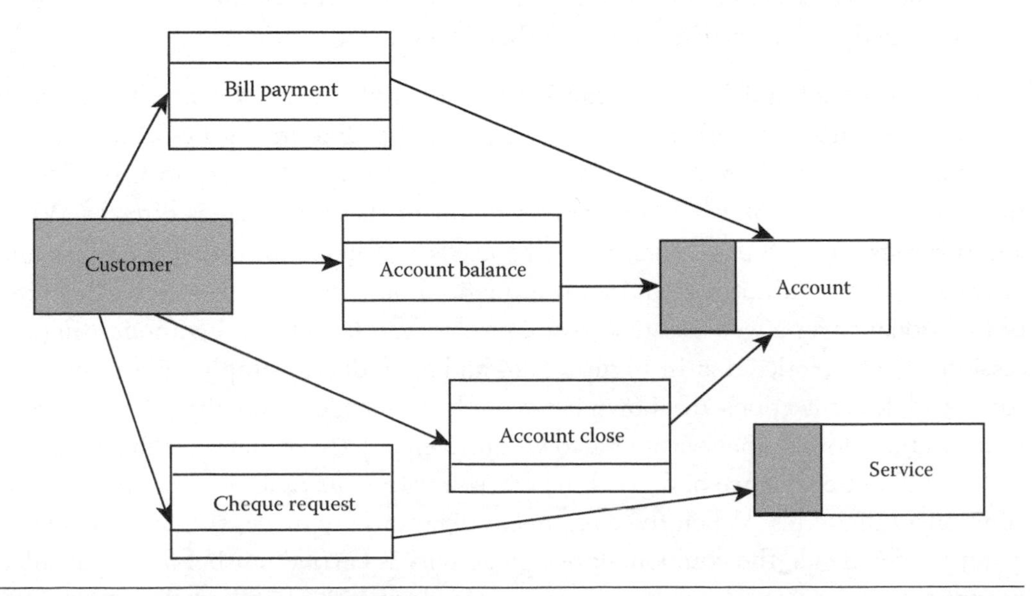


**Figure 5.26** Data flow diagram for OBAAS 1.1.

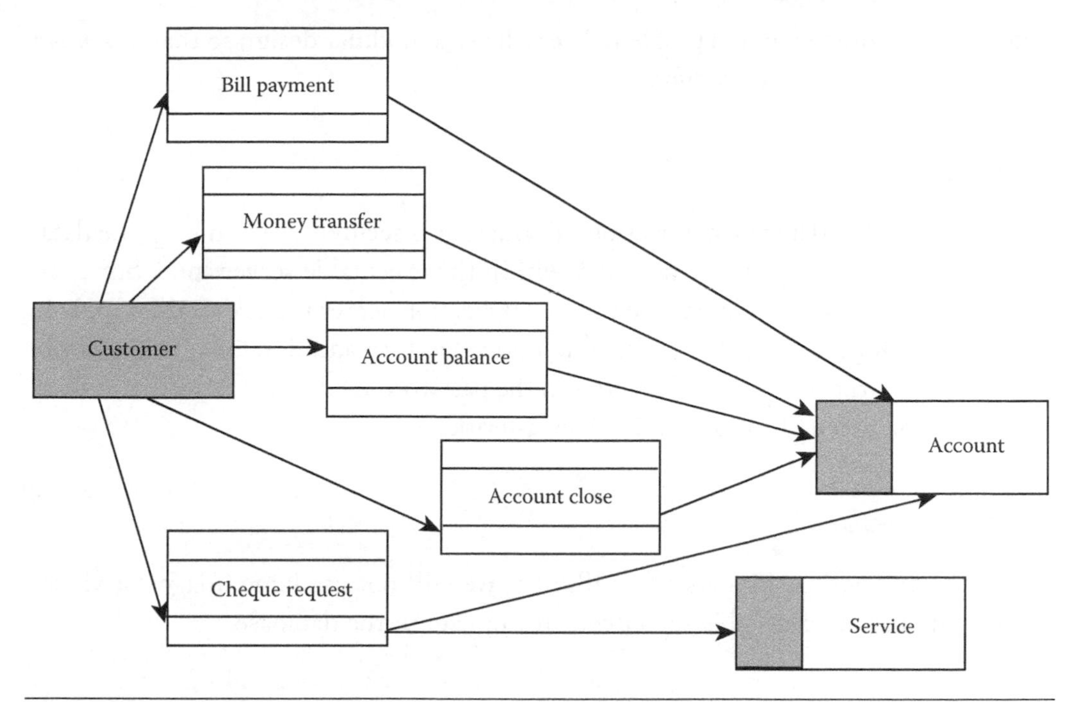


**Figure 5.27** Data flow diagram for OBAAS 1.2.

The account closing process updates the account data store. The cheque request process updates the service data store.

Figure 5.27 displays the data flow diagram for OBAAS 1.2. There are two differences in this diagram compared to that for OBAAS 1.1. For the cheque request process, both service and account data stores are updated. There is one additional process in OBAAS 1.2. A money transfer process has been added. This process updates the account data store.

### *5.17.7 Code Reuse*

For OBAAS 1.2, we will be using the same code base of OBAAS 1.1. The changed functionality of cheque request will be achieved by adapting the cheque request component using refactoring. An additional component for money transfer is added in OBAAS 1.2.

We will create a component for the database access definition. This component will be used to connect and manage all the database connection requirements for the entire OBAAS. Thus, the same code for the database connection management will be reused every time a connection with the database is required from any component of OBAAS.

### *5.17.8 Maintainability*

We will use an object-oriented programming language and create the classes that can be maintained. When the code base becomes larger, we will be able to refactor our code so that maintenance is a problem. We will use a modular design so that it is easier to maintain our software product.

*5.17.9 Security*

To ensure that the database connection information is secured, we will keep the database connection string information only within the executable server code. Since the executable server code is binary, even if the application server is hacked, this database connection information will be difficult to understand. In addition, as a basic security measure, when the user enters characters in the password field, OBAAS will not show the actual characters. Instead, it will show asterisks.

*5.17.10 Performance*

To guarantee good performance by OBAAS, we will not use large images or videos and we will use an executable server code to connect to the database.

*5.17.11 Methodology*

Our technology (i.e., three-tier architecture and object-oriented programming and design) allows us to work using any software engineering methodology. We can build OBAAS incrementally, if required. We can also cope with all product requirements and build a complete product under one project.

### 5.18 Chapter Summary

After identifying the requirements for building a software product and deciding to go ahead with the project, the project team has to brainstorm how to implement a software system that provides a solution to meet all the requirements. The result of this brainstorm is a high-level design of the project. What technologies are available and how these relevant technologies should be selected are some of the tasks to be performed at this level. You need to find out whether a web-based solution will be the best fit or a client–server architecture will suffice. You then need to determine whether a two-tier architecture is good enough or an *n*-tier architecture will be required. You need to find the right solutions for providing adequate security and performance.

Your software product design will depend on factors such as software product size (an estimate), availability of the technologies, methodology used, product type, and nonfunctional requirements.

The first computers were the mainframe type. In the 1970s, client–server computers became popular. After the advent of WWW, web architecture became popular. Because of the inherent benefits of web architecture, client–server architecture has become almost obsolete.

In essence, high-level design and modeling are carried out at three levels. At the top level, when we think of the entire software product and its parts, the software design is known as software architecture. At this level, the software design is broken into layers or tiers. You can use architecture patterns, which are industry best practices, available in the form of templates. For example, the three-tier architecture is an architecture pattern. You can implement this pattern to architect a software product.

At a lower level, the software architecture parts need to be further broken down into smaller parts so that each architecture part (e.g., a middle layer) can be designed in a modular way. At this level, we design the software components; thus, this level is known as software component design. At this level, the design involves creating component diagrams and data flow diagrams. Component diagrams depict the components, their relationships, dependencies, coupling (integration) points, and so on. Data flow diagrams depict the flow of data from the point of initiation (generally the user) to various components in response to some user action.

Software design patterns have been made to create a software design for a software product. These design patterns are the standard means to solve computation problems. There are many design patterns. They are divided into three categories: creational design patterns, behavioral design patterns, and structural design patterns.

Apart from architecture and design concerns, a high-level software design should also consider the limitations of the available programming languages and their associated technologies. Not all programming languages are equal. In fact, most programming languages were created to address some specific issues related to software development. For example, Java was created to achieve platform independence. MATLAB® was created to write software programs for heavy-duty engineering and mathematical computations. Programming language selection thus becomes a task that has to be addressed during high-level software design.

## QUESTIONS

1. Explain what a server-side script is.
2. Explain web architecture in detail.
3. Why does one need a high-level design?
4. What types of security mechanisms can you implement for your software product?
5. What are the benefits of a web-based architecture?
6. Why are platform-independent software products needed?
7. What is a client-side script?
8. What is a three-tier architecture?
9. What is a multitier (*n*-tier) architecture?
10. What measure(s) can you take to ensure good performance from a software product?
11. What is a software architecture pattern?

12. What is software architecture?
13. What are the software components in various software architecture layers?
14. What is a component diagram?
15. What is a data flow diagram?
16. What is a software design pattern?
17. Name some software design patterns.
18. What is a software component?

## Recommended Reading

Rod Ellis (1996), *Data Abstraction and Program Design: From Object Based to Object-Oriented*, 2nd Edition, Taylor & Francis, London.

Information Resources Management Association (USA) (2014), *Software Design and Development: Concepts, Methodologies, Tools and Technologies*, IGI Global, Hersey, PA.

Ivan Mistrik, Anthony Tang, Rami Bahsoon, Judith Stafford (Editors) (2013), *Aligning Enterprise, System, and Software Architectures*, IGI Global, Hershey, PA.

Katalin Popovici, Frederic Rousseau, Ahmad A. Jerraya, Marilyn Wolf (2010), *Embedded Software Design and Programming of Multiprocessor System-on-Chip: Simulink and System C Case Studies*, Springer, London.

Dev G. Raheja, Louis J. Gullo (Editors) (2012), *Design for Reliability*, John Wiley & Sons, New Jersey.

Ben Shneiderman, Catherine Plaisant, Maxine S. Cohen, Steven M. Jacobs (2009), *Designing the User Interface: Strategies for Effective Human-Computer Interaction*, 5th Edition, Pearson, New York.

# 6

# Software User Interface Design and Construction

**In Chapter 5, we learned**

- **What a software high-level design is**
- **What a software architecture pattern is**
- **What software architecture is**
- **What a software design pattern is**
- **What a software component design is**
- **What high-level software design technological considerations are**

**In Chapter 6, we will learn**

- **What a user interface is**
- **What the elements of a user interface are**
- **What HTML is**
- **What AJAX technology is**
- **What a Model–View–Controller design is**
- **What a client-side script is**

## 6.1 Introduction

In a software product, the user interface is the part that is visible to the users of that software product. The information processed at the business logic layer is displayed at the user interface (unless the software product is designed to send its output to an entity such as a program, a file, or hardware). The user interface is also a place where the user provides inputs for the software product to process. Thus, the user interface is used both for submitting the inputs and for getting the outputs.

Apart from the consideration of how to display the processed data, there is also a concern about what the user interface should look like. Aesthetics is an important aspect of user interfaces. The user interface should be interesting to the users for whom the software product is developed. At the same time, the user interface should allow the user to work productively. If the user needs to provide a lot of input that is not necessary to get the output, then the productivity of the user will be affected. Similarly, if the output from the product contains information that is not required by the user, then it may confuse him or her.

In the early days of computing, the user interface used to be only text based. However, after the advent of Windows and graphics, user interfaces have evolved to display even three-dimensional images.

Most users use a mouse and a keyboard to interact with the user interfaces of any software product. Some specialized software products also use special input devices such as sensors, electronic pens, and keypads.

The introduction of smartphones has made the development of user interfaces an exciting new area in software engineering. There are several challenges involved in developing user interfaces for these devices. These include different interface sizes, interactive screens, touch screens, and fulfilling the needs of different kinds of users, to name a few. Since the memory and the processor in these devices are smaller and less powerful than those of common personal computers, the user interfaces must consume less memory and processing resources than those of personal computers. Hence, providing good user interfaces for these devices is a real challenge.

There is another discipline called human–computer interaction (HCI) that is closely related to this chapter. As defined by the Special Interest Group on Human–Computer Interaction of the Association for Computing Machinery, "Human Computer Interaction is a discipline concerned with the design, evaluation and implementation of interactive computing systems for human use and with the study of the major phenomena surrounding them." HCI is a vast and mature discipline. Because of space limitations, HCI is not further discussed in this book. For more information, the readers are referred to any good book on HCI.

### 6.2 Graphical User Interface

The user interface of any software product is an important consideration. Most software products are used by the users (the only exception is software products such as device controllers for robots, where the user interface may not be a part of the software product). User interfaces enable the user to provide his or her input to a software product so that it can compute and display the appropriate output at the user interface.

In the early days, computers used to have text-based user interfaces. The text-based (also known as character-based) user interfaces would display output in the form of text only. In real life, most people communicate with each other through speech, images, multimedia, and text. So why should a computer interact with a user only through text? Communicating through graphics (images, videos, etc.) is more powerful than through text.

Usage of graphics in computer programs requires a lot of computer processing power. Early computers did not have the processing power to display graphics. However, when computer hardware technologies became advanced and powerful, computer output in graphic form became possible. This is when Windows-based software products started to be developed. Companies such as Microsoft, Apple, and

IBM started shipping personal computers that used to have software products and operating systems containing graphical user interfaces (GUIs). Indeed, GUIs revolutionized the software industry so much so that computers used by people in their office work could be used in their homes as well.

Early GUIs were used in the products that deal with operating systems and client–server applications. Later, with the advent of technologies such as the Internet and smart mobile phones, many other forms of GUIs emerged. Technologies such as thin client, thick client, and so on also helped introduce various kinds of GUIs.

### *6.2.1 Rich Windows-Based GUIs*

When client–server software products were built, the client part of the product used to contain a rich Windows-based GUI. The client part of the software product would be installed on all the computers that would be used to run as the client parts for the software product. They would be connected to the server part (installed on a computer that would work as a server) through computer networking.

Rich Windows-based GUIs are dependent on the operating system on which they are installed. Thus, a software product's client part, which needs to be installed on a user's computer, must be programmed in such a way that it can be installed on the operating system of that user's computer. For example, if the installation program is written for a Windows operating system, then this program will not be able to install the rich Windows-based GUI on a Linux operating system and vice versa. If you are developing a client–server software product, then you need to think about this aspect of client part and rich Windows-based GUI.

Windows forms provide GUI components to client–server applications. Several user interface developers work with Windows forms when a rich Windows-based GUI needs to be developed. A Windows form is specifically helpful if your goal is to build an application that uses the processing power of your local computer and performs processor-intensive tasks such as documents, databases, graphs, and digital video. The user interface developer first selects a Windows form and then chooses the user interface control elements (discussed in a later section) to put them on the Windows form to create a rich Windows-based GUI.

Some of the important characteristics of a rich Windows-based GUI include responsiveness, security, and access to local resources. Rich Windows-based GUIs run on a local computer. Hence, responsiveness of the user interface is fast compared to the user interfaces that do not run on a local computer (in such cases, they run on a server computer). Windows forms can have complete or partial access to the local computer resources. For example, you can control whether your Windows form can access the file system of your local computer when you need to upload a file from your local computer to a server (using a Windows form in a software product).

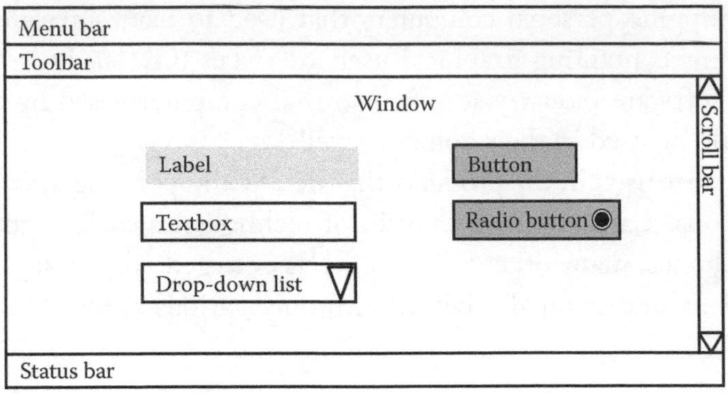

**Figure 6.1** Rich Windows-based GUI with various user screen control elements.

Windows forms are used to build client–server-based software products. Since Windows forms are part of a software product that is directly installed on any operating system, Windows forms can access any part of the operating system of the computer on which they are installed. Still, security is not a big issue with the software products that are built using Windows forms. This is because these software products are deployed behind a firewall and are not accessible outside the computer network inside which these software products are installed. The only security issue with such software products is ensuring the security of personal information inside the computer network in which they are installed. In contrast, web-based software products are exposed to all kinds of security threats because they are open to the entire World Wide Web (WWW).

Figure 6.1 depicts a rich Windows-based GUI with its various user control elements. We will learn about user control elements in a later section.

### *6.2.2 Web Browser*

A web browser is a GUI that was developed for web-based software products. Although a web browser looks very similar to a rich Windows-based GUI, it has many differences from a rich window. The browser itself is installed on the user computer. However, the content that runs inside a web browser actually runs on a web server (in the case of a rich Windows-based GUI, most of the processing takes place on the local machine). For example, if a user points his or her web browser to www.microsoft.com, then the content of this website actually runs on a web server maintained by Microsoft Corporation.

There are some technologies that allow the software programs to run locally for a web browser. These technologies are together known as client-side scripts. However, in most cases, the software programs and actually the entire software product run on a web server and it becomes available on a web browser of the user through Hyper Text Transfer Protocol (HTTP).

User interface developers develop a web application in the form of web pages. Each web page can contain a web form, which is also known as a Hyper Text Markup

Language (HTML) form. Interactive or dynamic graphic user control elements require the data to travel from the user interface to the web server and back for the update when they are used on web forms. This can affect the responsiveness of web-based software products. The responsiveness of web pages is slower (when compared to Windows-based GUIs) because all the content of a web page runs on a web server but not on the local computer. To increase the responsiveness, technologies such as client-side scripting, client-side validation, and so on are used. To connect to a web application (e.g., a website), a user computer needs to be connected to the Internet first. The Internet is a network of computers and any piece of data can flow among the Internet-connected computers. These pieces of data can also be malware. Malware can connect with a user computer even without any authorization. If this happens, then the malware can install itself on the user computer and damage the data on that computer. Most malware can enter a computer through a web browser. For this reason, all kinds of web browsers have some built-in security mechanisms that can prevent any dangerous computer programs from running on the web browser. This is why most web browsers cannot access the local computer resources on their own. Otherwise, some dangerous computer programs could access the operating system of the local computer (of the user) through the web browser and cause damage to the data residing on that local computer. Thus, browser security may prevent any computer application from accessing the resources on the local computer. If a user interface needs to access a local computer, then that interface needs to have a permission mechanism in its source code that may ask the user to allow that access. For example, if the software application has a file that can be downloaded to a computer, then the source code of that interface will ask the user (of that computer) to give permission to access the local computer for saving the downloaded file. This security mechanism ensures that no unauthorized file or computer program from any source on the Internet has access to the resources (files, programs, etc.) residing on the local computer.

Web forms are based on a model in which different components are loosely coupled with the front end. In general, these application components that are composed of web forms are invoked through HTTP. HTTP is an asynchronous and stateless model to transfer data from one computer to another over a network. If a web-based software product is deployed over the WWW, then the network is the Internet. HTTP is not the right model for environments in which a lot of interactions are needed from the users while the server is performing high-volume transactions. In addition, web forms are not suitable if high levels of concurrency control are needed while processing the databases.

However, these limitations (of web-based software products) are now being overcome using available hardware and software technologies. The hardware capabilities of computers nowadays are very high. Therefore, a website can easily support more than 10,000 simultaneous connections. Even at such high throughput, the performance of the websites does not fall below some acceptable limits. The security mechanisms have

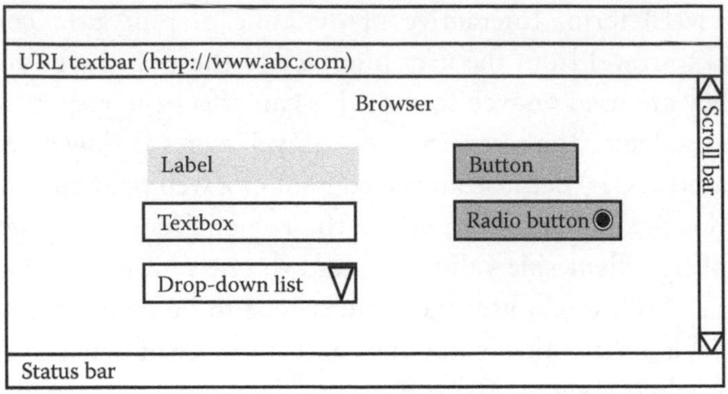

URL textbar (http://www.abc.com)
Browser
Label
Button
Textbox
Radio button
Drop-down list
Scroll bar
Status bar

**Figure 6.2** Web browser with user control elements.

also become sophisticated and users can safely visit websites and perform their transactions without fear of data theft or impersonation. The stateless nature of HTTP is also overcome using session objects. These session objects ensure that the user information is kept throughout the period when the user is browsing a website. We will learn about session objects in Chapter 7.

Figure 6.2 depicts a web browser with various user control elements. If you are designing a web application, then you need to consider the issues discussed in this section.

## 6.3 Graphic Control Elements

Nowadays, applications that use text (character)-based user interfaces are rarely developed. Predominantly, software products developed these days use GUI. Therefore, it is important to learn the details of GUI.

Whether you are developing a rich client application or using a browser as the user interface, most of the user control elements are the same. Let us understand some of the user control elements that are most commonly used.

Some of the commonly used user screen elements have been depicted in Figure 6.1. The window you see here is a rich Windows-based GUI. These rich Windows-based GUIs are used when the user interface is based on a common window format. The browser-based user interface and its elements (Figure 6.2) are very similar to the ones depicted here for a rich Windows-based GUI.

User screen elements can be broadly categorized as containers, selection and display elements, navigation elements, input elements, and output elements. We will discuss each of these now.

### *6.3.1 Containers*

Containers are the topmost elements in the user interface and they can hold all the user screen elements. In the typical Windows-based client–server software products, the top-level container is the Windows. For web-based applications, the container for the user interface is the browser window. Some commonly used container elements are explained in the next few paragraphs.

*6.3.1.1 Window* Windows are the topmost user interface elements. All other elements are contained within a window. Rich screens are used when the user interface is window based (Figure 6.1). In the case of Internet-based software products, the container is a browser window (Figure 6.2). Windows can be resized as well as placed anywhere on the user screen (desktop).

*6.3.1.2 Modal Window* The modal window is a variant of the common window. The modal window cannot be resized. It acts like a child window for the main window and the user must interact with the modal window before he or she can resume his or her interaction with the main window. The user also cannot change the focus away from a modal window until some action (from the user) closes this window. Generally, on a modal window, there are two buttons, Yes and No. The user has to click one of them. Once one of these buttons is clicked, the modal window is closed. Figure 6.3 depicts a modal window.

*6.3.1.3 Dialog Box* Dialog boxes are used in a software product when the user must make a decision by pressing a button on the dialog box. The user cannot change focus away from a dialog box until he or she chooses to dismiss the dialog box by selecting and pressing a button on it. Dialog boxes are always a type of modal window. Figure 6.3 depicts a dialog box inside a modal window.

*6.3.1.4 Frame* Frames are used to keep many user screen elements together. For example, if you have many groups of different screen elements that you want to keep separate from each other, then you can use frames to contain a group of similar user elements inside one frame, keep another group of similar elements inside another frame, and so on. Figure 6.4 depicts two frames that can be placed on the same window to segregate the user elements from each other.

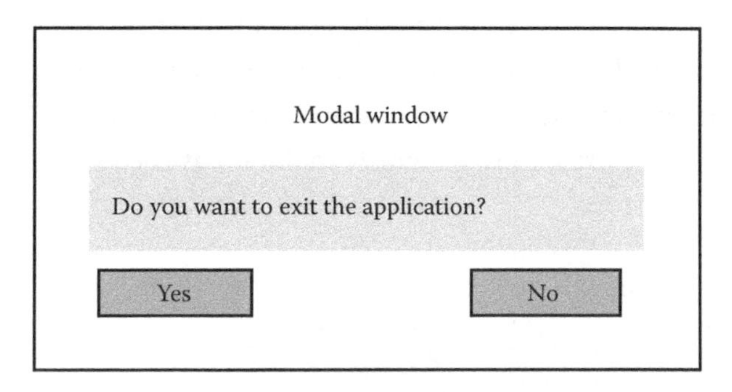


**Figure 6.3** Modal window.

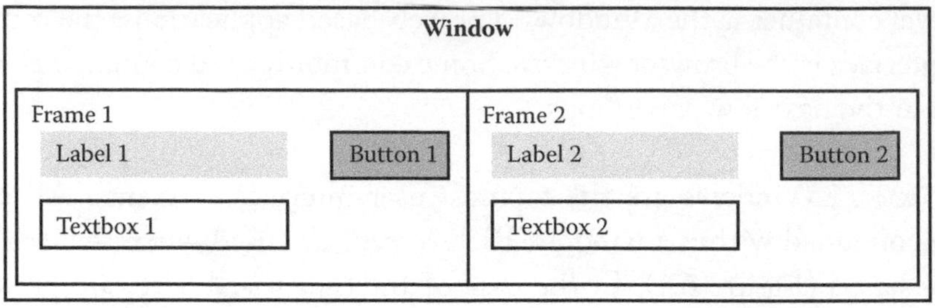

**Figure 6.4** Two frames on the same window containing various user control elements.

*6.3.1.5 Canvas* A canvas is similar to a frame. It contains palettes instead of user elements. For example, a canvas can be used to contain a palette of colors.

*6.3.1.6 Form* Forms are used on any user screen to keep all the user elements together. Forms are of two types: window forms and web forms. Window forms are used to create rich Windows-based software applications. Web forms are used to create web pages for a web-based software application. During processing, the forms take values (entered by the user) from the user elements to the business logic layer. Forms are especially useful on a web page. A web page may contain lots of information and user screen elements. However, for dynamic processing of the inputs provided by the user, only one form is provided on any specific web page. The user enters the values in the textboxes and other user elements (that are inside the form) and then presses a button commonly called the Submit button. The information provided by the user on the form is then sent to the business logic layer. Figure 6.3 displays a dialog box. This dialog box is made by putting a web form on a web page and then adding two buttons (Yes and No) and a label.

### *6.3.2 Selection and Display Elements*

There are a large number of user elements that can be used to display the contents and select the values and they are explained next.

*6.3.2.1 Button* Buttons are very commonly used on user screens. When a button is clicked, the program will make a decision (based on which button is clicked) and perform some calculations or take the user to some new window. For example, the Next page and Previous page buttons are generally included at the end or top of a web page. Figure 6.5 depicts a button.

**Figure 6.5** Button user control element.

Radio button

**Figure 6.6** Radio button user control element.

*6.3.2.2 Radio Button* The radio button has a clickable small circle. Each radio button has two states: on and off. Radio buttons are used along with the answers associated with a question. For a given question, there are multiple answers (options), with each answer preceded by a radio button. The user can select an answer by clicking the radio button that precedes that answer. Notice that it is not possible to select more than one radio button (answer) for a given question. On the basis of the radio button (that is representing the answer) that the user selected, the computer program will perform some calculations or navigations. For example, for the question that asks you to select your age group, you can select only one choice. Figure 6.6 depicts a radio button.

*6.3.2.3 Checkbox* Checkboxes are similar to radio buttons and each checkbox has two states: on and off. When a checkbox is checked, then its status is on, and when the checkbox is unchecked, then its status is off. The main difference between a radio button and a checkbox is that the user can select more than one checkbox to answer a given question. For example, for the question that asks you to select the names of the cities you visited, you can select more than one city. Figure 6.7 depicts a checkbox.

*6.3.2.4 Slider* Sliders are used to adjust the value of an element on the screen and the value is generally between 0 and 100. For example, if you want to adjust the speaker volume of a music program such as Winamp player, then you can do so by adjusting the speaker volume slider provided on the user interface of Winamp. Figure 6.8 depicts a slider.

*6.3.2.5 List Box* A list box contains a static list of names (options) and the user can select one or more of them. Each name in the list is in a separate line. Figure 6.9 depicts a list box.

**Figure 6.7** Checkbox user control element.

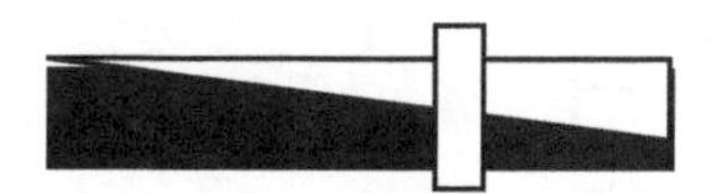

**Figure 6.8** Slider user control element.

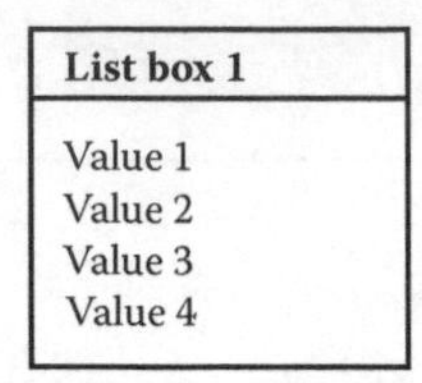


**Figure 6.9** List box user control element.

*6.3.2.6 Drop-Down List* Drop-down lists are similar to the list boxes just explained. The difference is that while the list box, by default, displays all the names inside the box, the drop-down list displays the list of names only when you click the drop-down list box. If you do not click it, it is in a collapsed state and occupies just one line. On the basis of the design of the drop-down list, you can select one or more names from it. Figure 6.10 depicts a drop-down list.

*6.3.2.7 Menu (Menu Item)* Menus are one of the fundamental elements that are invariably used in all software products to display different windows of the software product. Menus are also used to change the functionality of the software product, for example, showing or hiding a canvas or a dialog box. Menus sometimes consist of submenus, in which case, they are hierarchically organized. Figure 6.11 depicts many menus on a menu bar.

*6.3.2.8 Context Menu* Context menus are a variant of standard menus. These menus are generally hidden but displayed on the user screens depending on the context in which they are opened. For example, when the user clicks the right mouse button on a web browser, a context menu opens.

*6.3.2.9 Menu Bar* The menu bar is generally provided at the top of any user window. It contains all the menu items that are provided for the software product. Figure 6.11

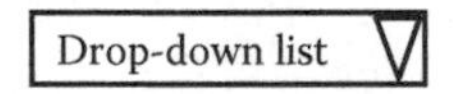


**Figure 6.10** Drop-down list user control element.

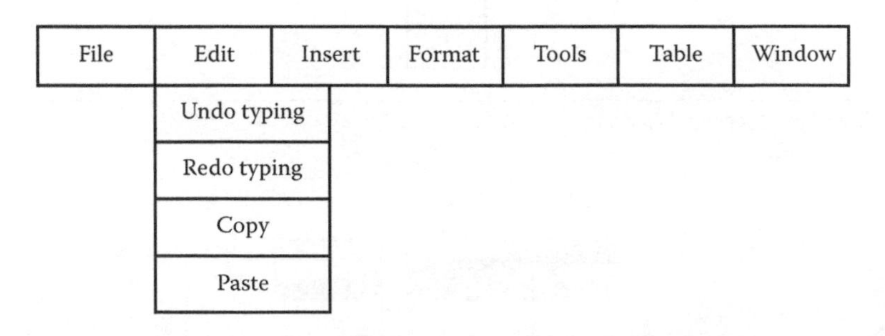


**Figure 6.11** Menu bar with main menus and drop-down menu user control element.

**Figure 6.12** Toolbar with many tool buttons.

depicts many menus on a menu bar. The figure also shows drop-down menus (Undo typing, Redo typing, Copy, and Paste) from the menu item Edit. These drop-down menus are submenus of the main menu.

*6.3.2.10 Toolbar* The toolbar is generally provided at the top of any user window just below the menu bar. It contains those tools that are most useful for perform common tasks inside the software product. Figure 6.12 depicts a toolbar with many tool buttons.

*6.3.2.11 Icon* Icons are small images that can be used to denote some useful information (to the user) inside the user screen. Icons are also used on buttons and in those cases they are known as icon buttons.

### *6.3.3 Navigation*

For navigation purposes, some user elements such as links, tabs, and scroll bars are used. Navigation user control elements are important because, by using them, users can navigate on a single window or web page from one user control to another. Navigation user elements also allow the users of any software product to navigate from one window to another or from one web page to another.

*6.3.3.1 Link* Links are extensively used on browsers to navigate the user from one web page to another. A link can appear anywhere in a web page. For example, http://www.microsoft.com is a web link. A click on this link will enable a web browser to navigate to the site of Microsoft Corporation.

*6.3.3.2 Tab* A tab is a user element that can be shown or hidden as the user clicks the tabs. Tabs are a good way of organizing a large number of windows inside the main window. Each inside window is actually a tab. When the user clicks a tab, all the contents of that tab will become visible on the user screen and, at the same time, the contents belonging to the other tabs that were closed will become invisible. Only one tab can be opened at a time. Figure 6.13 depicts tabs for browser windows. When one tab is clicked, a corresponding browser window is opened. When this new window opens, then the previous window is closed.

*6.3.3.3 Scroll Bar* Scroll bars are provided at the right-hand side of most browsers and rich windows for the users to scroll up or down to display the contents of the window. Figure 6.14 depicts a scroll bar.

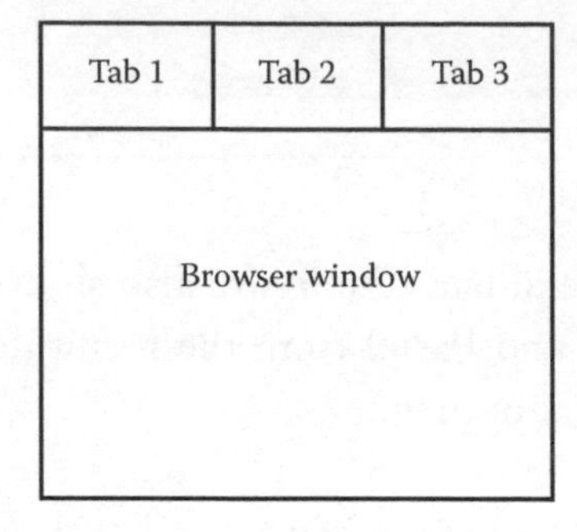


**Figure 6.13** Tabs for browser windows.

**Figure 6.14** Scroll bar.

*6.3.4 Value Input*

Value input elements are used to receive inputs from the users. These elements can be in the form of textboxes, checkboxes, radio buttons, and so on. Users need to either write inside these user elements (e.g., a textbox) or use the mouse to click an item (e.g., a checkbox) to provide inputs. When a Submit button on the user screen is clicked, the values entered or selected by the user will be sent to the middle layer of the software product for computation.

*6.3.4.1 Textbox* The most common element for entering user input is a textbox. A textbox allows the user to enter any text to be used by the software. Sometimes, a textbox may contain a validation program to prevent certain characters being entered into the textbox.

*6.3.4.2 Combo Box* Combo boxes are a combination of a drop-down list (Figure 6.10) and a textbox (Figure 6.15). Users can key in the values in the combo box and this entry becomes a selection value for searching the records. This selection value can be used to either navigate or carry out processing in the software product. If the value keyed in does not match any existing record connected, then it can be programmed to show an error page to the user, stating that no matching records were found for the searched word.

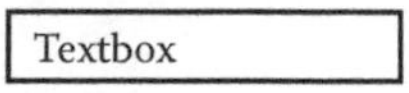

Textbox

**Figure 6.15** Textbox.

Combo boxes are used when searching for records is dynamic in nature. There are no prepopulated lists associated with a combo box. The best example of a combo box is the search words entry box of Google (http://google.com).

### *6.3.5 Output*

Output user elements are used to hold the values that are generated by the software. They to display those values to the user on a user screen.

*6.3.5.1 Label* Labels are the most commonly used user elements to display the values that are calculated by any software product. They are used for displaying the data coming (due to a computation) from the middle layer. These data cannot be edited by the user because they are "read only." Labels do not allow editing of text or data during the runtime of the software product.

Another use of labels is to provide identification to other user elements. Usually, users do not know what input needs to be entered in a textbox on a form (web form or window form). For example, suppose you have a textbox that is supposed to be used by a user to provide the username in that textbox. If additional information is not provided to the user, then the user will never know the kind of input he or she needs to enter in that textbox. If an "enter username" label is placed against that textbox, then the user will know that this textbox is where he or she should enter his or her username. Labels are most often used for this purpose.

Figure 6.16 depicts a label.

*6.3.5.2 Tooltip* Tooltips are help systems that are generally hidden on any user screen. When the user hovers over some text or any other user element that contains the tooltips, only then are they visible in the form of balloons or call boxes. Figure 6.17 depicts a tooltip.

Label

**Figure 6.16** Label.

Spelling and Grammar Status

**Figure 6.17** Tooltip.

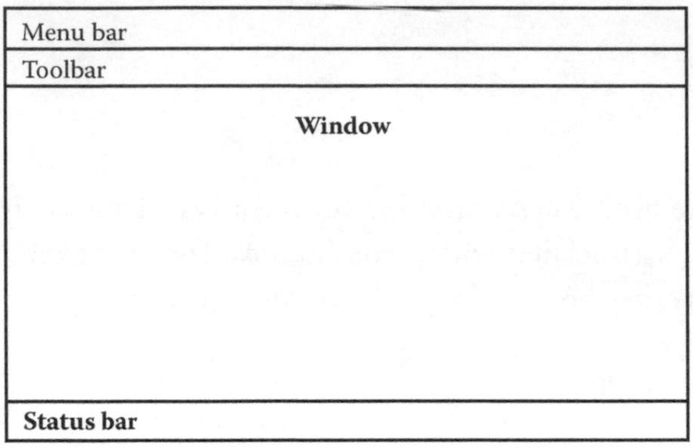

Menu bar
Toolbar
**Window**
**Status bar**

**Figure 6.18** Status bar at the bottom of a rich window.

*6.3.5.3 Status Bar* Status bars are generally provided at the bottom of any user screen. Status bars can contain valuable information (about the software product) that might be helpful to the user. For example, on any Microsoft Word program, the status bar provides information such as page number, number of words in the document, and language used in the document. Figure 6.18 depicts a status bar in a window.

*6.3.5.4 Progress Bar* Progress bars are used to provide information to the users about the progress of any task. For example, when you download a document or a program from the Internet, using the progress bar, the browser may show the progress of the download.

## 6.4 Hyper Text Markup Language

HTML is a language that is used to format web pages. It is not a general-purpose programming language. It is a standard format to display web pages inside a web browser. Web pages can be written in many programming languages including Java Server Pages (JSP), Active Server Pages (ASP), and pure HTML. However, a web browser is only capable of displaying the content that is in HTML format. When a web page is written in a programming language such as ASP or JSP, the programming code is first sent to the application server. The application server runs the programming code and generates HTML code. This content is in pure HTML format and can be displayed inside a web browser.

HTML is characterized by a pair of formatting tags. All HTML tags are enclosed within angle brackets < >. For example, "<H1> This is in very large fonts. </H1>" will convert the letters inside the angle brackets into letters with very large fonts. Most HTML tags are used for formatting. However, there are also tags that are used for other purposes. For example, <href http://www.abc.com> will make the http://www.abc.com into a hyper link. The user can click this link to visit the website at http://www.abc.com. Some tags are used as empty tags. They do not have the corresponding ending tag. For example, <img> is an empty tag used for embedding an image in the web page.

Each web page is saved as an HTML file. A web browser can read the HTML files and compose them into visible web pages. The browser will hide the HTML tags and only show the formatted text.

HTML can embed images and other media files. It is also possible to embed programming scripts inside HTML pages. These scripts make the HTML pages dynamic and interactive.

### 6.5 Cascading Style Sheets

Cascading Style Sheets (CSSs) are used to design web pages. While the web page itself is composed of HTML, the CSS builds a formatting layer on top of HTML. If a web page is created to be displayed on a 15-inch computer screen, then the same web page can be resized to be displayed on a mobile phone screen 4 inches in size. This is possible using CSSs. Mobile phone screens are too small and cannot correctly display all the elements of a web page meant for a 15-inch computer screen. Using CSSs, it is possible to specify which elements can appear on the small screen. Generally, CSS formatting commands are stored in a file. This file name is then saved in a web page. The formatting commands from that file will apply when that web page is loaded in a browser (when you run that web page). You can write several CSS files and save them on the computer on which the browser will run. You can change the CSS file name inside the web page. Depending on which CSS file was applied for a web page, the web page will look differently although the HTML content of the web page remains the same.

There are some methods that can be used to format HTML pages. In one method, you can mix CSS commands with HTML tags in the page. For example, you can write <h1 style = "color:red"> Chapter 1 </h1>. Here, the HTML tag <h1> … </h1> is used to make the letters "Chapter 1" red. This way, you can use CSSs directly with the HTML contents on a web page. A better method is to store all the CSS commands in a CSS file and save it. Later, you can call the CSS file inside the HTML page to get the formatting done. You can use a statement such as <link href = "..path/file.css" rel = "stylesheet"> to call the CSS file in the HTML page. By using a separate CSS file, you can achieve good-looking and similar web pages for a website. This effect is achieved by using the same CSS commands over all of the web pages of a website.

### 6.6 Client-Side Scripting

Scripting is basically a procedural programming language. Scripting is used to create dynamic web pages. A dynamic web page contains content that is generated dynamically using a scripting language. There are two types of scripts for building web pages: the client-side script and the server-side script. A server-side script is run on an application server. The output of this script, which is in the form of HTML, is sent to a web server. Then, the web browser is able to display this HTML content when connected to that web server via the HTTP protocol. Server-side scripts belong to the middle tier of any web-based software application. They are discussed in Chapter 7. Client-side scripts, on the other hand, belong to the user interface side. Hence, they are discussed here.

A client-side script resides on a web server. When a web browser is connected to the web server (via HTTP protocol), this client-side script starts running on the web browser. The output of the script is in HTML format; therefore, it can be easily displayed on any web browser. When a web page containing the client-side scripts is loaded inside a web browser, the client-side script runs in the web browser and generates the HTML content. This content becomes visible on the web browser and the user can see it. Some client-side scripts are used for validation purposes as well. In such cases, the script is only doing the user input validation and there could be no visible output from the script.

Although the web page validations can be done either by client-side scripting or by server-side scripting, there is a difference between the two. The difference is that client-side scripting makes these tasks faster. This is natural because if a validation task is to be done using a server-side script, then a round-trip of data needs to be done (data from the client to the server and then back to that client) and this will take some time. Things such as web page validation will result in slower responses from the web pages if server-side scripting is used. Thus, client-side scripts are better in such cases.

User interfaces of web applications are composed of HTML elements. Although HTML is great because of its platform independence (any HTML page can run on any computer platform), it has a drawback. HTML is totally static. You cannot have some part of the user interface be composed dynamically during runtime if the user interface is entirely composed of HTML elements. This is a major drawback because some tasks such as providing user preferences or automatic services to the user (such as saving passwords and filling textboxes automatically) are not possible with static HTML content. You need client-side scripts to do these tasks.

Client-side scripts can be directly written on the web page along with the HTML elements or you can save the scripts in a file and call it in the web page.

Some additional uses for client-side scripts are the following:

- Some websites do not want the users to download the images that are embedded on those web pages. This functionality can be achieved by writing the scripts on the client side to disable the right-click functionality (i.e., right click of the mouse). Generally, when the right-click functionality is enabled by default, it opens a context menu that has an option for the users to save the image on a local disk. When the right-click functionality is disabled, users will not be able to save the images embedded in the web page.

- For security reasons, some websites do not want the users to see the source code of their web pages. This functionality can be achieved by writing the scripts on the client side to disable the right-click functionality. Generally, when the right-click functionality is enabled by default, it opens a context menu that has an option for the users to view the source code of the web page. When the right-click functionality is disabled, the users will not be able to see the source code of the web page.
- If validation of user input is needed on a website, then it can be done using a client-side script. For example, if the length of a password field should be in the range of four to eight characters then, by using a client-side script, the users can be provided with a message box if the user input has either less than four or more than eight characters. In that case, the user has to dismiss that message box and reenter the password. This process is repeated until the password is entered correctly. This will ensure that the length of the password is always between four and eight characters.

### 6.7 Asynchronous JavaScript and XML

Whenever a user needs to supply some information to the server, the user submits it by filling in a form on the web page. When the user clicks a Submit button on the form, the information (provided by the user) that is inside various textboxes and other user input fields is sent to the server for processing. Sometimes, a form creates problems for the user. This problem may be related to the user input validation. For example, if a user chooses a username when creating a new account on a website, the business logic checks if this username has already been taken. This validation is done by checking the existing data in the database for the usernames. If the username has already been taken, then the user is informed that a different username should be chosen. On the user registration form, there is a lot of information (required about the user) such as name, age, sex, preferences, and address. This information needs to be provided by the user along with the username. When the user submits all this information and if it is found that the username supplied by the user has already been taken, then all this information is lost during the registration processing. This happens because the web page is reloaded after processing on the server. Reloading of the web page resets all the user elements it; thus, all the user information entered by the user is lost. The user has to input all this information again while trying with another username. If the supplied username is found to already be taken, then the user has to again fill in all those boxes with the required information. This cycle is repeated until the username supplied by the user is one that has not been already taken. You can easily imagine the frustration a user has to go through for this kind of registration.

Let us see an example where the previous issue is tackled. Suppose a user wants to register on Twitter (a popular social media website) so that he or she can become a

Twitter user. When the user points his or her web browser to http://www.twitter.com, the default web page appears on the user screen. On this screen, there are two options. If the user is already registered, then he or she can log into the Twitter account by providing his or her username and password in the corresponding textboxes. However, if the user has no account with Twitter, then he or she can create one by clicking on the "Create account" link. The fun begins on the next page. When the user enters a username in the username textbox, within a few seconds, a label appears next to the textbox informing the user that the username has already been taken and he or she needs to provide a new username. This is the case when a matching record is found in the back-end database. If no matching record exists, then the username provided by the user will be accepted and a label will appear next to the username textbox stating that the username is fine.

How was this kind of functionality achieved? How come the username was checked and verified even without the user submitting the registration form? Has not this kind of functionality spared the user from the headache caused by repeatedly submitting the registration form?

This kind of functionality is achieved using a technology known as Asynchronous JavaScript and XML (AJAX). Using this technology, a partial web form can be validated on a server without reloading the entire web page. Thus, if some information is entered by the user on a web page, then this information is not lost because the web page is not reloaded.

### *6.7.1 AJAX Details*

The scenario described earlier can be implemented using AJAX. AJAX consists of many elements such as JavaScript and XML. Initially, HTML, JavaScript, and XML were used to create AJAX components that could be used to carry out client-side formatting and validation. Later, many alternative technologies started to be used to create AJAX components. Now, it is possible to use AJAX to even check the values in the user elements by comparing them with the values stored in the database. This is how the username checking scenario described above works.

Figure 6.19 depicts the earliest implementation of AJAX technology with XML file validation. The XML file resides not at the back of the business logic layer but at its very front. Thus, the client-side script can access it without doing a round-trip via

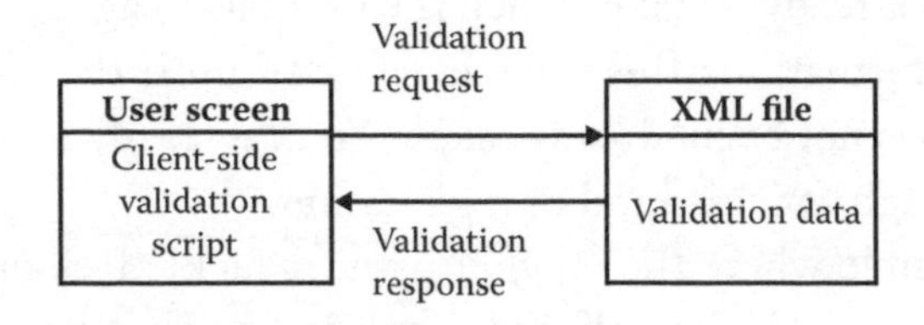


**Figure 6.19** Schematic view of the earliest implementation of AJAX.

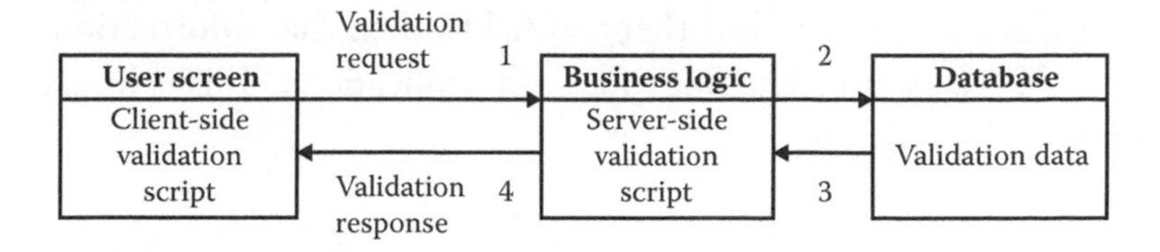


**Figure 6.20** Schematic view of AJAX implementation with database checking.

the business logic to access the XML file, and the client-side script residing at the user screen can directly validate the user input against the XML file.

Now, it is possible to validate the user input using a database, as depicted in Figure 6.20. Here, when a validation request is made from the user screen, the request is passed on to the business logic. This is Step 1 as depicted in the figure. Next, the business logic accesses the database to check whether the validation request is valid (Step 2). The validation data are compared after getting the data from the database (Step 3). The result of the validation is finally sent to the user screen (Step 4).

### *6.7.2 AJAX Implementation*

Let us see how we can implement AJAX in the username checking scenario. First, you can create a web page containing a form with fields such as username, password, age, sex, and address. Later, you will have to read the input (in the fields such as username) that needs to be checked using some JavaScript or any other scripting language.

There is a mechanism that can be used here to check and validate the data filled in the username field although the user has not invoked any event (such as clicking the Submit button). On a web page, automatic events such as move over and move focus are available. These events can occur even without the user invoking any event explicitly. We can use these events instead of using an event (e.g., a Submit button) invoked by the user. Special functions have been created in the scripting languages to invoke a trigger after automatic events such as move focus happen. For example, in JavaScript, you can write a method that will run when an event such as move focus happens. What could be an event to use here? Well, when the user enters the username in the username textbox, he or she moves to fill the next textbox (e.g., a password field). Thus, we can use an event such as move focus. As soon as the user leaves the username textbox field, this move focus event will fire. To implement that event, you can write your script for checking and validating the username field by comparing it with the stored values in the database. If a stored value matches the username entered by the user, then the program will give a message such as "username already exists." If the username provided by the user is not found in the database, then the program can give a message such as "username is available." When this validation script is fired, reloading of the web page does not happen. Thus, if the user has already entered any information such as age, sex, and address in the other text fields, then that information will not be lost. Thus, AJAX is a useful technology for such applications. This is how you can use AJAX technology in such cases.

### 6.8 Simple (Model–View) User Interface

For simple programs that do not have a large number of user screens, a simple user interface can be designed. In this design, you can put some static elements and dynamic content (dynamic content comes from the business logic layer) on the user screens. In this arrangement, any dynamic content processed by the business logic (model) will be displayed on the user screen (view) if the processing involves a display of content to the user. Figure 6.21 depicts this simple model–view user interface schema.

In this arrangement, you create the required user interfaces. Then, you write the business logic separately. You link the business logic to the user interface through some buttons or links. When the user presses those buttons or clicks those links, the business logic layer is triggered. If a database is connected to the business logic layer, then the business logic may connect to the database and read/write the data from/to the database. The business logic layer will then process these data again and display the processed data on the user screen. For example, suppose the user interface is a login web page. The user enters the username and password on the user screen and then presses the OK button. The middle layer is then triggered. It will collect the data entered in the textboxes for username and password and then connect to the database. From the database, it will fetch the data for the username and password. If the username and password data from the database match the data supplied from the user screen, then the middle layer (business logic layer) will compose a welcome message and send this message to the user screen. This message will then be displayed on the user screen. If the data pair from the user screen does not match the data from the database, then the middle layer will compose a warning message and send it to the user screen. This warning message is displayed on the user screen.

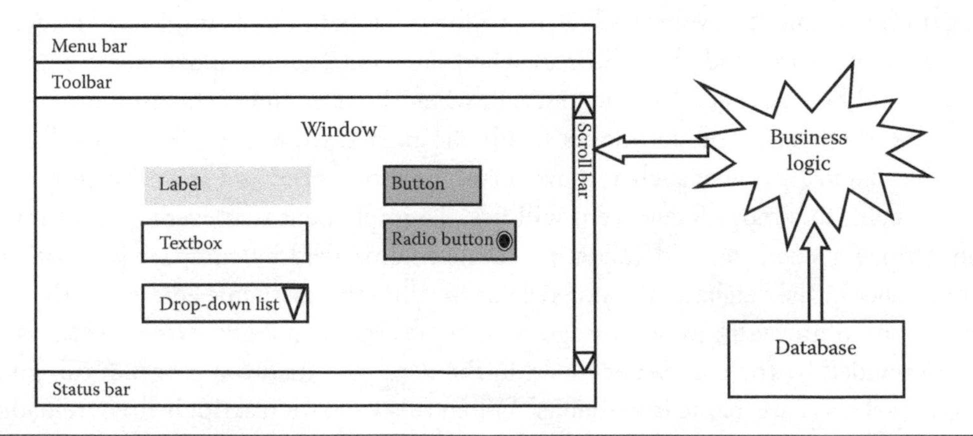


**Figure 6.21** Simple user interface connected to business logic layer.

In this model–view architecture, you can see that for each action from the user, there is a corresponding response from the software product. This means that there is a corresponding model for each view. Whenever a user clicks on the user screen to request the software product to do some computation, the other layers (business logic and database) perform the computation and bring the required view to the user.

### *6.8.1 When to Use*

A simple (model–view) user interface with business logic is easy to implement. It does not require sophisticated programming. This architecture can be used for small and simple websites.

## 6.9 Model–View–Controller

When a user interacts with a software product, the interaction is carried out in many ways. Sometimes, the user may want immediate results for his or her queries. In this case, when the user presses a suitable button on the user screen, the software product does some processing and displays the results on the user screen. The results of whatever processing is done by the software product are immediately displayed on the user screen. For example, if you are a manager in a bank and want to see a report on all the cash deposits made for that day, then you can get the report immediately.

Let us see one more scenario. A clerk at a point-of-sale (POS) terminal may want to do a purchase transaction by receiving money from a customer. The clerk reads the barcode label on a product using the barcode scanner. Then, he or she reads the display on the screen to see if scanning was correctly done. The user screen at this point will show the item description and price. If the quantity of the same product is more than one, then the clerk may manually enter the quantity. Then, the clerk proceeds to process other items in the customer's trolley in the same manner. When all the items are entered, the clerk computes the total amount the customer has to pay for the purchase. When the customer makes the payment, the clerk presses the complete transaction button and then the transaction is completed. In this scenario, the transaction is not complete until the clerk processes all the purchased items for the customer and then presses the complete transaction button after receiving the payment.

In many instances, human–computer interaction does not happen in a straightforward way as explained in the previous scenario. Let us take one more example. On a web-based bookstore, a customer browses the books listed there. When he or she sees a book that looks interesting, he or she may press a button on the website to get more details about the book. If the customer finds the book interesting, he or she can press a Buy button on the web page. Next, the customer is shown a web page that asks him or her if he or she has finished and wants to check out or wants to continue shopping for more books. If the customer wants to continue shopping and presses the Continue button, then he or she is taken to the main page of the website where the books are listed. The customer can select additional books. The customer can also delete any book from the shopping cart. Once the customer has finished browsing the books and wants to check out, he or she presses the Buy button. Now, on the next page, the customer is shown a list of all the books that he or she is buying. Here, a total amount to be paid is also shown. Now, the customer can pay the amount and the purchase transaction will be completed. In this example, what we see is that the customer browses the books and keeps the interesting books in his or her shopping cart. Finally, at the checkout counter, the customer pays for the books and completes the transaction.

If you compare the POS example and the web-based bookstore example, you will see a difference. In the POS example, the computer program took note of the purchased goods only at the checkout. However, in the bookstore example, the computer program stored the information about the books that were in the shopping cart throughout the time when the customer was browsing. When the customer finally checked out with the selected books, this shopping cart information was used for making the transaction.

From the POS and bookstore examples, we can see the difference in direct and indirect rendering of information on the user screen. In the case of POS, rendering was direct, but in the case of the bookstore, rendering was indirect. In indirect rendering, some information was kept in the memory and recalled after the user pressed the Checkout button. You can surmise here that in indirect rendering, some information was kept outside the user screen and, in fact, the response from the computer was not commensurate with the input provided by the user. Now, the question arises, how does one implement such an indirect rendering behavior in a software product? Let us discuss it here.

There is a technology known as Model–View–Controller (MVC). There is a component known as a controller that determines what output is to be provided to the user for each input provided by the user. This way, there is no direct link between the user input and the computer output. This important feature (controller) is used in many situations where the computer output is not directly related to any of the user inputs, as we saw in the bookstore shopping cart example.

### *6.9.1 MVC Description*

In MVC technology, the view (what the user sees on the screen) is separated from the computer output through two entities known as model and controller. When the user provides some input through the user screen, the information is passed on to a program known as controller. The controller then initiates a program, known as model, to do some computation. The model does the computation and sends the computed results back to the controller. The controller then finds what information needs to be sent to a program known as view. When the information becomes available to the view, only then does it become visible on the user screen, as shown

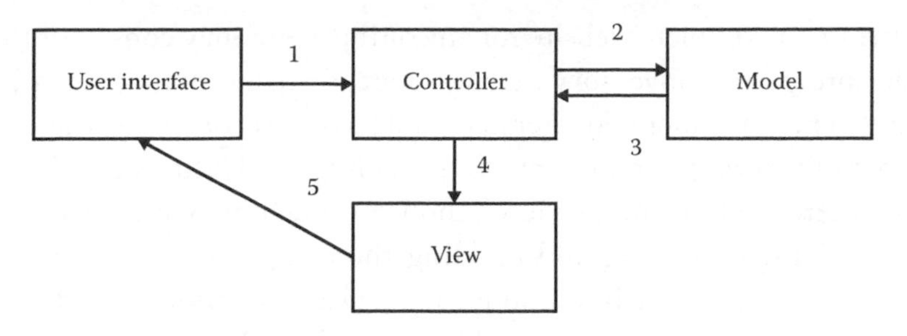


**Figure 6.22** Schematic diagram of MVC.

in Figure 6.22. How data flow for the MVC arrangement is shown by arrows. The numbers (1 to 5) next to the arrows show the order in which the flow of data takes place.

From this description, you can see that the controller plays the most important role in sending and receiving the information with the model and the view. The most important feature of MVC is that it can hold the view until the user specifically wants to see it. In the case of the bookstore example discussed earlier, the user keeps on adding or deleting the books from the shopping cart and the application just keeps on updating the shopping cart information transparently to the user. When the user specifically presses the Checkout button, only then will the contents of the shopping cart be displayed to the user.

A fundamental aspect of computing is to save the information into a database whenever some update happens in the model. Using MVC, this behavior is controlled. Although the model is updated whenever the user adds or deletes some items in the shopping cart, this updated information is never passed on to the database. Instead, the updated information is sent to the controller and the controller keeps this information in the memory. Only when the controller needs to save some information to the database will the controller instruct the model to save these data to the database. From Figure 6.22, you can see that the model and the controller communicate both ways so that all the updates in this model are sent back to the controller. This is in contrast to the traditional systems where any update in the model would be sent to the view.

### *6.9.2 MVC Implementation*

The web-based bookstore example can be implemented using MVC. Let us recapture the scenario. The user browses the bookstore and saves some books in the shopping cart. The user can also delete some books from the shopping cart at any time. After performing all the browsing and book selection, the user finally checks out.

Here is the complete scenario on what each component (controller, model, and view) is doing.

**Assumption:** A complete website for any online store may consist of functionality for order processing, inventory management, financial management, shopping cart management, customer management, and so on. Accordingly, there could be a large number of software components for providing the features for all these functions. Thus, there could be many views, models, and controllers for taking care of all these functions. Designing and building the complete software product for a website for the online store will definitely require a large effort. The design of such a software product would also be large. Understanding through such a large design is relatively difficult. If the design of only one functionality is described, then it is easy to understand each aspect of the MVC architecture. This is why we only focused on the shopping cart scenario and will describe its design here.

The components and methods of the MVC architecture for the shopping cart are shown in Table 6.1. Now, let us consider one operation and see how the data flow and command flow take place. We will take the "Save books in shopping cart" operation of the user (see Table 6.1).

The user screen shows view—"Save books in shopping cart" (this can be done using a button or a checkbox on the user screen). The moment this button is clicked or the checkbox is checked, the command goes to the controller where the action "Ask model to save books in shopping cart" is done. The command goes to the model where the method "Save books in shopping cart" is executed. At the end of the method, the command goes back to the controller. The controller then creates a

**Table 6.1** Components and Methods of MVC Architecture for a Shopping Cart

| Controller | Model | View | User |
|---|---|---|---|
| Receive request from user () | | | Show shopping cart () |
| Ask model to display shopping cart () | Display shopping cart method () | Display shopping cart () | |
| Ask model to save books in shopping cart () | Save books in shopping cart method () | | Save books in shopping cart () |
| Ask model to delete books from shopping cart () | Delete books from shopping cart method () | | Delete books from shopping cart () |
| Ask model to checkout () | Checkout method () | | |
| Send saved books list (after addition/deletion) in the shopping cart to view () | | | |
| Send checkout to view () | | Display checkout () | Checkout () |

view of the updated shopping cart. This view is then sent to the user screen. When the user presses the Checkout button, the Checkout method is activated and the checkout screen displayed (by activating the Display checkout method in view) to the user with all the books that were saved in the shopping cart.

In Figure 6.23, a data flow view is provided for the operation of selecting a book and saving it in the shopping cart. When the user selects a book on his or her screen, data flows happen from User screen to Controller, from Controller to Model, again from Model to Controller, then from Controller to View, and finally from View to User screen.

In Figure 6.24, the data flow view is provided when the user presses the Checkout button at any stage. Similar kinds of data flow views are possible for each of the operations in Table 6.1 when the user presses a button or checks a checkbox.

We will learn about MVC implementation in detail in Chapter 7. Implementation of MVC is mostly related to the development of the middle tier when we write the business logic. Controller, model, and view are all part of the middle tier. Only their manifestation is visible to the user in the form of user screens. However, their discussion in this chapter is important because the system architecture involving these elements (controller, model, and view) affects the user in the way he or she interacts with the application.

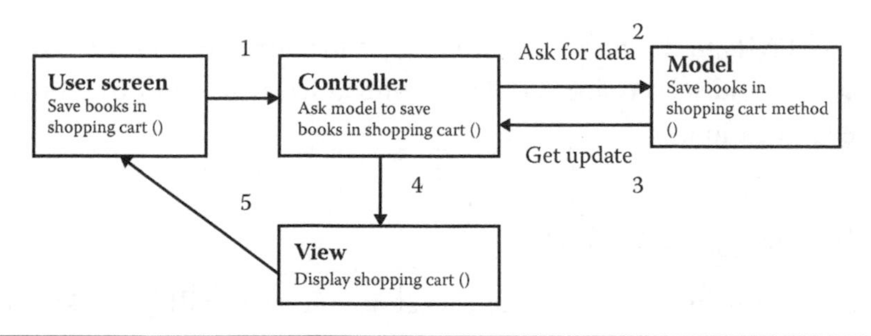


**Figure 6.23** Shopping cart data flow view for save books in shopping cart.

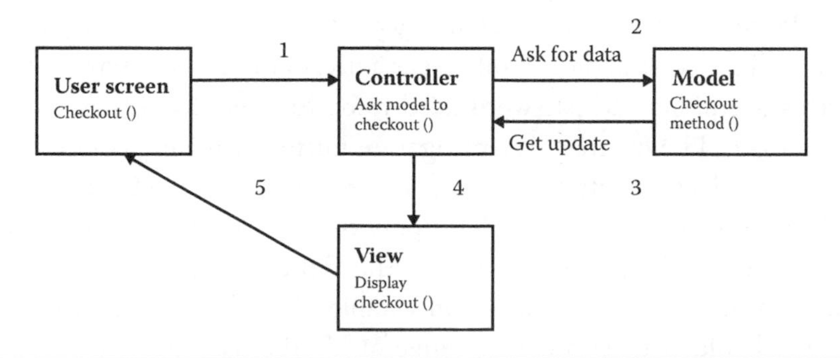


**Figure 6.24** Shopping cart data flow view for checkout.

## 6.10 Case Study

Both versions of our online bank account access system, OBAAS 1.1 and 1.2, use the standard user control elements for the browsers. The application can be accessed using any browser available in the market. We will use some client-side scripts for user data validation and page formatting. Note that Figures 6.25 through 6.35 belong to OBAAS 1.1 and Figures 6.36 through 6.39 belong to OBAAS 1.2.

### *6.10.1 Client-Side Scripts*

We will use client-side scripts for these purposes:

- Validation: We will check the user input for a range of values entered. For each textbox, we will check if it has been left empty. If so, we will raise an error in a dialog box. For the username field, we will raise the following message in a dialog box if the username field is left blank: "Username must not be left blank. Username should also be in characters." For the account number field, we will raise a message, "Account number must not be left blank. Account number must be in numerals," in a dialog if it is left blank.
- Style sheet: We will use a style sheet to format our web pages.

### *6.10.2 Hyper Text Markup Language*

We will use HTML to design all the web pages for both versions of OBAAS. Both client-side scripts and server-side scripts are embedded inside HTML on all the web pages of the application.

### *6.10.3 Mockup Screens for OBAAS 1.1*

We will use mockup screens to represent what user screens will look like for OBAAS 1.1. In the mockup screens, you can see some user control elements used for building the user interface. These user elements are connected to the middle tier using server-side scripts.

Figure 6.25 depicts the Create account page. This process starts when a user asks the bank for online access and the bank issues a username and a password to that user. However, this username and password need to be registered in OBAAS. Only then can a user access it. Hence, there is a navigation button, Create account, in OBAAS. When the user clicks this button, he or she is presented with the Create account page shown in Figure 6.25. The user needs to provide the Username, Password, Retype password, Phone number, and Address in the fields given on this page. There is another field named Category that is noneditable. This field is used for the internal purpose of the bank. It takes a default value of U. The user is not required to fill in this field. Once the information is submitted for Username, Password, and Retype

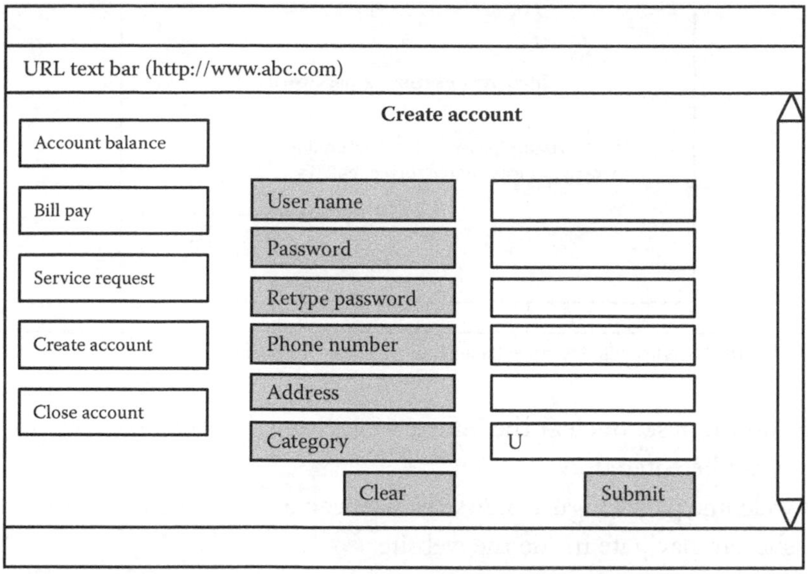

**Figure 6.25** Create account page for OBAAS 1.1.

password, information is entered for other fields, and the Submit button is pressed, the user is presented with the welcome page along with the "account number" that is assigned to that user, as shown in Figure 6.26. If the information related to the password and retype password does not match, then a dialog box similar to the one in Figure 6.27 appears with a message: "Entries in the password and retype password fields do not match. Please ensure password and retype password are the same." After pressing the OK button on this dialog box, the user can again enter the information on the Create account page (Figure 6.25). There is also a Clear button on this user screen. If a textbox was left empty and the user clicks the Submit button, then a dialog box informs the user to fill in that empty textbox. If the user presses this Clear button, then any information entered in the textboxes is cleared. This button can be

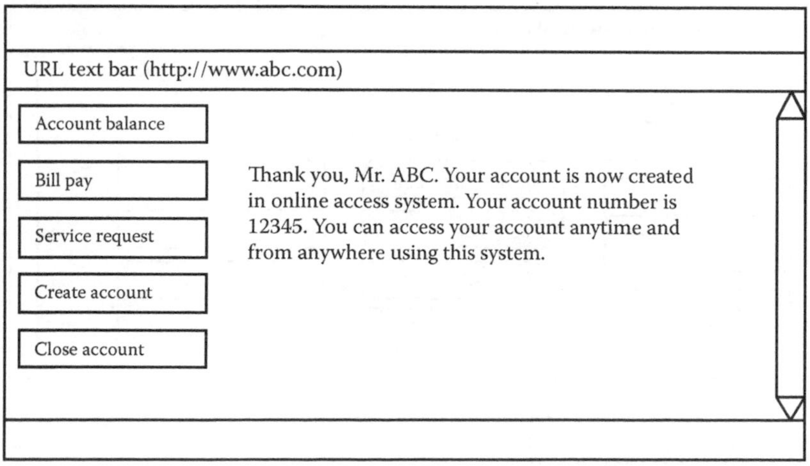

**Figure 6.26** Create account response page for OBAAS 1.1.

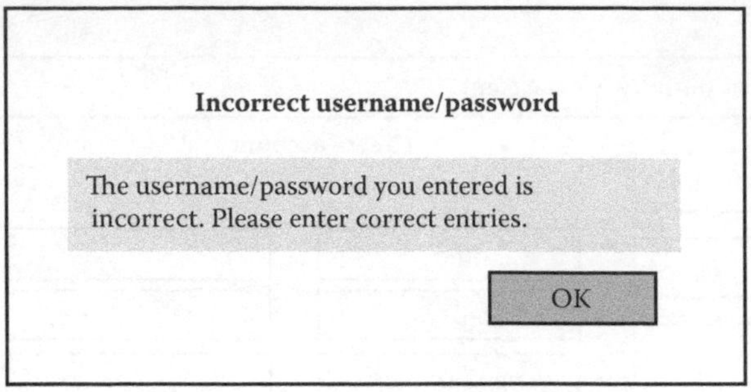

**Figure 6.27** Incorrect username/password entry information dialog box in OBAAS 1.1.

used by the user to reset or clear the information entered in the textboxes so that fresh information can be entered.

On the welcome page (Figure 6.26), the user can also see the navigation buttons so that he or she can navigate inside the website.

When the user clicks the Account balance button (e.g., as given in Figure 6.26), the Account balance page is displayed as depicted in Figure 6.28. The user enters the Account number, Username, and Password and presses the Submit button. If the user enters correct input for these fields, then he or she is presented with the account balance response page as depicted in Figure 6.29. If a wrong account number is entered, then a dialog box appears similar to the one given in Figure 6.27. There is also a Clear button on the user screen of Figure 6.28.

When a user clicks the Bill pay button (e.g., as given in Figure 6.28), he or she is presented with the bill payment page as depicted in Figure 6.30. The user has to select a biller from the drop-down list. Currently, there are two different billers and the user can select one of them. Then, the user has to enter the amount to be paid in the Amount textbox. The user also needs to provide the inputs for Account

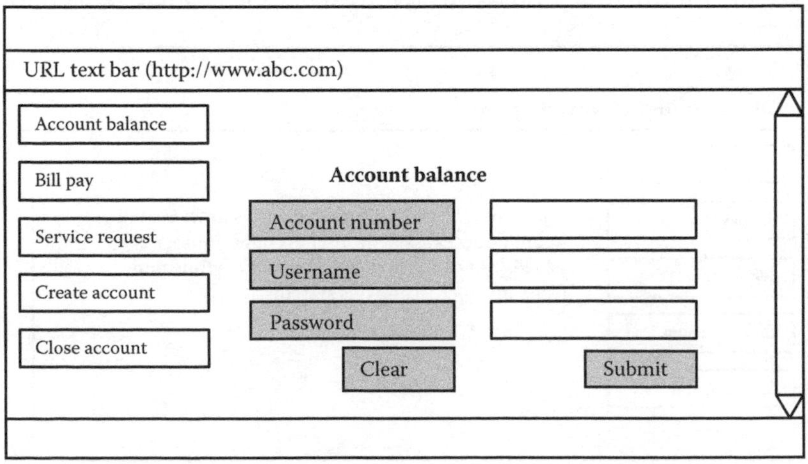

**Figure 6.28** Account balance page for OBAAS 1.1.

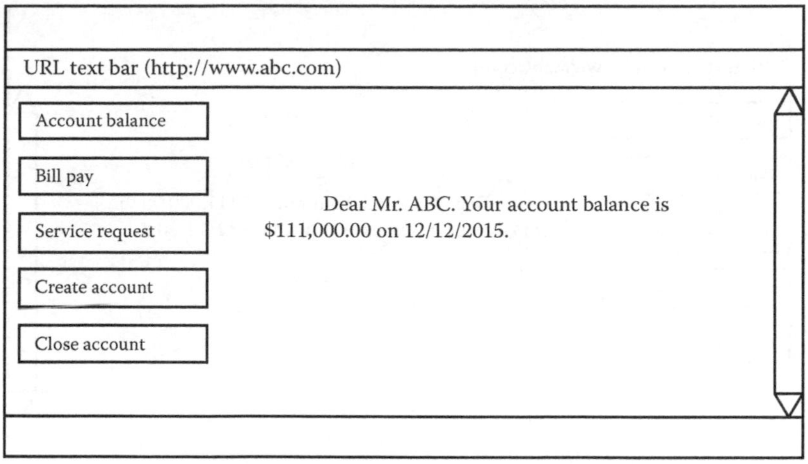

**Figure 6.29** Account balance response page for OBAAS 1.1.

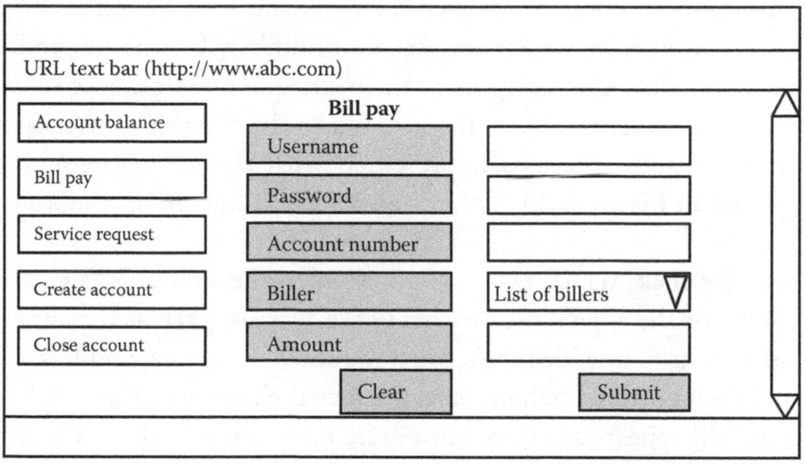

**Figure 6.30** Bill pay page for OBAAS 1.1.

number, Username, and Password in the respective fields. Once the user clicks the Submit button, the given amount will be paid in the account (maintained in the same bank) of the biller. This amount is deducted from the account of the user. If the user leaves the Amount field blank or does not select a biller from the drop-down list or forgets to provide the input in the Username, Password, or Account number fields, then a dialog box informs the user to do what is needed. If the available amount of money in the user account is less than the amount requested for bill payment, then an error message is displayed. After a successful payment to the biller, a response page appears (Figure 6.31). There is also a Clear button on the user screen for Bill payment (Figure 6.30).

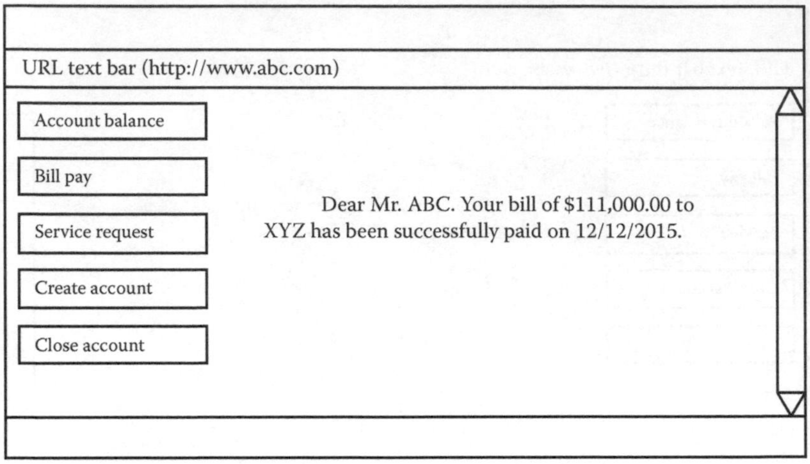

**Figure 6.31** Bill pay response page for OBAAS 1.1.

When the user clicks the Service request button (e.g., as shown in Figure 6.26), he or she is presented with the Service request page as depicted in Figure 6.32. The user has to provide inputs for the Account number, Username, and Password in the respective fields. Currently, only a new cheque book request is available in the service request function. Once the user clicks the Submit button, after selecting the service request, the user is presented with the service request response page as depicted in Figure 6.33. There is also a Clear button on the user screen of Figure 6.32.

When the user clicks the Close account button on the web page (e.g., as shown in Figure 6.26), he or she is presented with a Close account page as depicted in Figure 6.34. The user needs to provide the Account number, Username, and Password in the textboxes and then press the Submit button. Once the user does this, a Close account response page will appear as depicted in Figure 6.35. There is also a Clear button on the user screen in Figure 6.34.

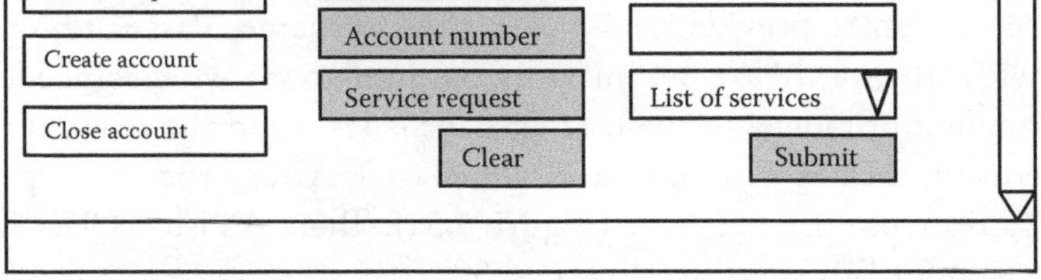

**Figure 6.32** Service request page for OBAAS 1.1.

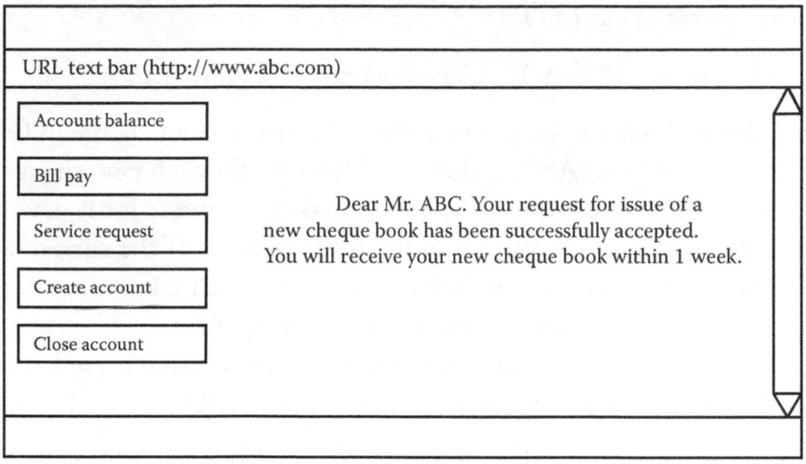

URL text bar (http://www.abc.com)

Account balance
Bill pay
Service request
Create account
Close account

Dear Mr. ABC. Your request for issue of a new cheque book has been successfully accepted. You will receive your new cheque book within 1 week.

**Figure 6.33** Service request response page for OBAAS 1.1.

URL text bar (http://www.abc.com)

Account balance
Bill pay
Service request
Create account
Close account

**Close account**

Username
Password
Account number
Clear
Submit

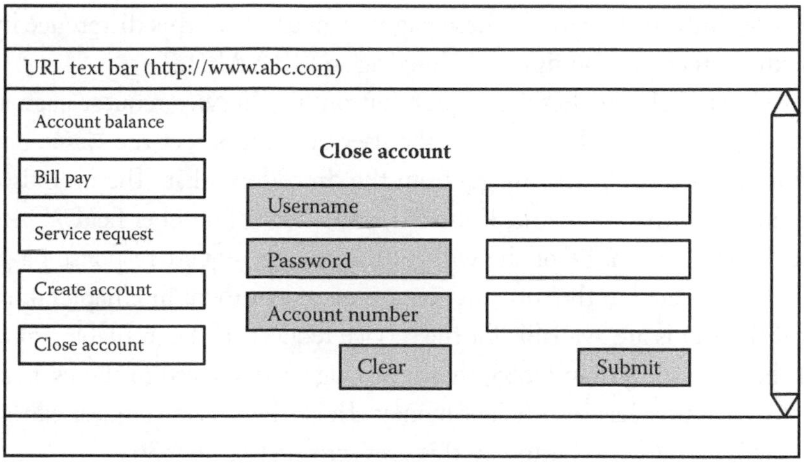

**Figure 6.34** Close account page for OBAAS 1.1.

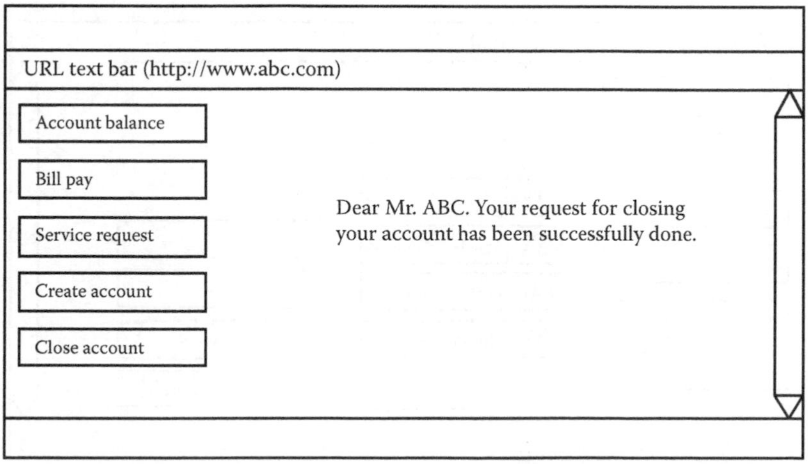

URL text bar (http://www.abc.com)

Account balance
Bill pay
Service request
Create account
Close account

Dear Mr. ABC. Your request for closing your account has been successfully done.

**Figure 6.35** Close account response page for OBAAS 1.1.

*6.10.4 Mockup Screens for OBAAS 1.2*

There are two changes in OBAAS 1.2 compared to OBAAS 1.1:

1. A new cheque book can be issued to the customers free of charge in OBAAS 1.1. However, in OBAAS 1.2, there is a US$5 fee for each new cheque book. Thus, in OBAAS 1.2, when a customer makes a request for a new cheque book, US$5 is deducted from his or her bank account. If the current balance in the customer's bank account is less than US$5, then a warning message is issued to that customer and the request is not accepted.
2. In OBAAS 1.2, a customer can make a money transfer from his or her account to another one if the other account is also with OBAAS.

Mockup screens are provided for these two functionalities for OBAAS 1.2. Mockup screens for other functionalities are not provided because they are the same as those of OBAAS 1.1 except for the additional navigation button for money transfer, and this Money transfer button appears on these pages. You can find this difference by looking at the mockup screens in the figures belonging to OBAAS 1.2.

When the user clicks the Service request button, the Service request page appears as depicted in Figure 6.36. In Figure 6.36, the user needs to select the "issue new cheque book" option (not shown in the image) from the drop-down list. The user also needs to provide the inputs in the Username, Password, and Account number fields. Once the user presses the Submit button, he or she will get the service request response page as given in Figure 6.37. As stated in the software requirement document in Chapter 4, the system will check if the funds are available for the service request. If the available amount in the account of the user is less than US$5, then a message is displayed to the user stating that the account of the user has insufficient money. Thus, the service request cannot be performed. There is also a Clear button on this user screen (Figure 6.36).

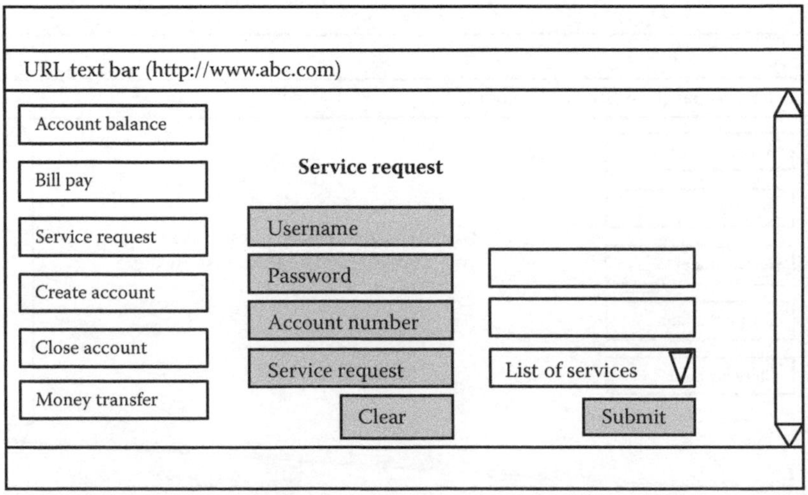

URL text bar (http://www.abc.com)

Account balance
Bill pay
Service request
Create account
Close account
Money transfer

**Service request**

Username
Password
Account number
Service request
List of services
Clear
Submit

**Figure 6.36** Service request page for OBAAS 1.2.

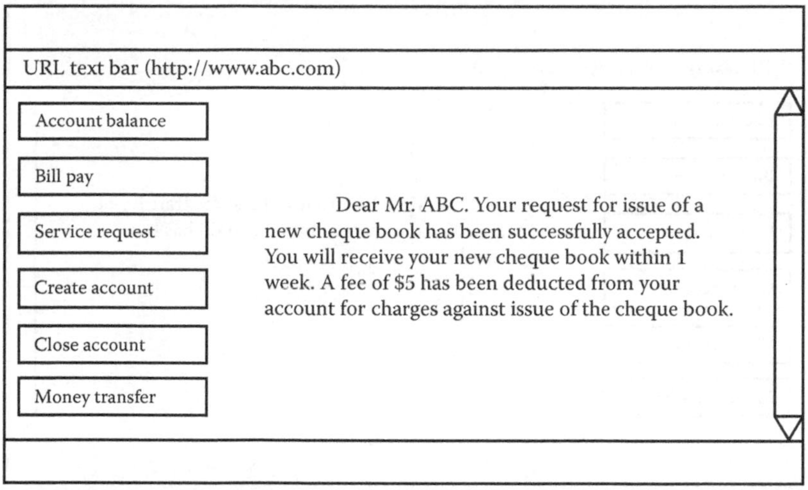

**Figure 6.37** Service response page for OBAAS 1.2.

When the user clicks the Money transfer request button on any web page, then the Money transfer page appears as depicted in Figure 6.38. In Figure 6.38, the user needs to provide the inputs to Username, Password, From Account (user account number), and Amount. In addition, the user needs to provide input to the To account field that represents the account number of the party to which the money needs to be transferred. Once the user provides correct information and clicks the Submit button, the money transfer response page is displayed as given in Figure 6.39. As stated in the software requirement document in Chapter 4, the system will check if the funds are available for money transfer. If the amount of funds in the account of the user is less than the amount that needs to be transferred, then a message is displayed to the user stating that the account has insufficient money, in which case the money transfer cannot be performed. There is also a Clear button on the user screen (Figure 6.38).

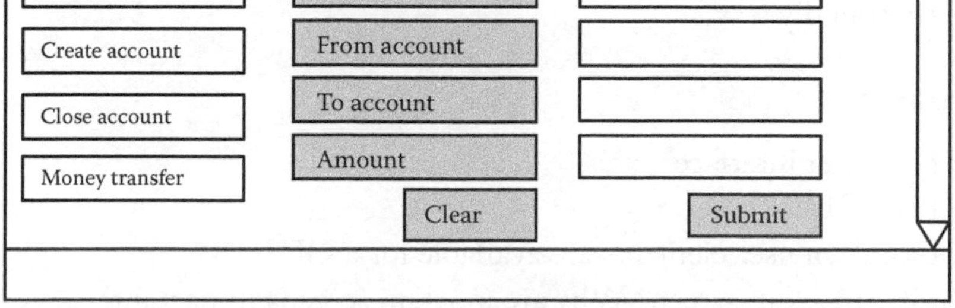

**Figure 6.38** Money transfer page for OBAAS 1.2.

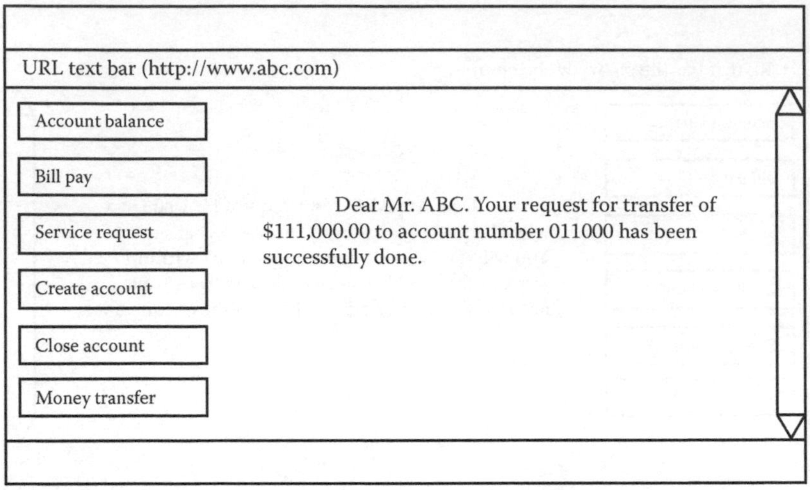

**Figure 6.39** Money transfer response page for OBAAS 1.2.

### 6.11 Chapter Summary

User interfaces are the only visual part of any software product. What the software product does and how the user interacts with the software product are important considerations for building user interfaces.

In the early days of computing, user interfaces used to be character based. However, after the advent of Windows and GUIs, user interfaces provide many ways in which users can interact with the software product. GUIs consist of user control elements. These elements are of many types such as containers, user input elements, output and elements.

For web-based applications, the user interface is a browser. Any web page can be displayed on a browser using HTML. The limitation of HTML is that it supports only static data. If you want to display data that are processed by the business logic, then you will have to use server-side scripts. Similarly, to enhance user experience, you need to provide client-side scripts. These scripts can be easily embedded inside HTML tags. You can also create separate script files and embed them in the HTML page. During runtime, these files can be called and their compiled content can be displayed in HTML format on the web page. Some technologies such as AJAX and MVC help in user input validation and in tasks such as displaying the views to the user in a user-friendly way.

## QUESTIONS

1. What is a user interface?
2. What is a GUI?
3. What kinds of user elements are available for a GUI?
4. What is a mockup screen? Why are mockup screens important?
5. What is AJAX? How has AJAX technology evolved over the years?

6. What is MVC?
7. What is a model in MVC?
8. What is a controller in MVC?
9. What is a view in MVC?
10. For what purpose are the client-side scripts used?
11. What is a style sheet?
12. What is HTML?

## Recommended Reading

Wilbert O. Galitz (2007), *The Essential Guide to User Interface Design*, 3rd Edition, Wiley Publications, Indianapolis, IN.

Ben Shneiderman, Catherine Plaisant, Maxine S. Cohen, Steven M. Jacobs (2009), *Designing the User Interface: Strategies for Effective Human-Computer Interaction*, 5th Edition, Pearson, New York.

Joel Spolsky (2001), *User Interface Design for Programmers*, 1st Edition, Apress, New York.

Jennifer Tidwell (2010), *Designing Interfaces*, 2nd Edition, O'Reilly, Seabastopol, CA.

Larry E. Wood (Editor) (1997), *User Interface Design: Bridging the Gap from User Requirements to Design*, 1st Edition, CRC Press, Boca Raton, FL.

# 7

# SOFTWARE MIDDLE LAYER DESIGN AND CONSTRUCTION

**In Chapter 6, we learned**

- **What a user interface is**
- **What the elements of a user interface are**
- **What HTML is**
- **What AJAX technology is**
- **What a Model–View–Controller design is**
- **What a client-side script is**

**In Chapter 7, we will learn**

- **What a software detailed design is**
- **What a software middle layer is**
- **What object-oriented programming is**
- **What procedural programming is**
- **What a class is**
- **What an object is**
- **How to build a web-based software product**
- **What database programming is**
- **What refactoring is**

## 7.1 Introduction

Software design and construction is the heart of any software development project. This activity accounts for a major portion of the total effort required to build any software product. Because of this, software design and construction activities are divided into many parts. In Chapter 5, we saw that a good way to divide these tasks is to use layers. Thus, we have a user interface layer (presentation layer), a middle layer (business logic layer), and a database layer. We discussed the user interface layer in Chapter 6; here, we discuss the middle layer. The database layer is discussed in Chapter 8.

The user interface layer uses more design and less software construction (coding or programming). Not much effort is required for database design and implementation. In contrast, the middle layer involves a lot of effort for both construction and design. The entire business logic is implemented in the middle layer.

The middle layer sits between the user interface layer and the database layer. The middle layer is responsible for connecting the user interface layer and the database layer. Thus, you not only have to build the middle layer, but also need to connect it with the other two layers.

Software programming can be done using any old-fashioned procedural programming language such as PASCAL, Fortran, and Basic. However, most software projects are made using the latest object-oriented programming languages such as Java, C++, and C#, to name a few, because of the inherent advantages of object-oriented programming compared to procedural programming. There is always a debate as to which programming language can best implement customer requirements. The reality is that each programming language has its own strengths and weaknesses, and it all depends on what kind of software product your team is going to develop. For example, the best approach to develop the middle layer of a web-based software product is to use business logic implemented by any object-oriented programming language. This is because object-oriented programming languages have matured and provide techniques such as code reuse and better quality when compared to procedural programming languages. Object-oriented programming languages also provide the benefit of faster software development.

You also need to use server-side scripting to connect the business logic to the user interface. Even database programming involves procedural programming. Client-side scripts are also a type of procedural programming language. The reason for using procedural programming for these tasks is that they are all procedural in nature. Although the database and the user interface consist of objects, connecting these objects can be done only through procedures. For example, when we need to access a database to manipulate a record, we can do so only at the record (row) level. Access is also sequential in nature. The records are read from top to bottom. This kind of functionality is not possible with object orientation. A Structured Query Language (SQL) is used for this purpose (we will learn about SQL in Chapter 8). Thus, we generally end up using both object-oriented programming and procedural programming to develop software products.

Before you start developing a product, a design and programming methodology must be chosen. The design team and the programming team must be skilled in that chosen methodology. This will allow the work to proceed in a streamlined and systematic way. Each team member should be thoroughly skilled to understand and work with concepts such as objects, inheritance, and polymorphism for the object-oriented methodology. These features greatly improve the design and development process.

In this chapter, we have two tasks to accomplish. First, we need to understand some object-oriented programming and design concepts and then apply these concepts to solve real-world problems. We will also learn some concepts in procedural programming to build client-side and server-side scripts as well as scripts related to database programming.

We will cover the areas related to what a good software product design and implementation should be. The source code based on a good design generally results in a better software product that contains the least number of software bugs. In this chapter, we discuss the design and construction of the middle layer, that is, the business logic layer, using object-oriented programming as well as some procedural programming concepts.

This chapter is structured as follows:

- Object-oriented design and programming.
- Object-oriented design and programming to build middle layer components.
- Database programming pieces inside the business logic components. Although the database layer is discussed in Chapter 8, it is important to understand how database scripts are embedded inside the business logic components. Thus, this portion of databases is covered in this chapter.
- Learning procedural programming for building client-side and server-side scripts.

**Usage notes:** This chapter is more about learning object-oriented design and programming. The discussion provided in it will help readers understand the benefits of object-oriented programming languages compared to procedural programming languages. At the same time, some procedural programming language concepts are provided to implement client-side and server-side scripts.

### 7.2 Software Design and Implementation and Software Engineering Methodology

In the Waterfall model of the software development life cycle, the complete software design is created by software engineers skilled in the chosen design methodology. The design includes all the details about methods and classes and how they interact with each other. The implementers or software programmers are required to just write the source code on the basis of that design. In circumstances where writing the code based on the logic of the design would incur a problem or conflict, the designer must be notified to make design changes. Thus, there is clear demarcation between design jobs and programming jobs if the Waterfall model is used.

In the case of agile methodologies, things have become blurred. This is because software product design is never considered more than just some high-level detail when the product development starts. The project team has no knowledge of the complete software product. The customer may keep on providing additional requirements or product features to be developed in short iteration cycles. Once these product features are implemented, the customer will provide the next set of product features to be developed. In this environment, the project team should enhance the software design with each new feature. When the development team receives a set of product features to be developed, they only have to think about how to build these features. The other issue for the development team is building such new features on top of the features that are already built. The project team thinks in terms of incremental design and programming.

In agile environments, the design and construction of a software product are often done by the same people. There is no demarcation between a software designer and a software developer. They have to determine how many classes and methods they will have to create to build the required features of the product. Since most agile projects build the software products incrementally, these people also have to think about reorganizing their existing code so that the new development can be done on top of the existing code. Reorganizing the existing code is required for reasons such as maintaining suitable naming conventions for the classes and class members, code reuse, and maintaining a balanced structure of packages.

In agile methodologies, the complete architecture of the product is not built at the beginning of the project. This means that after some time, when the code base has become large, the software structure starts to become brittle and the software product needs some redesign. Redesign is done through refactoring (explained later in this chapter). Therefore, software design activities take place both before and after programming.

Most of the projects these days are done using agile methodologies. Therefore, there is also a debate on how software design is separated from programming to perform the project activity. In agile methodologies, design goes hand in hand with programming activities; therefore, separation of these two activities is difficult.

In a nutshell, software engineering methodologies greatly affect the tasks involved in software design and programming.

### 7.3 Procedural Programming: A Brief Introduction

Some procedural programming concepts will help in learning object-oriented programming. We will also use procedural programming to perform scripting tasks such as client-side scripting and server-side scripting.

The basic concept in procedural programming is a procedure call. Each procedure (also known as a method, subroutine, or a function) contains a sequence of steps to be carried out. A procedure can be called anytime during program execution. Procedural programming is imperative because it makes explicit references to the state of the environment where the program is executed. As procedural languages evolve, they start having constructs such as assignment statements, loops, decision trees, and switch statements. Languages such as C, Pascal, and Fortran are procedural programming languages.

In this chapter, we will learn most of the constructs used in object-oriented programming languages. Some of these constructs are applicable to procedural programming languages as well. These constructs include variables, operators, decision trees, and loops. The difference between these two languages starts from methods. From class constructs afterward, only object-oriented programming is involved. The details are discussed in the latter parts of this chapter.

The scripts used for writing client-side and server-side gluing are also a type of procedural programming language. Some examples of these scripts include JavaScript, VBscript, and Python. We will learn about these scripting languages in a later section of this chapter.

### 7.4 Object-Oriented Programming

The foundation of object-oriented programming languages is objects. Once you write the source code for some objects and run them (i.e., when the byte code is executed), these objects will be created in the software product. These virtual objects are supposed to behave in the same way as their real-world counterparts. These objects interact with each other to give you the desired output. For example, when you want to shop on a website, a shopping cart object is created. This virtual shopping cart behaves exactly as a real-world shopping cart. Similarly, you can browse the web store in the same way you browse a real brick-and-mortar store. You can look and see things in the web store in the same manner you do in a brick-and-mortar store. You can pick up some merchandise and put it in your virtual shopping cart as you do in the real world. When you find better merchandise, you can remove the earlier merchandise from your virtual shopping cart in the same way you do with your real-world shopping cart. Finally, when you check out at the point-of-sale counter, you can pay exactly as you do in the real world.

You can do the same things with procedural programming languages as with any object-oriented programming language. However, object-oriented programming languages have become very powerful compared to procedural programming languages. They have acquired many features that the procedural programming languages lack. Object-oriented programming languages can be compared with civil engineering. If you have prefabricated building blocks, then you can construct a building very quickly. However, if you do not have prefabricated blocks, then it will take more time to construct the building. There is one more advantage with prefabricated blocks. Prefabricated blocks lead to a better finish of the building when compared to wet mortar construction. If you compare procedural programming languages and object-oriented programming languages, you will see that this analogy is very appropriate. Object-oriented programming languages have well-defined building blocks. This helps in designing and building better software products when compared to those with procedural programming languages.

### 7.5 Basics of Programming Languages

Programming languages are not as easy to learn as natural languages. However, after some effort, humans can understand a programming language. As you can see in Figure 7.1, humans can understand both natural languages (such as English and French) and programming languages (such as Pascal and C#). However, as depicted in Figure 7.2, computers understand only native machine code. This machine code is different for different operating systems. For example, machine code for a Windows operating system is different from that for Macintosh computers.

Programmers write computer programs in a variety of programming languages. These computer programs are then compiled using compilers. A compiler converts the source code into a machine code (Figure 7.2). The computer understands the machine code. The machine code consists of just 0's and 1's. A loader program assigns the machine code in an area of RAM (i.e., primary memory of the computer) and the machine code is now ready to be executed (i.e., run). When the machine code is run, the computer performs the tasks (i.e., computations) based on the instructions written in the program.

Some programming languages are interpreted. Here, instead of a compiler, an interpreter interprets the programming language into byte codes. A runtime environment converts the byte codes into the machine code. A loader program can then load the machine code into the RAM of the computer and the machine code can run. The process of generating a byte code and then converting that byte code into a machine code is depicted in Figure 7.3. Examples of programming languages that are interpreted include Java and C#.

A machine code is specific to an operating system, whereas a byte code is platform independent. This means that you can take a byte code generated on a computer and run it on any other computer that may have a different operating system and you can still run the program that was in the form of byte code. This is why programming languages such as Java and C# are truly platform independent. Platform independence

**Figure 7.1** Humans can understand natural languages as well as programming languages.

**Figure 7.2** Computers can only understand machine code.

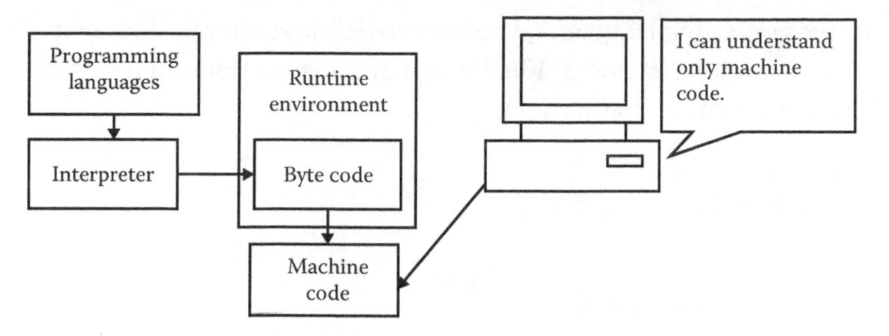


**Figure 7.3** Platform independence of byte code.

is of great significance. When a software product is made, the owner of the software product would like it to run on all operating systems so that any customer can buy the product irrespective of the operating system. In cases where a software product is made using a programming language that does not support platform independence, the owner may have to build many versions of that software product for each operating system. This is a time-consuming and costly matter.

Most programming languages define variable data types, computational operators, and methods to write computer programs. You can create variables to assign and hold some values and then perform some computations using the available operators; the result of these computations is stored in some variables and can be displayed or used in some other computation. A function or a method in a program is a process that performs a logical task. A function or a method is often named according to the task performed. You can also pass values to a method. This property is used to break the computation task into many methods. It has many advantages. We will see these advantages in later sections of this chapter.

In object-oriented programming, we use classes and objects. We will learn about classes and objects in later in this chapter.

### 7.6 Variables and Variable Types

Variables are the most basic elements in all programming languages and they are used to hold some values. You can create variables and then assign some values to them. After that, you can do some computation using such variables.

Let us look at an example. There are five people and each person has six candies. You want to find out how many total candies there are among all five people. In this case, you want the computer to compute and return the result of 5 × 6. First, let us understand how a human will perform this task. A human sees 5 multiplied by 6. Using a multiplication table, a human computes and returns the result as 30. Now, the question is how to tell this to a computer in a language that it can understand for it to give you the results.

In most programming languages, you have to define some variables first. Then, you create the computation statement. Finally, you give instructions on how to display the result of the computation. Let us do it.

```
Create variable a;
Create variable b;
Assign a = 5;
Assign b = 6;
Compute a multiplied by b;
Display a multiplied by b on the user screen;
```

Depending on the programming language you choose, the earlier statements may vary. We have used the computation expression in pseudocode. However, essentially, all the above steps need to be written in a programming language. Now, you need to compile the program (in this book, the word *compile* is used in a generic sense to represent both *compile* and *interpret*). This will turn your program code into a machine code. When you run the machine code, the computer will promptly display 30 on your screen.

For variables to perform various kinds of computations, they are defined in different ways. For example, if we need to do a computation involving mathematical expressions like we did in the earlier example, we must define the variable data type and name. Some common types are integer, float, and character. In our example, variables *a* and *b* are integer types. If you need to do a computation involving characters or texts, then you can have variable types such as "Character" and "String." For example, if you need to find the number of letters in "December," then we can define "December" as a string variable and count the number of letters. Therefore, you can see that each variable type is associated with a data type. There are many other variable types used in different programming languages. They have been created to do certain operations. In many cases, you need to evaluate a condition to find out if it is true or false. In these cases, you can create a variable that can hold only one of the two values: true or false. In most programming languages, the variable type "Boolean" is used to take care of this need. You can learn about variable types defined for a specific programming language by reading any programming language book.

Variables can exist in some other forms. For example, parameters are also variables. Parameter variables are used to pass data from one method to another. Depending on where a variable is declared, it can belong to either a method or a class. Class variables behave differently from the variables declared within the methods. We will learn about these aspects in Section 7.10.

In the earlier example, we have not declared the data type. Generally, in most of the programming languages, the syntax for declaring a variable is given below:

```
<data type> variable name;
```

There is also one more qualification we need to add when we declare a variable. It is the scope or accessibility level of the variable in your program. Thus, the general declaration of a variable can be like this:

```
<modifier> <data type> variable name;
```

The modifier determines the scope or accessibility level of the variable. If the variable is declared inside a method and is declared private, then this variable will be visible only inside that method. Thus, this variable can be used only inside that method. Let us pay attention to some differences here. If you are using a variable in object-oriented programming, then you have some scope limitations. You cannot use a public variable inside a method in object-oriented programming because it is not required; therefore, compilers will raise errors when you do so. In object-oriented programming, there is ample scope to define a public variable at class level and thus, at the method level, a public variable is definitely not needed. We will discuss why this is so when we discuss methods.

### *7.6.1 Variables in Object-Oriented Programming versus Procedural Programming*

Figure 7.4 shows the scope of the variables in procedural languages. Procedural programming languages have limited space to define the variables. Variables are contained inside the methods; thus, both public and private variables have to be declared inside the methods. When the variables need to be accessed from outside the method, they are declared public. When the variables need to be accessed from inside a method, they are declared private inside that method.

In contrast, object-oriented programming languages provide more flexibility because they have more space for defining the variables. Figure 7.5 shows the scope of the variables in object-oriented languages. Variables can be defined inside the methods or at the class level. If a variable needs to be accessed from outside a class, then it should be declared public. Thus, there is no need for any variable to be declared public inside the methods. We will further discuss the scope of the variables when we discuss methods and classes later in this chapter.

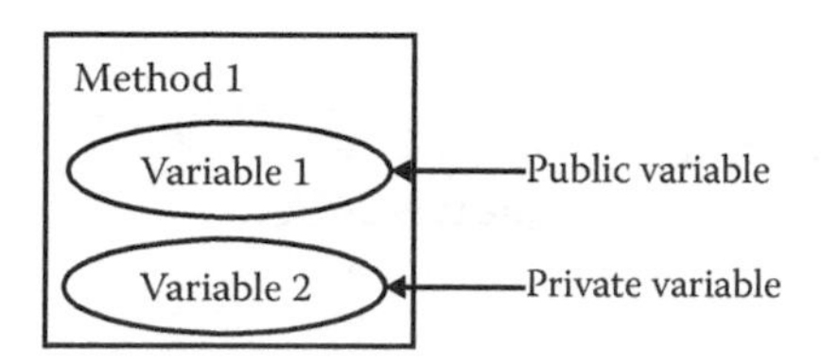


**Figure 7.4** Scope of variables in procedural programming languages.

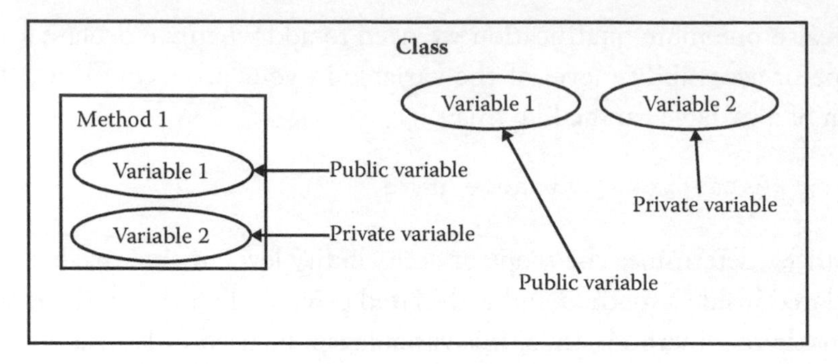


**Figure 7.5** Scope of variables in object-oriented programming languages.

### 7.7 Operators

Operators are used to perform computations on variables. The most common operator is the assign operator. You can assign a value to a variable using this operator. In Java and most programming languages, it is denoted as "=" (equals sign). You can see that we have used this assign operator in the example given in the previous section. In Java and most programming languages, the operator for testing the equality is "= =". Some special operators are also created for various programming languages.

One can look at the documentation of a programming language to learn about the operators in that language.

### 7.8 Decision Trees and Loops

Suppose the problem statement is the same as the one given in Section 7.6 but with a condition. The condition is that there is a leader among those five people and you need to compute only when that leader is at least 30 years old. Let us write the computation program now.

```
Create variable a;
Create variable b;
Create variable c;
Assign a = 5;
Assign b = 6;
Assign c = 34;
If c > = 30 then
Compute a multiplied by b;
Display a multiplied by b on user screen;
End if;
```

Now, we are telling the computer to compute 5 multiplied by 6 only when *c* (age of the leader) is at least 30. Figure 7.6 presents how an if–then decision tree works.

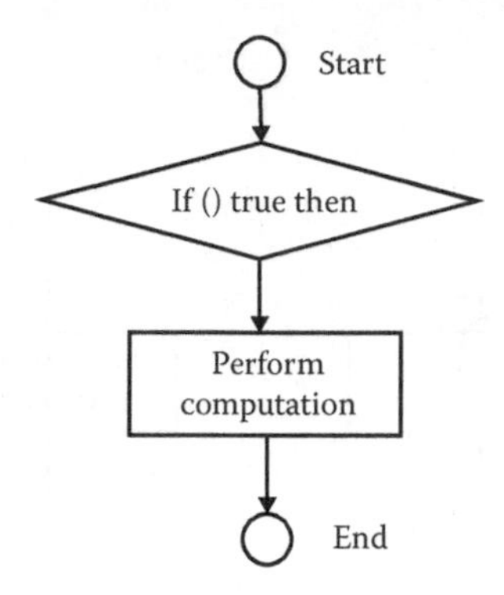


**Figure 7.6** If–then decision tree diagram.

When we need to evaluate a condition, we can use "If <condition> then do something." Generally, in most programming languages, the statement is closed by a corresponding "End if" clause.

Now, it is possible that the leader may be below 30 years of age. Hence, the problem statement now says to compute the total number of candies owned by the people when their leader is above the age of 30, and if the leader is below the age of 30, then display the message "age of the leader is below 30 and computation is not possible." Let us write the program now.

```
Create variable a;
Create variable b;
Create variable c;
Assign a = 5;
Assign b = 6;
Assign c = 26;
If c > = 30 then
Compute a multiplied by b;
Display a multiplied by b on user screen;
Else
Display a message "age of the leader is below 30 and computation is
not possible";
End if;
```

Now we are telling the computer to compute 5 multiplied by 6 only when *c* (age of the leader) is at least 30. If the leader's age is below 30, then the computer will display the message "age of the leader is below 30 and computation is not possible."

Here, we are testing a condition and telling the computer what to do if a condition is not met. The "Else" clause does this task. Figure 7.7 depicts this idea to test the condition. If a condition is found to be true, then perform one set of computations; else, perform another set of computations. You can also notice that we have used the condition evaluation result in terms of true and false. If a condition is met, then the condition results in true, and if the condition is not met, then it results in false.

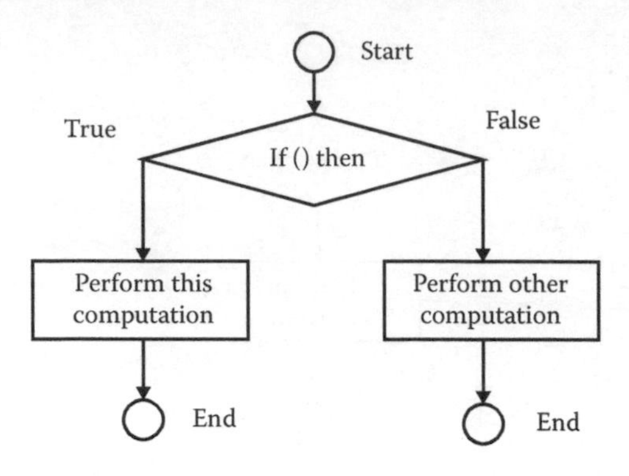


**Figure 7.7** If–else decision tree diagram.

There are many types of decision trees used in programming languages. There are switch statements, loop statements, and so on. In switch statements, many conditions can be checked one after the other. For example, in the previous example, if we need to perform different computations for each age group to which the leader belongs, then we can use a switch statement. The loops are used for testing a condition for a range of test values. For example, suppose we need to find the age of each person in the group holding the candies and perform some computation; then we will need a loop. First, we may need to find the age of each person and then do some computation if the age matches some specified value. Figure 7.8 depicts a while loop. As long as a condition is true, some computation is done in each cycle of the loop. After running each cycle of the loop, we may need to increment or decrement the condition and then check the condition again with the increased or decreased value of the condition statement. When the condition is no longer true (evaluates to false), the computation will be stopped and we can exit the loop. The common syntax for a while loop is as follows:

```
Do while <condition>
i = i + 1;
do some computation;
End;
```

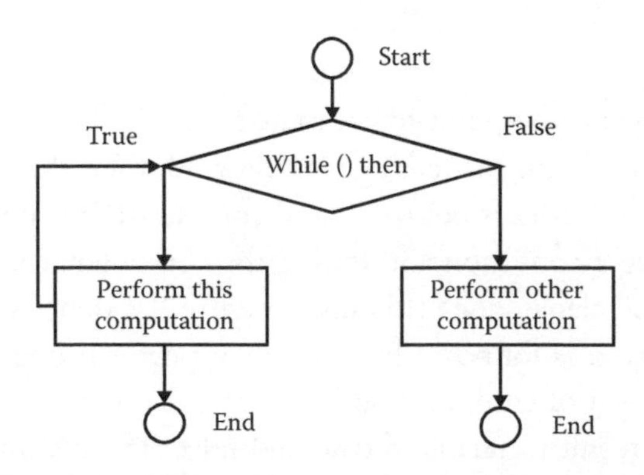


**Figure 7.8** While loop diagram.

When creating a while loop, be careful to set the condition and the increment. If your condition for exit is such that the code may keep evaluating the condition for exit but may not be able to meet that condition, then you are creating an infinite loop.

### 7.9 Methods

When you have created a program to do some data processing, you need to save it with some unique name. There are also some other aspects in data processing that you need to think about. An aspect arises when you have many data processing tasks to be done as part of one project. Each piece of code or program is wrapped under a structure element known as a method. In procedural programming, these methods are also known as procedures or functions. In object-oriented programming languages, these methods are just known as methods, and they do a lot of things. We will discuss them in a later section.

Suppose, in the example we gave earlier, we save our piece of code under a method name "Candy_computation." Generally, in programming languages, there are some programming conventions for naming the variables, methods, and other programming elements. We have used an underscore so that we have one word for the method name yet the method name is clear as to what it is supposed to do. Let us wrap our code under the method name now.

```
Candy_computation()
  Create variable a;
  Create variable b;
  Create variable b;
  Assign a = 5;
  Assign b = 6;
  Assign c = 26;

  If c > = 30 then
    Compute a multiplied by b;
    Display a multiplied by b on user screen;
  Else
    Display a message "age of the leader is below 30 and
    computation is not possible";
  End if;
```

Like variables, methods may have some data type, and it depends on what values the methods may return. The Candy_computation method is not returning a value. Therefore, it has no data type. In such cases, in most programming languages, the word *void* is used in place of returning data type. For example, our method Candy_computation will have the following declaration:

```
void Candy_computation(){
– same code as earlier
}
```

Methods are used to do some computations and pass some values to other methods. When you are building a large software product, you do a small computation in each of your methods and then pass some values (from each of these methods) to some other methods. This way, you will be able to create a software product with small components. One of the goals of software design is to break down the entire design into small components so that the design complexity can be reduced. The Candy_computation method can be partitioned into two methods. One method will do some computation and pass some value to the other method. This way, the two methods can do some work that will be divided over these two methods.

We can divide the calculations in the Candy_computation method into two methods. Let us try doing it.

```
Pass_candies (a, b)
  Create variable x;
  Assign x = a multiplied by b;
  Return x;

Candy_computation()
  Create variable a;
  Create variable b;
  Create variable y;
  Create variable c;
  Assign a = 5;
  Assign b = 6;
  Assign c = 35;
  Assign y = Pass_candies (a, b);
  If c > = 30 then
    Display y on user screen;
  Else
    Display a message "age of the leader is below 30 and
    computation is not possible";
  End if;
```

You can see some changes we have made to our original code of the Candy_computation method. We are no longer doing the multiplication of *a* and *b* inside this method. Instead, we are doing this multiplication in the new method Pass_candies. We are calling the Pass_candies method inside our Candy_computation to do the multiplication. We have also shown that a method can be called inside another method using parameter passing. Parameters are the variables that are defined outside of a method but can be called inside of that method.

Even without parameter passing, we can call the methods inside another method, and this other method will do some computation. This other method has no parameters and there is some return type. Let us see what happens when we have this type of method.

```
Public class my_class{
  My_method add_integers x(){
  a = 10;
  b = 5;
  y = a + b;
  Return y;
  }
}
```

**Analysis:** This method is doing the computation of adding two variables and then returning the value of this computation. Is this method designed well? No, it is not. Why? It is doing the calculation using two variables, but you cannot change the values of these variables from outside of the method. Notice that any method created in your program must interact with other methods. This is a big design issue. This design issue is violated in this example method. We will learn about these aspects later.

You can "call" one method from another method. In order to use this feature, it is better to create a separate method to do some computation and then call that method in the other method. This separation of methods will provide good clarity to your code. This concept is illustrated in Figure 7.9.

What is the use of creating two methods when we could do the same thing in one method? Several important things can be done when we create two methods instead of one. First, we are delegating a task to be done to another method. This, itself, makes it possible to divide the programming work among many people at the same time: one team can start constructing one method and another team can simultaneously start constructing the other method. The design of the software product becomes easier (i.e., design complexity will be reduced) when we divide the design into many parts. The method that is being called (i.e., method A in Figure 7.9) can later be modified (e.g., to optimize the code or make that code execute faster), without touching the other method, as long as those modifications do not change the functionality of method A. Similarly, any modifications that are performed on the calling method (i.e., method B in the figure) will not affect method A, as long as method B is calling method A using the same set of parameters as it did earlier. Therefore, maintenance will be easier by having two separate methods than having a single method in that place.

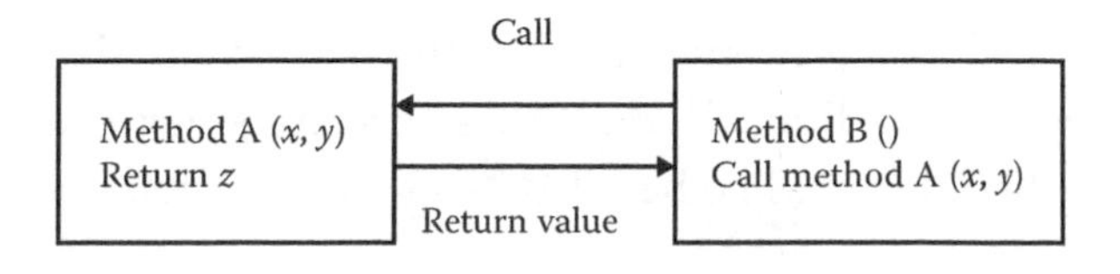


**Figure 7.9** Call of method A ($x$, $y$) in method B ().

*7.9.1 Method Definition*

Methods are used for many purposes in object-oriented programming. The main purpose of the methods is to provide a wrapper for some code. However, there are special cases in which the methods are used. For example, methods are used for method overloading, to define a method with the same name as the class in which that method is defined (such a method is called a constructor), and for method overriding (method overriding is explained later). Let us see some examples.

First, let us see how methods are declared in object-oriented programming.

```
<Modifier> <data type> method name (<data type> parameter 1, <data
type> parameter 2,...)
{method body
Exception list
Exception handling
}
```

Let us look at different parts of the method definition.

*7.9.1.1 Modifier* Modifier defines the access level of the method. For example, if the modifier is defined as "public," then the method is accessible to outside classes in addition to the class in which it is defined. If the modifier is "private," then the method is accessible only inside the class in which it is defined.

*7.9.1.2 Data Type* The data type of a method declares the type of data (e.g., integer or string.) of the return result of the computation done by that method. If the method is not returning any values, then in most programming languages, the keyword *void* is used in the placeholder for data type.

*7.9.1.3 Method Name* Method name is the name by which the method is known inside or outside the class. You can use the method name to call it anywhere in your code.

*7.9.1.4 Parameter List* Parameter list is the list of variables (parameters) that are passed to the method, from some other method, while calling that method. Each parameter has a data type.

*7.9.1.5 Exception List* During computation, if an error arises, then the error can be trapped using exception handling. To do this, an exception list is declared, and later, if exceptions arise, then they will be handled inside the method body.

*7.9.1.6 Method Body* This is where all the computation code of the method is written.

*7.9.1.7 Method Signature* Method signature is a declaration by which a method can be called in the source code. For example, if a method is declared as follows:

```
Public Integer my_method (integer x)
{
}
```

then this method signature will be my_method (integer x).

### *7.9.2 Scope of Methods*

To create the methods, we use identifiers, return value, and data types, among others. For example, we can use an identifier to indicate whether a method is accessible outside a class (we will learn about classes in a later section). For example, in Java language, identifiers such as private, public, and static are used for this purpose.

When you define your method, you have to think about whether that method should be accessible only from inside the class in which it is created or if it can be accessible from any other classes as well. This is one of the most important considerations for designing your code. In object-oriented programming, you need to make your methods private in some situations so that they are not accessible from outside of the class. In other situations, you must make your method accessible from outside of the class. The same is true for the variables that you create. Let us first discuss the accessibility level for the variables and then discuss the same for methods.

In Figure 7.10, you can see that we have created variables at the class level as well as inside the methods. Class-level variables can be both public and private. These variables are also known as class variables. If they are public, then they can be accessed from both outside and inside that class. If they are private, then they can be accessed from within that class only. This accessibility level has profound effects. If any variable is declared public at the class level, then it can be modified by any class. This will lead to bad coding because it may allow some defects to creep in through this free calling (of the public variable) from any class. Therefore, these variables should be private. However, if they are private, then what is the use of declaring them at the class level? If you think about

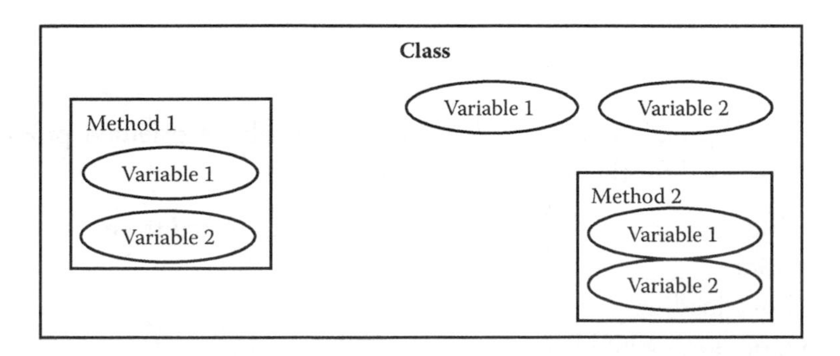


**Figure 7.10** Scope and accessibility level of the methods and variables in an object-oriented language.

it, then it is one of the fundamental principles of object-oriented programming. It is known as encapsulation. We will discuss encapsulation in Section 7.10.

It is still possible to use the variables in a class (even if those variables are private) from outside that class. It is done through what is known as accessor methods. One accessor method is used to get the values that are stored in a private variable. Another accessor method is used to set the values to be stored in a private variable. Together, these two methods are used as accessor methods. In Java, there are "get" and "set" methods. Other languages have their own way of implementing these accessor methods. This way, the variables are not directly accessible from outside of that class yet they can be read or assigned values from outside of that class in a roundabout way using get and set methods. The beauty of this approach is that you have total control of whether the values need to be read only from outside or they can be assigned values from outside. We will learn more about accessor methods in the next section.

Most computations that need to be done locally should be done inside private methods. There is no need to make them public. On the other hand, in most object-oriented programming languages, a type of method known as "constructor" is always declared a public method. We will see constructor methods later.

### *7.9.3 Accessor Methods*

As mentioned in Section 7.9.2, public variables are bad programming practice and should be avoided. Then, how will the objects (when classes are instantiated and the objects are created in memory) interact with each other during runtime?

In practice, public variables used to be created so that the methods from other classes could use them for interaction with the class where the public variables were defined. The problem with public variables has already been discussed in Section 7.9.2.

A technique that is used in object-oriented programming is the use of accessor methods. We create a get method to read the value of a private variable that is declared inside the object in which that get method resides and a set method to write a value to the private variable that is declared inside the object in which that set method resides.

Accessor methods essentially encapsulate the data of classes. We will learn more about accessor methods in Section 7.10.4.

### *7.9.4 Constructor*

Constructors are special constructs that are used in object-oriented programming. Constructors are used for the following purposes:

- Identifying the name of the objects that need to be created from classes (object initialization)
- Method overloading
- Copying or cloning an object
- Assigning the value of one object to another

**Object initialization:** When we need to create objects from a class, we use a default method of that class. For example, we can have the statement

```
getset person = new getset ("Natasha");
```

This statement uses the default method of the class getset. Here, the method name is also getset. This default method is also known as constructor.

**Method overloading:** Method overloading is used to define several methods with the same name but different signatures. This helps in reusing that same (method) name for performing different tasks. Inside a class, you can define several methods with the same name but different signatures. When another method wants to use any of these methods for performing a specific computation it can use the right method out of these methods having the same name. Method overloading is useful when the same class (that contains the methods with the same name but different signatures) is used to perform different tasks. For example, you can use method overloading to draw various sizes of rectangles as follows:

```
public class rectangle {
  private integer length;
  private integer height;
  private integer x;
  private integer y;

  public rectangle() {
     this.x = 15;
     this.y = 20;
     this.length = 20;
     this.height = 25;
  }

  public rectangle (integer length1, integer height1) {
     this.length = length1;
     this.height = height1;
     this.x = 1;
     this.y = 2;
  }

  public rectangle (integer length2, integer height2, integer x,
  integer y) {
     this.length = length2;
     this.height = height2;
     this.x = x;
     this.y = y;
  }
}
```

You can see that although the method name, rectangle, is the same for all three methods, the signatures of these three methods are mutually different. The benefit with method overloading is that, by creating just one method and by passing different parameters, we are able to perform many types of computations with the same name. Otherwise, we would end up defining many methods with different names where each method performs a different computation.

Let us see how the same constructor name can be used to create various shapes:

```
Rectangle rectangle1 = new rectangle();
Rectangle rectangle2 = new rectangle(30, 35);
Rectangle rectangle3 = new rectangle(40, 50, 5, 10);
```

When you test your code, you will see that three objects are created: rectangle1 with length 20, height 25, and center at (15, 20); rectangle2 with length 30, height 35, and center at (1, 2); and rectangle3 with length 40, height 50, and center at (5, 10).

Maintaining many different methods with different method names is a time-consuming task. In contrast, many methods with the same name save time. By using the same name, you can easily remember what a method does rather than remembering different names for the same job. Therefore, to perform a task, you call the same method (i.e., same name) without worrying about the type of data you are passing to that method.

The drawback of method overloading is that it can result in errors. Generally, to enforce better programming practices, for each entity, whether it is a variable or a method, its name should be descriptive so that we know what it is supposed to do in the program. In the previous example, if the methods' names are small_rectangle, medium_rectangle, and large_rectangle then these descriptive names easily identify that the first method is to draw a small rectangle, the next one is to draw a medium size rectangle, and so on. When using the same name for more than one method, we are sacrificing this clarity in the code.

The technical limitation of method overloading is that you cannot define two (or more) methods with the same name if the number of parameters in each of these methods and the data type of the corresponding parameters in these two (or more) methods are exactly the same. This is because the compiler will get confused, even if the return data types of such methods are mutually different. The bottom line is that the method signature should always be different.

There is another concept known as method overriding. People often get confused between these two concepts, namely, method overloading and method overriding. We will learn about method overriding in Section 7.10.

## 7.10 Classes

In object-oriented programming, we wrap the methods and variables inside another programming element that is known as a class. The class concept is powerful and many object-oriented concepts are built around classes. We present some of them here.

Why is there a need for classes? In procedural languages, all programming code is saved inside the methods and you can build the entire software product by creating various methods to do various required computations. From this point of view, classes look superficial and redundant.

Let us think about classes from another angle. One of the primary requirements while building a software product is to make it platform independent. The entire product is supposed to run on any operating system once the product is developed. To achieve this goal, the programming language should provide a rich and complete environment where the programmers do not have to worry about the operating system. This means that the people who developed the programming language should provide the necessary infrastructure on which the programmers can build their own products. This infrastructure comes in the form of extensive libraries that the programmers can use in their source code.

When a programmer writes source code, he or she can use the infrastructure provided by the operating systems. This infrastructure is available through the operating system's Application Programming Interface (API). For example, if the programmer needs to create a user screen, then he or she can use an operating system's API. However, in this case, the source code written by the programmer will run only on that operating system. If the programmer has to make his or her source code run on some other operating system, then he or she will have to write the source code separately for it.

To facilitate the programmer's work of writing the source code that can run on all platforms, the developers of some programming languages have provided some libraries that are specific to those programming languages. In such cases, a programmer will not use the operating system's API and instead will use the libraries available in the programming language in which he or she is writing the program. For example, Java language has libraries like Java.awt.* that take care of the infrastructure required for building the user interfaces. When this library is available to the programmer, he or she does not need to rely on the APIs provided by the operating systems. The programmer will use these libraries and the code will run on all the platforms. In this way, the software product will be truly platform independent.

The programming language libraries are built using classes. Several classes are then combined under one package. Several packages are in turn kept under one library. Classes greatly help build these libraries. Accessing the code inside these libraries is also easy. The programmers have to just include these libraries inside the source code of their product using an include statement. When the source code of that product is run, the code inside these libraries also runs.

Imagine if all the libraries provided in a programming language were just methods. Managing and accessing the methods would be extremely difficult. Therefore, building the software products using the classes makes the source code manageable and accessible. Packages (explained later) include your classes. Classes, of course, include their own methods and data. For example, if you have created a package named "my_package" and have included a class named "my_class" in that package, then you can call your class "my_package.my_class" anywhere in your source code. If you have included a method named "my_method()" in this class, then you can call this method using a statement "my_package.my_class.my_method()" anywhere in your source code.

The other advantage of a class is that it acts as a template for creating objects. When you run your source code, objects that hold their own data are created. The important thing about objects is that they are created when they are needed and are destroyed when they are not needed. This helps memory management. When an object is destroyed, the memory occupied by that object is freed and can be used by new objects.

The concept of class was conceived because of the accessibility of variables and methods. Naturally, methods should always be declared public in procedural languages. If they are declared private, then they cannot be called anywhere in your program and this results in dead code (i.e., code that never executes). In many instances, private methods are required. Generally, in procedural programming languages, this concept is difficult to implement. To circumvent this problem, procedural programming languages devised some structures that somehow made some methods private. For example, in VB.Net, there is a structure known as module. Inside these modules, you can keep private methods. These methods are accessible only inside that module (in which those methods are declared) and cannot be called from outside. Classes in object-oriented programming allow you to declare private methods universally. You can create private methods anywhere in any class. These methods will be accessible only inside that class and not from outside.

Compare Figures 7.10 and 7.11. In Figure 7.10, we have depicted a class in an object-oriented language, whereas in Figure 7.11, we have depicted a method in a procedural language. In procedural programming, a method is the top-level entity, whereas in object-oriented programming, a class is the top-level entity. Classes represent a better way of managing your source code. We have also been discussing public variables and the issues associated with them and have seen that this problem is taken care of in object-oriented programming.

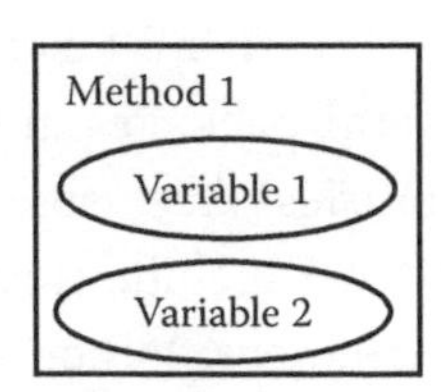


**Figure 7.11** A method and its variables in a procedural language.

Now that we have learned about classes, let us build and learn how they work. Remember that we have made methods to do computation for candies. We will now build a class using the same methods to compute the candies.

```
<modifier> class my_class {
Pass_candies (a, b)
  Create variable x;
  Assign x = a multiplied by b;
  Return x;

Candy_computation()
  Create variable a;
  Create variable b;
  Create variable y;
  Create variable c;
  Assign a = 5;
  Assign b = 6;
  Assign c = 34;
  Assign y = Pass_candies (a, b);
  If c > = 30 then
    Display y (a multiplied by b) on user screen;
  Else
    Display a message "age of the leader is below 30 and
computation is not possible";
  End if;
}
```

What we have created here is the class, my_class, with two methods. Now, if we want to access any of these two methods, then we can use my_class.Pass_candies (*a*, *b*) or my_class.Candy_computation() if these two methods are declared public.

Let us further discuss classes.

### *7.10.1 Class Variables*

Variables in an object-oriented programming language can be of many types depending on where they are declared. When a variable is declared inside a method, it is known as a local variable. When the variable is declared outside the methods, it is naturally declared inside a class. These variables are known as class variables. Class variables behave differently from local variables. Let us declare some class variables.

```
Public class my_class {
  Public integer a;
  Private integer b;
```

```
  Public void my_method()
  {
  }
}
```

In this example, we have declared two class variables and one method in the class, my_class. Declaration of a class variable is done in the same way it is done for a local variable.

When you declare your class variables as public, they are accessible from outside that class. These variables are known as instance variables. This is because you can create many instances of the same variable inside all the objects you create from that same class. Each instance of that same variable can hold a different value. In case of private class variables, you can use them through get and set methods. Get and set methods were discussed earlier in this chapter.

There is another type of variable that acts more like a constant. In Java programming language, these are known as static variables. Static variables are never instantiated. It means that there is just one copy of this variable when you run your code. This means that the value stored in an instance variable is a constant value for the entire life of an object. In contrast, we have seen instance variables and they have as many copies as the number of objects.

### *7.10.2 Inheritance*

Once you define a class, you can define some child classes based on that parent class. One advantage of doing this is that you can reuse the code, which you have written for the parent class, in those child classes. Another advantage is that you can build a hierarchy of classes that provides a useful structure for your code.

Let us see how we can define a parent class (also known as super class) and a child class that inherits the properties and behavior of that parent class.

```
Public class parent {
  public integer age;
  public string language;
  age = 44;
  language = "English"
}

Public class child extends parent {
  age = 20;
  language = "French";
}
```

We have defined a child class "child," which is a child class of "parent." If you have defined class variables and methods in the parent class, then they are inherited in its child class. You can see that the variables “age” and “language” are defined in that parent class, “parent.” Since we have declared that the class, “child,” is a child class of “parent” through the keyword “extends,” the class members of the parent class become the class members of the child class. You do not have to declare the variables “age” and “language” in the “child” class again. You can directly assign values to these variables in the child class. Inheritance is useful because you can create similar classes (to create similar objects) without having to write too much code to create all these classes separately each time you need them.

### *7.10.3 Interfaces*

Interfaces are the templates for creating classes. A template defines a common structure that can be used to create many classes with similar structure. Because an interface is just a template of a structure, it cannot be used in the implementation directly. Classes that use this structure template (provided by the interface) will implement that structure.

```
Public interface my_interface {
  Public void yellow();
}
```

Now you can implement a class based on this interface as follows:

```
Public class my_class implements my_interface {
  Public void yellow(){
    String color = "light yellow";
    Print to user screen (color);
    }
}
```

You can create many classes that have a structure similar to that of the interface by implementing this interface. The benefit of creating an interface first and then creating many classes on the basis of that interface makes the source code reusable. Reuse of the source code is extremely desirable because it leads to a better-structured source code and increases the productivity of the software developers.

### *7.10.4 Encapsulation*

In object-oriented programming languages, you can hide the data that are in a class from other classes. You can do this using get and set methods (also known as accessor and setter methods), which we briefly discussed earlier in this chapter. You can declare your variables as private so that they are completely invisible from outside. Later, you can use the get and set methods to initialize and provide values to these variables.

```
public class my_encapsulate{
  private String name;
  private integer age;
  public integer getAge(){
    return age;
  }
  public String getName(){
    return name;
  }
  public void setAge (integer newAge){
    age = newAge;
  }
  public void setName (String newName){
    name = newName;
  }
}
```

In the previous example, we have created two private variables, name and age, in the class my_encapsulate. These variables cannot be accessed from any other class because they are declared private. Now, if we want to access them, we can only do so through the methods getName() and getAge(). If we want to set values to them, then we can only do so through the methods setName() and setAge(). Although it looks like we have to write an unnecessary code for accessing these variables and setting the values to these variables, these private variables and this code have advantages. In procedural programming languages, we could access a variable only if it is publicly accessible. However, the values of the variables that are publicly accessible can be changed by any code. Managing this kind of code is extremely difficult. Imagine if you have a publicly declared variable and it is accessed or modified by some pieces of code located at 10 different places. Finding which of these 10 pieces of code accessed and changed the value of that variable will be a nightmare. Debugging will also be extremely difficult in this situation. However, this nightmare will not happen with my_encapsulate class.

```
public class test{
  public static void main() {
    my_encapsulate encap = new my_encapsulate();
    encap.setName ("Rebecca");
    encap.setAge (20);
    print to user screen ("Name:" + encap.getName() +
    "Age:" + encap.getAge());
  }
}
```

As you can see in the above code, we are setting the values of the variables by using the set methods. The values of the variables are displayed on the screen after retrieving them using the get methods.

In the above code, you will also notice that we have introduced a new method of type "main". Main methods are used in one class to run the code of that class as well as the code of all the classes that are called inside that class.

Public variables are the most dangerous things in programming. Most software defects occur because of these public variables. Avoiding public variables is one of the best practices to secure your code from defects. Usage of get and set methods, instead of public variables, is one of the best practices in object-oriented programming.

There is a warning to the use of accessor methods. These methods still create problems. The point is that a good design should never allow too many calls from one method to another. All the computation that is necessary to be done in a class should be done inside that class itself. A class should not be dependent on another class for computing. Otherwise, a vital objective of maintenance performance is compromised. If a class is connected with too many classes and this class is changed in the future, then it will affect the other classes and you may need to make changes in all those connected classes. For example, even if you only need to change the data type of a variable that is shared among many classes, then you will have to make changes in all the classes that are dependent on that variable.

Calls to other classes should be made only when required. Unnecessary calls will always make your source code difficult to maintain.

#### *7.10.5 Method Overriding*

Method overriding is used when a method in a child class performs its computation differently from the same method in the parent class of that child class. Hence, although the method remains the same in both classes, for the child class, the computation will be different from that for the parent class.

Let us look at an example:

```
class Animal{
void eat(){
System.out.println("Animal is eating");}
}

class Cat extends Animal{
void eat(){
System.out.println("Cat is eating mouse");}
public static void main(){
Cat obj = new Cat();
obj.eat();
}
}
```

In this example, Animal is the parent class and Cat is its child. When you run your code, you will get Cat is eating mouse as the output because of the definition of "eat" in the Cat class.

### *7.10.6 Import Statement*

In most object-oriented programming languages, you may encounter the import statement. Import statements are used to import the source code of a class or a package into another class. In Java, the keyword for import is "import," while in C#, the keyword is "include." In some other object-oriented languages, a different keyword may be used for import. In Java, here is the syntax of import:

```
Import <package>.<sub-package>.class_name
```

There could be many subpackages nested inside a package and it depends on the size of the package and its branching into subpackages.

An import imports the source code from one class (say, Class 1) into another (say, Class 2). Therefore, all the source code of Class 1 becomes available to Class 2 and the compiler treats it as if all the source code (from Class 1 and Class 2 together) is actually written inside Class 2.

Importing one class into another is different from creating objects. By importing, you are just getting the source code of a class into the other class. However, you instantiate a class to create an object.

### *7.10.7 Class Diagram*

We have learned some high-level design concepts including architecture patterns, component diagrams, and data flow diagrams in Chapter 5. These high-level designs are used to design the architecture of the software. Now, we will learn software designing at a class level. Class-level designs are low- or detailed-level designs. In class-level designs, we provide most of the implementation details for a software component or for the complete software product.

While designing a system, several classes are identified and grouped together to represent the static relation between them. A class diagram represents the classes and their relationships. Each class consists of a class name and the members of that class (as depicted in Figure 7.12). Members of a class can be methods or class variables. A method can be declared with return type if it is returning any values. If a method takes any parameters, then the parameters can also be mentioned. A class variable should be declared (inside that class) with its data type.

**<<Interface>>**
**Class name**
Method name 1(): return type
Method name 2 (): return type
Variable name 1: integer

**Figure 7.12** Class diagram.

*7.10.7.1 How to Create a Class Diagram* Declaring a single class is not of much use. Declaration of all the classes and the relationship among those classes (e.g., the classes and their relationships form the middle layer or part of the middle layer of the software product) is required when we create a class diagram. When this kind of declaration (in the form of a class diagram) is done, it can become the static design of the software product.

There is an approach to model the detailed design for software systems and it is known as the mental model. Here, the nouns and verbs used in the requirement specifications are used to create classes (classes are the templates to create the objects) and their behaviors. The idea behind this approach is that a model that is built this way will closely resemble the way the things are in the real world and how a human mind perceives real objects and their interactions. However, there is a limitation to this idea. When you actually start building a software system this way, you realize that the source code becomes difficult to reuse. For example, if you built two classes representing a credit card and an ATM machine, then many attributes and behavior such as customer information, bank information, and data related to a transaction will be common to both. Thus, you end up creating two classes that have many things in common. This will result in unnecessary duplication of the source code. This is a violation of the principles of code reuse. To counter this challenge, the mental model approach tries to remodel the classes by using refactoring. The common behavior and attributes are then shifted from many classes to new classes that will be built.

We can start modeling our software product from the mental model. After refactoring, a group of refactored classes may be used to build the components. You have to think of a group (or a set) of classes. Software modeling is more about abstraction and refinement than anything else. You need to find ways to aggregate the classes first. Classes with similar attributes need to be aggregated first, and this aggregation will form a group of classes with similar attributes. Thus, we will end up with many groups of classes. Each group of classes will have similar attributes. This is achieved through abstraction. Later, these aggregated classes need to be refined so that the classes with very similar attributes can be categorized together inside this group of classes.

For example, suppose we found that we initially need five classes (Classes 1–5) to build a software product. After some analysis, we found that Classes 1 and 2 have similar methods. After further investigation, we found that Classes 3–5 also have some similar methods. Hence, we have two groups (of classes). For each group, we need to do refactoring. Let us take the first group and discuss this refactoring. The code reuse between the classes in this first group can be done by eliminating the repeating source code (which is in the form of methods that have a similar source code) between the classes in that first group. How can we eliminate this repeating source code? This is done by moving the similar (or common) methods (from the classes in that first group) to one place either by creating a new class and placing these similar methods inside this new class or by removing a similar method from one class and modifying the source code of the method in the other class to take care of the needs of the other class from where the method was removed. Once this refactoring is done for the first group of classes, we repeat the same process for the second group of classes. This way, we can completely avoid the repeating source code and, at the same time, reuse the source code.

The details of the classes and their members can be easily derived from the mental model. The mental model finds all the verbs and these verbs will form the methods of a class. The adjectives will form the class variables.

From this discussion, you can see that the mental model is a low-level activity that is concerned with the internal structuring of a single class. When you have individual classes built, you can think of aggregating similar classes. At the aggregate level, a group of classes can correspond to a component. Later, you can do refinement to categorize the classes inside each group of classes. Refinement is the process of separating dissimilar things from a common group. Let us find out how refinement can be done by considering the same example (five classes) that we provided earlier. Suppose, after refactoring, we have ended up having three classes instead of the original five. Now, in one of these classes, we found that there is a method that performs the tasks of database connection management and the calculation of taxes. Separation of these two tasks is better because they are very dissimilar to each other. Using refactoring, the original method (that is doing these two dissimilar tasks) can be divided into two separate methods. A new method can be created for one of these tasks (e.g., tax calculation) and the relevant piece of source code for tax calculation can be removed from the original method and put inside this new method.

You can see that refinement is opposite of the concept of abstraction. While abstraction aggregates things, refinement segregates things.

We have already seen in Chapter 5 how to draw the component diagram from use cases as well as an example this. From the component diagram, you can later draw your class diagrams.

Another point to note here is that we have already separated the user interface layer and database layer in our software architecture and these parts will be developed by the people working in those specific areas. In this book, these areas are covered in Chapters 6 and 8, respectively. Thus, we are only concerned with designing the business logic in this chapter. Therefore, the class diagram we design in this chapter is meant only for the business logic layer.

*7.10.7.2 Class Diagram Example* We will see how a class diagram can be drawn from a component diagram.

Figure 7.13 depicts a class diagram for an order management system for a restaurant. The component diagram for this system is discussed in Chapter 5. Let us look at the required functionality. First, let us see what functions we have here.

- Login
- Customer registration
- Delivery

- Payment
- Take order
- Confirm order
- Menu search
- Menu
- Logout

Now, how many actors do we have here? We have two actors: Customer and Sales Representative. One large entity in the component diagram with many subentities is Order. Here, we have entities such as menu, confirm, take order, menu search, and so on that are actually part of order. Sales Representative is represented in the component diagram through login and logout functions. The Customer entity is represented in the diagram through payment and delivery components.

By this analysis, it turns out that we can have classes for Customer, Sales Representative, and Order. We also need to think about the relationship among these classes. The Customer and Order classes have a relationship because each order belongs to a customer. The Sales Representative is also directly related to the customer and the order because the sales representative takes the order from a customer.

If you observe the component diagram carefully (provided in Chapter 5), the types of orders have not been taken care of. This is because they have to be taken care of at a detailed level and not at the aggregate level. Hence, apart from other details from the component diagram, we will take care of this aspect as well when we design our class diagram.

The Dine-in Order and Take-out Order classes are child classes for the Order class. These child classes will inherit properties and behavior such as close(), confirm(),

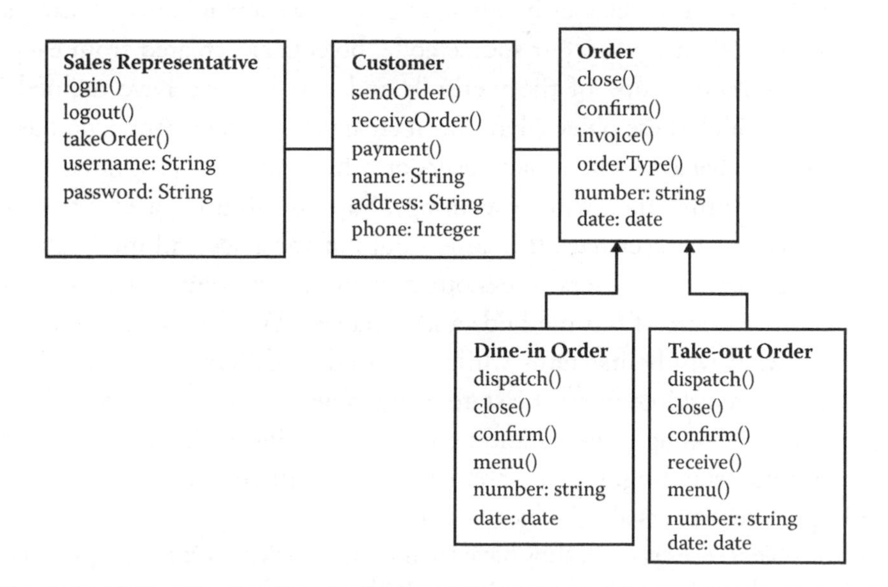


**Figure 7.13** Class diagram for order management system.

number, and date of the parent class. Then, each of the child classes can have their own properties and behavior that are not part of the parent class.

Now, by taking everything together, our class diagram is given in Figure 7.13. The class members have been included based on the use cases. For example, login, take-Order, and logout methods are needed by the Sales Representative. He or she also needs a username and a password to access the system. The Customer needs to have sendOrder, receiveOrder, payment, name, phone number, and address members. The Order entity needs to have close, confirm, invoice, number, and date members.

Unlike other Unified Modeling Language (UML) diagrams, a class diagram can be directly mapped to a program written in an object-oriented language. The class diagram clearly shows the elements that can directly be used to create the programs in an object-oriented language. For example, a class, its methods, and class variables shown in a class diagram can be used in an object-oriented programming language to create the source code. The relationships among the classes help define the interfaces of the classes through which the data will flow from one class to another. The class and its members become the building blocks on which the entire source code will be written.

In summary, class diagrams do the following:

- Describe the construction of systems using an object-oriented language
- Represent the associations or relations among the elements in static view
- Represent different functions provided by the systems

### 7.11 Objects and Object-Oriented Programming

No discussion on object-oriented programming would be complete if objects are not discussed. When we create classes in our source code, we are actually creating a template for objects. When we run our source code, objects are created from each class. These objects contain a copy of the methods and variables we have defined inside that class (from which these objects have been created). The most important aspect of an object is how it behaves when many users or other computer programs access our program during runtime (i.e., while executing). Suppose 10 users access the runtime concurrently. Will each user have the same values of variables and methods we have defined in that class? If the users access our runtime at different times, what will be the impact on the values of our methods and variables? What is the current status of an object and what will be its status while it is used by different users over time? In other words, we are asking about the state of the objects at various times. To answer these questions, a thorough understanding of the behavior of the objects is required. Without this understanding, your component design will go wrong.

Some important aspects about objects follow.

When the objects are created, they have their own data type. Objects can be assigned to a variable just like the methods or other variables. You can create an object by using the "new" keyword (used in most programming languages for creating objects). If you have created a class, my_class, then here are the statements to create some objects from this class:

```
my_class My_object1 = new my_class;
my_class My_object2 = new my_class;
my_class My_object3 = new my_class;
```

Here, we have created three objects from my_class. You can create as many objects from a class as you require. Once you create your objects, you can assign them to other objects or variables in the same way you make assignments for variables. However, there is a major difference between the assignments of variables and the assignments of objects. When you assign variables, they are assigned using values. However, when you assign objects, they are assigned by reference. We will understand the difference after looking at the following example.

First, let us see instance variables.

```
class Echo {
  integer count = 0;
}
```

Now, let us create an object of our class Echo and call it in another class.

```
public class my_class {
public static void main() {
  Echo e1 = new Echo();
  Echo e2 = new Echo();

  Integer x = 4;
  e1.count = x;
  e2.count = x;
  Print on user screen (e1.count + e2.count);
  }
}
```

**Explanation:** We have created two objects, e1 and e2, from the class Echo. There is an instance variable "count" in this class. The value of this variable will be computed when we create and call the objects from another class, my_class. We created a variable *x* and assigned a value 4 to this variable. Then, we assigned the value of *x* to the variable "count" of the objects e1 and e2. In the print to user screen, we asked to print the value of the addition of e1.count and e2.count. The result, 8, should be shown on the user screen because both e1.count and e2.count hold 4.

Now, let us change some assignments.

```
public class my_class {
public static void main() {
  Echo e1 = new Echo();
  Echo e2 = new Echo();
```

```
  Integer x = 4;
  e1.count = x;
  e2 = e1;
  e2.count = e2.count + 1;
  Print on user screen (e1.count + e2.count);
  }
}
```

**Explanation:** This time, we have assigned the value of e1 to e2. In fact, it is not only the value but also the whole object e2 being referenced to the same memory location of e1. In the next operation, we are increasing the value contained in e2 by 1. What value will e2 hold now? It should be 5. Why? Because when we put e2 at the same memory location as e1 (through the assignment e2 = e1), the value of e2 became 4 (same as e1). Now if you run your program on your screen, you will see 10 (5 + 5). Surprised? You may have expected that e1.count = 4 and e2.count = 5, but this is wrong, and the correct answer is e1.count = 5 and e2.count = 5.

The result may seem outrageous. How come e1.count is 5 and not 4? The answer is this: after you assigned e1 to e2, they were now pointing to the same location in the computer memory (RAM) and the value at this location is 5. If you try doing the same assignment with variables instead of objects, you will get 9 as the result. This is because when you define two variables, even if you assign one variable to another, only the value contained in one variable is assigned to the other variable but the two variables will still point to the respective memory locations of the objects that were created.

Now, you can understand that the objects are assigned through reference, whereas variables are assigned through value. Thus, always be careful about the assignment operations for objects.

### *7.11.1 Objects and Data Structures*

When you create variables or methods with a return type, they can hold some scalar data once you assign values to these variables or to the parameters of the methods. It means that they can hold just one value of any data type. In contrast, a class contains a structure with many values of different data types. For example, if you have created a class with two class variables of different data types and one method with a return type, then this class will have a structure and can contain three values after assignment. Of course, this assignment will come after you create the objects based on that class.

Let us revisit procedural programming. In procedural programming, we have variables and we use them to create methods. The left-hand side of Figure 7.14 represents this situation. These methods, in turn, do some computation. Many such methods together perform some computation and form a complete software product. When it comes to object-oriented programming, instead of variables (more specifically public variables; we still need to use private variables inside the methods), we have objects that we use inside the methods to do some computation. Many such methods, in turn, can form a complete software product. The right-hand side of Figure 7.14 represents this situation.

Now, you may ask, why do you need to go this long way to create classes and then build the objects from these classes and use the objects to build the methods that in turn will help build the complete software product? Although it is a long process, it has some advantages. The first one is that we can completely avoid using public variables (public variables are dangerous). The other advantage is that we build a structure inside a class that becomes available to its objects. This promotes reuse. As we have seen previously, defining space in procedural programming is rather limited, and this issue is addressed in object-oriented programming through providing a larger space. This helps provide flexibility to write our code in an object-oriented language.

Another point to be discussed is comparing arrays and objects. Arrays consist of a range of variables that can hold values or data. However, there is a limitation with arrays. Arrays can hold variables of the same data type. An array cannot hold variables of different data types, which is not true for objects. Objects can hold variables of any mixed data types. There is no limitation whatsoever in this regard.

Figure 7.15 depicts the difference between arrays and objects. Array 1 is an integer data type and can hold only integer values. Array 2 is of string data type and can hold only string values. However, an object has no such restriction. It can hold integer, string, Boolean, or any other data type. This indeed is a strong concept.

One of the goals of object-oriented programming is to bridge the gap between relational databases and the programming paradigm. As you now probably understand, procedural programming is not able to provide basic structures that could hold any values of any data type. In contrast, a record in a database can contain the values of any data type. This goal has not been achieved yet by object-oriented programming, but object-oriented programming is moving in that direction. If you are able to create

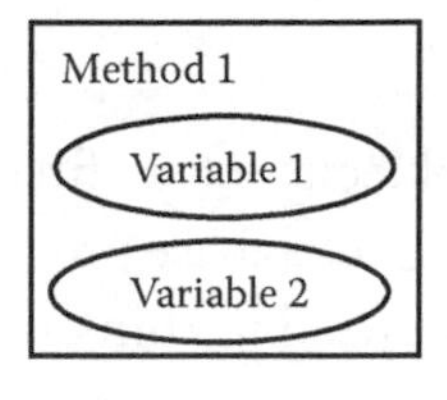


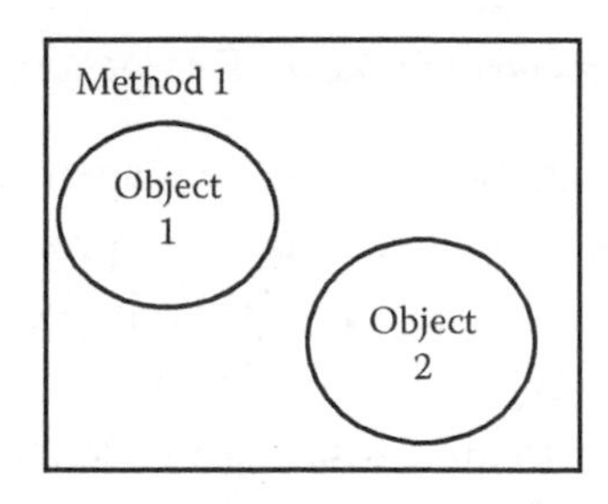


**Figure 7.14** Variables and objects: the figure on the left-hand side represents a situation in procedural programming; the figure on the right-hand side represents a situation in object-oriented programming.

| Array 1 (integer) [5] | Array 2 (string) [5] | Object |
|---|---|---|
| Values: 56, 23, 3, 5, 6 | Values: duck, rabbit, cow, horse, banana | Values: 56, banana, cow, 5, rabbit |

**Figure 7.15** Difference between arrays and objects.

a class that can provide a structure for its objects that is exactly the same as those of a record in a database, then you will have a true relationship between a class and a table in a database. In that case, you just create an object, and when you try to save the data (that belong to that object) in a database table, you do not need to do some extra processing to match the data structure of the object and the data structure of the table in the database.

#### *7.11.2 Object State and Behavior*

When you create a class, you define some class variables and methods in that class. The methods do some computation and give some output based on the inputs provided. These methods are known as the behavior of the object (the object that is instantiated from the class you have created) because, for different inputs, the method provides different outputs.

However, the class variables behave differently. You do not provide any input to them. You only assign some value to them and they keep this value throughout their life unless they are reassigned some different value later. At the same time, it is possible that the values of the class variables are changed from some other class(es). The change in the value of a class variable is known as state change for the object (when the class is instantiated).

If a class variable is declared public, then any part of your software application will have access to this variable; thus, its value can be changed by any of the classes residing in your software application. This is dangerous. Accessor methods are used to make sure that this should not happen. The variable itself is declared private and its value can be set or accessed using the get and set methods. In Section 7.9.3, we discussed how private class variables can be accessed using the get and set methods.

#### *7.11.3 Object State Management in Web Applications*

Before we think about the state management of objects, let us first understand the fundamental tasks involved in database programming. Database programming aspects are provided in a later section of this chapter, but we need to think of databases when we talk about object states. In database programming, a user can insert, delete, or change some information residing inside a database. To perform any of these operations, the business logic layer of the system receives requests from the user and does the requested changes in the database. More specifically, to perform any of these operations, the variables defined in the business logic hold some data (provided by the user through the user interface) and then process those data and use those processed data to insert, delete, or modify the database.

*7.11.3.1 Scenario in Client–Server Application* Consider Figure 7.16. A simple schema is provided to understand what happens when a user provides some data and wants the data to be saved in the database or wants to modify some data in the database.

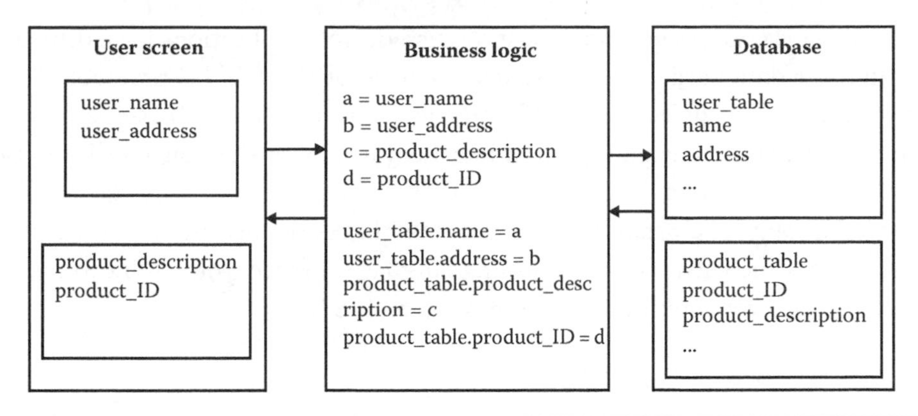


**Figure 7.16** Data taken from the user screen are converted into the database format by the business logic layer.

The business logic performs the task (requested by the user) through some written code. This code converts the user inputs and then performs some addition or modification in the database. To perform these transactions, the business logic holds the data (supplied by the user) in some variables. It means that the data are in computer memory. Once the business transaction is completed and the data have been written into the database, all the values of the variables in the business logic (i.e., in the memory) are reset. Now, these variables do not keep any values and are ready for the next transaction.

*7.11.3.2 Scenario in Web Applications* Let us consider a web application. The user clicks a link on a web page. This link leads to another web page. The information about the user will be lost because the old web page is closed. This is the standard behavior of web applications because they use Hyper Text Transfer Protocol (HTTP) and this protocol is stateless. How should the user information be stored so that it is available on the new web page? This question is important in situations where the response page must keep the user information to be used for reasons such as setting user preferences, keeping the information selected by the user in memory, and so on for the entire session until the user finishes browsing the website. One way is to use a session object that keeps the user data. As long as the user is browsing on a website (including its web pages), the user data are available in the session object.

A session object is created when the user first visits a website. This object is kept alive as long as the user is on that website.

A problem faced by this behavior of web-based applications is that the user information is lost during navigation on a website. For example, a user may browse a website and use the shopping cart to place some goods. After putting the items in the shopping cart, the user may browse the website, but when he comes back to the shopping cart, the goods that he has put in the shopping cart should be available to him. How can this functionality be achieved? As pointed out earlier, a solution is to use a stateful session object to record all the activities of the user.

*7.11.3.3 Session Management* As we just discussed, web applications have difficulty in managing user data because of the stateless nature of the HTTP protocol. To solve this problem, session objects are used and they store user data. These data are then available for the entire navigation a user may do during a session (on a website). Let us see how it is implemented.

The session object is instantiated from a class that is defined in one of the prebuilt packages that come with the programming language. Hence, you will need to import this package into your source code before you can use it.

```
import programminglanguage.http.*;
import programminglanguage.util.*;

//Extend Httpsession class. This class is found in the http package
that we imported

public class SessionManage extends Httpsession {

//the doGet method uses request and response parameters to get and
set the session object

//properties and methods

public void doGet(HttpSessionRequest request,
      HttpSessionResponse response)
{
//Create a session object if it is not yet created.
HttpSession session = request.getSession(true);
String page_title = "Welcome! You are back to this website";
Integer count_visit = keep track of count_visit;
String count_visit_key = keep track of count_visit_key;
String userID_key = keep track of userID_key;
String userID = keep track of userID;

//When a new person visits the web page the following message to be
displayed
if (session.isNew()){
  title = "Welcome! Thanks for your visit to this website";
  session.setAttribute(userID_key, userID);
} else {
  //keep the visits count to be displayed on the web page
  count_visit = (Integer)session.getAttribute(count_visit_key);
  count_visit = count_visit + 1;
  userID = (String)session.getAttribute(userID_key);
}
session.setAttribute(count_visit_key, count_visit);
}
}
```

If you observe the source code given above, you will find that there are two methods, getAttribute and setAttribute, which are associated with a session object. For example, the value for userID_key is set in session.setAttribute(userID_key). Once this value is set, you can fetch the value stored in userID_key by using session.getAttribute(userID_key). When the session object is initialized during web page loading, the values stored in these variables can be displayed on the web page. You only need to provide the fields on the web page and link them with the defined and assigned variables.

When you refresh this page, the visit count counter will increase by 1 and will be displayed on the web page. For each visit (or refresh) of the web page, this visit counter will continue to increase. When a user loads this web page on his or her browser for the first time, the "Welcome! Thanks for your visit to this website" message will be displayed. For any subsequent visit, the message "Welcome! You are back to this website" will be displayed.

You can see that the session object stores the user information; thus, it is able to use this information across all the navigations on the website for a session.

### *7.11.4 Object Diagram*

Objects in a running software product go through many states. For example, in an order management system, an object representing an order can have an initial state when the order is being taken. When an order is created, then the order object can have an order value, say US$10. At the same time, another order object may be created that may have different values from the first order object. In fact, if we consider the five classes, namely, Customer, Order, Take-out Order, Dine-in Order, and Sales Representative, that are given in Figure 7.13, then we can create many objects from them. Each object will have its own values (state) for the properties associated with that object.

Representing all the states that an object goes through is difficult. The best approach is to take a snapshot of an object at any given time and present its state (at that moment) in a diagram.

In the object diagram in Figure 7.17, we can see that there are three instances of Order object, three instances of Customer object, two instances of Take-out Order object, and one instance of Dine-in Order object. This snapshot of the status of all these objects was taken just after the orders were confirmed by the sales representative. You can see the customer numbers, order numbers, and so on in the figure.

Figure 7.17 shows the state of each of the objects and their relationships. The objects, their states, and their relationships are all dynamic because they change over time. Object diagrams are used only for showing the states of all the related objects at any given point of time. This is the limitation of object diagrams.

In Figure 7.17, if the Order number 14 is changed from Take-out Order to Dine-in Order, then instance 3 of the Take-out Order object will no longer exist and a new

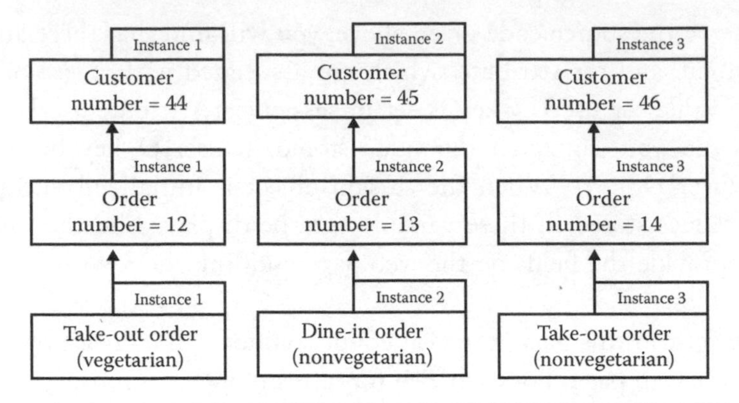


**Figure 7.17** Object diagram showing the objects and their relationships at a particular time.

instance of the Dine-in Order object will be created instead. We will need to create another object diagram to represent the relationships between the objects and their states for this new scenario.

### *7.11.5 Sequence Diagrams*

The relationship among the objects during runtime can be captured in many ways. The object diagram has already been discussed in Section 7.11.4. The object diagram displays the snapshot of all the objects during runtime at a specific time. How the objects interact with each other can be captured in interaction diagrams or sequence diagrams.

Figure 7.18 depicts the message calls starting from the first object to the last object. These calls are depicted in a sequence in time and that is why these diagrams are known as sequence diagrams. You can see that the Sales Representative object invokes the message to the Order object using the method takeOrder(). Notice that takeOrder() is a method of the Sales Representative object (see Figure 7.13 for details). The Order object in turn makes a message (using the confirm() method) to the Take-out Order object if the customer has chosen a take-out order. The next message is a self-message by the Take-out Order object on itself using the method dispatch(). You can also see that each of these methods returns some value and the return is also depicted in Figure 7.18.

When the restaurant order management system is running, the objects and their methods actually pass the messages as per the sequence just described.

In Figure 7.18, apart from the Sales Representative, Order, and Take-out Order objects, there is a Dine-in Order object as well. This object is not part of the message flow because it is not part of the business logic (used to process the message flow) when the Take-out Order is processed. However, the Dine-in Order has been depicted in this figure just to highlight the fact that other objects (which are not part of the business logic) may be present when some message processing is being carried out (when the software product is running).

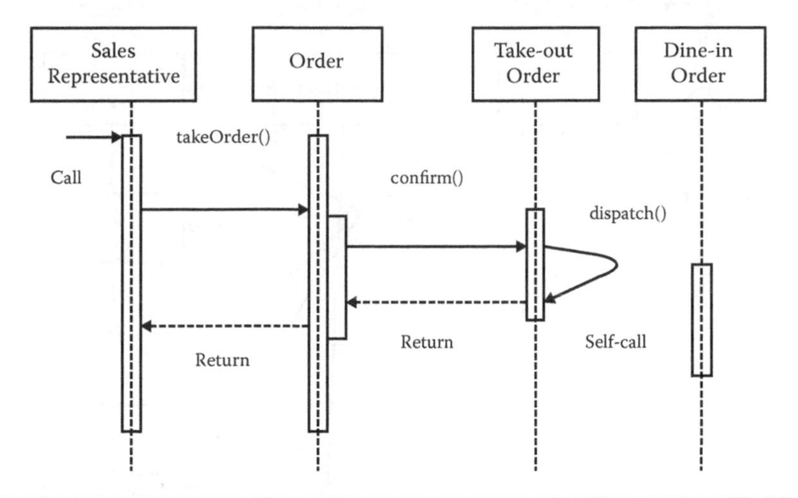


**Figure 7.18** Sequence diagram for the method calls (messages) sequence among the objects.

When some other business logic is run and you draw a sequence diagram, some other objects will take part in the message flow. This will be as per the message flow defined in the business logic. For example, when the sales representative takes a Dine-in Order, the Take-out Order object will not be part of the message flow.

### *7.11.6 Statechart Diagrams*

In order to represent the objects and their states (*states* means the values of the members contained in the objects), you need to understand how the objects will behave during runtime. Each object goes through different state transitions, starting from initialization (instantiation) to destroy. Capturing the state transition of each object is done through statechart diagrams.

Statechart diagrams depict the life of an object from its initialization state to its final state. The object can interact with other objects during its lifetime. The state change of an object can be caused by any event. These events can be either internally generated or from the outside. For example, Figure 7.19 represents the state of the Order object when external events occur. All these events are triggered by the user who initializes this object. In our restaurant management system, the user is the sales representative. The user takes the order (using takeOrder() method), selects the order type (using orderType() method), selects the menu (using menu() method), confirms the order (using confirm() method), dispatches the order (using dispatch() method), and closes the order (using close() method).

Statechart diagrams are extremely useful because they depict the life cycle of an object. This helps us to understand the behavior and state changes of an object during its lifetime.

As you may know, the state of an object is based on the values of the class variables that are in the class from which that object is instantiated. Similarly, the behavior of an object is based on the methods defined in the class from which that object is

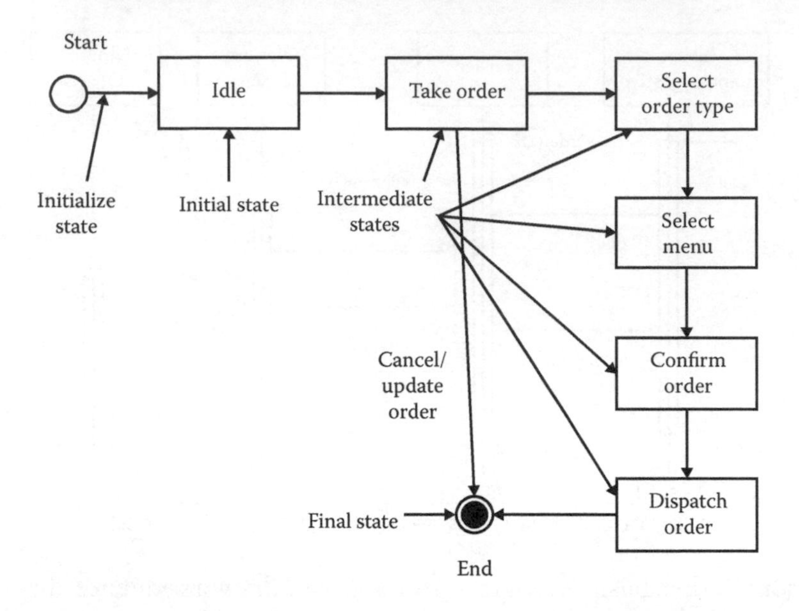


**Figure 7.19** Statechart diagram for the Order object, from the start state to the end state.

instantiated. The statechart diagram thus displays at what events the methods defined in the object get triggered.

## 7.12 Packages

Packages are used in object-oriented programming to hold and manage the classes. Packages themselves have no role in any programming but they are used for management purposes only. The role of packages is similar to the role of folders in an operating system. Files are organized in folders and it is easy to work with this arrangement. Folders generally form a tree structure. In each branch of the folder, there may be many files as well as subfolders.

Packages are also organized in a tree structure. In a package, there can be many class files as well as subpackages. In Java programming, class files are saved with the extension .java. These .java files are source code files. The packages for Java files keep these .java files. For a small software product, you may need just one package. However, for a large software product, which may consist of hundreds of class files, you may need many packages and subpackages. You will create packages to manage all those hundreds of class files.

Suppose you have created a package named my_package and have created a class file my_class and put this class file into this package. Now, if you want to refer to this class file in some other class file, then you can use my_package.my_class. The libraries that come with the programming languages also consist of packages.

If you are using a good Integrated Development Environment (IDE), then, through Intellisense, you can find the classes and methods inside a package. IDEs (e.g., Microsoft Visual Studio) are the platforms that the software development teams use to design and build the software products. Intellisense is a feature available in most text editors that are included in the IDEs. If you type a package name, then Intellisense will come up with a list of all the classes inside that package. From that list, you can choose the class you want. Therefore, you do not need to remember all the classes inside a package.

### 7.13 Database Programming

Most commercial software products use databases in some way. Databases are specially built for software products that hold a huge quantity of data. Some popular databases in the market include Oracle (from Oracle Corporation) and SQL Server (from Microsoft). The database is organized into database tables and each table can have many columns. Apart from the tables, a database can also have database triggers, sequences, functions, and so on. You can modify or delete the data residing inside a database or can create and store new data inside a database. For doing these operations, a query language is used. This query language is commonly known as Structured Query Language (SQL). Each database engine understands its own version of SQL. However, SQL syntax is not very different from one version to another. You can create a script consisting of SQL statements and execute it after connecting to the database engine.

Let us see some examples of SQL statements:

- Select statement: Select <column name 1, column name *n*> from <table name> where <condition>. This statement will fetch columns 1 and *n*, from the table "table name," in which the "condition" is satisfied.
- Create statement: Create table <table name> with <column name 1, column name *n*>.
- Delete statement: Delete from table <table name> where <condition>.
- Insert statement: Insert into table <table name> <column name 1, column name 2, …> values (1, 2, …).

You can find more details about databases and database programming in Chapter 8 or you may refer to a database programming book. In this chapter, we introduce database programming because database queries need to be embedded in the source code for developing the middle tier. Hence, some knowledge of database queries and databases is essential.

As a simple example on the usage of databases, we have a candy seller who keeps an account of all his sales. Whenever he makes a sale, he updates his account book. He keeps information about the date of the sale, quantity, price of each, and total sales for the day. He is using a software system that has a user interface, a business logic layer, and a database to store the sales data. On his user screen, he has an icon or a button to start this program. Inside this program, there is a user screen where he can enter the quantity of the sales done. When he clicks the save button, the program runs and the sales data are saved in the database.

If you want a software program to execute SQL queries, then you need to embed SQL statements inside your software program. Once you write such a program and execute it on the database, the database engine will run and execute the required operation (via the middle layer) on the database.

Here are the steps you need to take for database programming:

- Import database packages: Most object-oriented languages have their own libraries that create necessary infrastructure to connect to the database. You need to import these libraries into your source code. For example, in Java, the database connection management libraries include java.sql.*.
- JDBC/ODBC driver registration: You need to register the correct database driver so that you will be able to connect to the driver. For example, if your database is mySQL, then you need to register a driver related to connecting to a mySQL database engine.
- Open a connection: In your code, you need to explicitly make a statement to open a connection to the database.
- Execute a query: You can write an SQL query embedded in your source code. When your source code runs, this query also runs and executes the query on the database and then the database is updated.
- Cleaning up: Once your query is executed, you need to close the connection to the database.

Let us look at a sample code here:

```
import languagename.sql.*;

public class database_insert {
    static final String JDBC_DRIVER = "com.mysql.jdbc.Driver";
    static final String DB_URL = "jdbc:mysql://localhost/
    database_name";

String USER = "username";
String PASS = "passcode";
public static void main() {
Connection con_database = null;
//connect to database and execute query to insert a record
Statement get_stmt = null;
  Class.forName ("com.mysql.jdbc.Driver");
  con_database = DriverManager.getConnection (DB_URL, USER, PASS);
  get_stmt = con_database.createStatement();
  String sql = "INSERT INTO <table name>" +
"VALUES (100, x, y, z)";
  get_stmt.executeUpdate (sql);
    if (get_stmt ! = null)
      con_database.close();
}
}
```

Let us see how the steps mentioned about database programming are implemented.

- Import database packages: We have imported database packages in the import languagename.sql.* statement.
- JDBC/ODBC driver registration: We have registered the database driver to a mySQL database engine through JDBC_DRIVER, DB_URL, USER, and PASS.
- Open a connection: We have opened a connection to the database using the connection string con_database.
- Execute a query: We have executed a query to make an insert operation in the database.
- Cleaning up: con_database.close(); has closed the connection to the database after inserting the values to the database.

### 7.14 Model–View–Controller Revisited

In the Model–View–Controller (MVC) architecture, the view is the user interface. The model and controller are the middle layer components. We discussed MVC in Chapter 6. Let us recapture what MVC is all about:

User—provides inputs such as entering some text and pushing the Submit button on the user screen (view).

Controller—takes the user inputs and instructs the model to do the tasks such as saving the inputs (or some values) to the database. The controller also creates the views that are then displayed on the user screen.

Model—does the work such as saving inputs to the database and providing feedback to the controller.

View—the view that the user can see on his or her screen. The view is constructed by the controller.

Now, let us see how MVC can be implemented. For demonstration purposes, let us create a software program where a sales representative can see the menu information (menu name and menu number). We will create a small program with just a few classes. We can have one class each for the controller, view, model, and user. To run the complete program, we can create one additional class. Let us think about what methods and classes we need to create.

**Model:** We need information about menu number and menu name. We can create a class with get and set methods to get and set the menu number and menu name.

**View:** We need to display the information about the menu number and menu name. Here, we can create a class that will display this information. Inside that class, we will have a method to do this task.

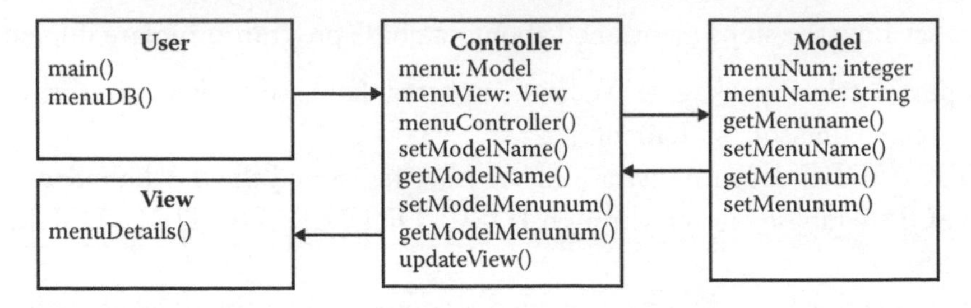


**Figure 7.20** Classes for the MVC product.

**Controller:** We will have a Controller class that will instruct the model to get and set the menu number and menu name. Once the controller gets updated from the model, it will transmit a view based on each of the updates that were received from the model. The Controller class has two special structure types: menu and menuView. They are of model and view types. The controller will get and set the properties of the objects being created for the model and will update the view object accordingly. The controller uses menuController() and updateView() methods to carry out these get and set activities.

**User:** When the user clicks the View button, the current menu information should be displayed. There are two methods in the user class: main() and menuDB(). The main() method will call the updateView() method of the Controller class to get the updated view after the user has entered some data in the user screen. The menuDB() method can be used to connect with a database and get the menu data and display it.

The class diagrams are shown in Figure 7.20. Once you have all the classes and methods for your MVC application, you can implement it easily.

This example depicts both the MVC architecture and the model–view architecture. The update from the controller method (updateView()), called in the main() method in the User class, shows a view coming from the controller, but the menuDB() method shows a view directly coming from the model. There is no involvement of the controller in the view coming from the menuDB() method.

The other interesting part in this example is that each element of the model is called in a method in the controller. For example, the accessor method getMenuname() of Model is called in Controller by the getModelName() method. Thus, whatever changes happen (by getting or setting the values) to the properties of the model will come to the controller. Later, by using the updateView method of the controller, these changes are reflected in the view (because main() method in User calls updateView()).

### 7.15 Refactoring

Refactoring is the task of redesigning the existing source code in such a way that some new source code can be integrated with that existing source code. Refactoring may involve performing one or more of the following tasks on the existing source code: renaming the classes and methods, moving the classes and methods, removing duplicate code by reusing one piece of common source code across many classes, and introducing polymorphism by using interfaces.

### *7.15.1 Refactoring Scenarios*

Let us see in what scenarios refactoring is needed.

*7.15.1.1 Rename Classes and Methods* Some classes and class members may need to be renamed when their usage is changed and their old names are no longer appropriate. If you assign appropriate names to the classes and class members, then it leads to more clarity. For example, you may have a class named issue_chequebook. Now, the same class is going to be used for issuing debit cards as well. It will be appropriate if you change this class name to something such as issue_chequebook_debitcard. When you rename a class or any class member, you will have to change that name everywhere in the entire package or in the entire application. Automated refactoring tools can be used to carry out renaming because they can do this task extremely quickly.

*7.15.1.2 Move Class to Another Package* In some cases, a class may need to be moved from one package to another because that class will be used extensively by the classes belonging to the other package, in which case you will have to change the package name of that class wherever that class is referenced. You also need to update the source control (configuration management) system in that case. For example, suppose you have a package named "calendar" and most of the classes related to calendars were put in this package. Later, you found out that a class named "account_summary" was put in this package although this class was not using any features of the calendar. In this case, it is better to move that class to a package that is related to accounts.

*7.15.1.3 Divide and Extract Method* When a long method needs to be divided into two or more methods to enhance its readability and maintainability, then you have to do refactoring. You can break a long method by taking out a piece of code from it if that piece can itself form a new method. You can then insert that taken out piece of code inside a newly created method. By giving clear and descriptive names to the original larger method (which is now shortened) as well as the newly created method, it becomes simpler to understand both methods. By extracting a method out of a large method, you can reuse that extracted method in other places. This reuse of the extracted method was not possible when it was part of the larger method. For example, we can have a method to create the contact details of customers. The contact details contain the name, address, e-mail, phone number, and so on. Now, it is possible that the phone number is used for some other purpose such as personal verification. If the phone number can be extracted and a new method is created to hold the phone number, then that new method can now also be used for the personal verification process where the phone number is required.

*7.15.1.4 Introduce Super Class* In some cases, the functionality provided by an existing class needs to be changed. A new class can be introduced as the parent or super class of an existing class, in which case the common behavior between that new parent class and the child (or the existing) class is moved into that new parent class. The clients of that existing class are also modified so that they now reference the newly created parent class.

Let us look at an example:

```
Public class order {
  Public void menu_veg(){
  }
  Public void menu_nonveg() {
  }
  Public void take_out_order() {
  }
  Public void dine_in_order() {
  }
}
```

In the above class, there are two kinds of orders (take-out and dine-in). One method for each kind of order is defined in that class. Each kind of order will have many functions such as dispatching order, menu selection, and confirm and close order to perform. (We have seen the class diagram for a restaurant management system in Section 7.10.7.2. There, we have seen all these functions defined inside the classes for those two kinds of orders.) Performing all these functions inside a method will make the code difficult to read and maintain. Hence, it is better to define a common "order" class and then create two child classes, one for the take-out order and another for the dine-in order. This will make the code more readable and thus maintainable.

```
Public class order {
  Public void menu_veg(){
  }
  Public void menu_nonveg() {
  }
}

Public class take_out_order {
  Public void menu_veg(){
  }
  Public void menu_nonveg() {
  }
  Public void take_out_order() {
  }
}
```

```
Public class dine_in_order {
  Public void menu_veg(){
  }
  Public void menu_nonveg() {
  }
  Public void dine_in_order() {
  }
}
```

Now, this is better. Your take-out order class can contain all the information related to the take-out type of order (but no information related to the dine-in order). The dine-in order will keep all the information related to the dine-in order (but no information related to the take-out order).

#### *7.15.2 When Should You Consider Refactoring?*

Here are some pointers for considering refactoring:

- Duplicate code: If you find that a duplicate code is present, then refactoring is a must to eliminate that duplicate code.
- Long methods: If you find that there are long methods in a class, then that class is a good candidate for refactoring. Make many small methods instead of one long method.
- Big classes: If a class contains many methods and class variables, then it is better to partition the class and make many smaller classes instead of one large class.
- Big switch or if–else statements: If your code contains many switches or if–else statements, then it is better to partition your code and remove excess statements. Move them to another method.
- Long navigations: If your packages contain many long navigations such as x.y.z.a.b.method(), then it is better to repackage your code. Finding a class and its members inside very deep package statements leads to slower running of your code. Refactor such packages so that you should be able to find your class within two or three deep subpackage structures. For example, limit your package structure so that you can have a class at x.y.method().
- Empty classes: Sometimes, you may have defined a class but, for some reason, you could not put much code inside it. In such cases, it is better to take the code out and put it in some other class.

The real benefit of refactoring is achieved long term, although there may be some immediate benefits. The benefits include the following:

- Substantially reducing the work on debugging and maintenance: For example, suppose you have a class that has a method to compute the weight of natural gas stored in a gas tanker. The weight of natural gas depends on many factors such as temperature, pressure at which it is stored, chemical composition, and volume. The method obtains the values for temperature, pressure, and chemical composition of the natural gas from various devices and instruments. From this description, it is clear that this method is exposed to many interfaces with external electronic devices. If any problem occurs in any of these devices, then it will be difficult to find out which device has the problem or why there are errors in weight measurements. This kind of functionality may require writing lots of decision tree codes (logical paths) in the method. Debugging a method containing so many logical paths will be a nightmare. It will be better to refactor this method and divide it into many methods so that testing and debugging will be a lot easier. It will also become easier to maintain it.

- Improving the extensibility of the product: When a new feature is required to be added to an existing software product, the existing code should allow it to be done. For example, suppose a restaurant has an order management system. If the restaurant decides to provide additional menu offerings, then the system should support these additional features. If the code in the existing system is completely refactored, then these additional features will not create any problem to the system.
- Making the code robust: A good refactored software product is more robust because the chances of defects creeping into the product are low.
- Reducing the duplication of code: When new features are added to a software product, developers add some new code without checking if some existing code could be reused. This results in adding new code pieces that are duplicates of some existing code pieces, which results in duplicate code at many places in the software product. Using refactoring, all the duplicate code can be eliminated. Identifying the duplicate codes and then merging them together will always create a robust and maintainable code.
- Reducing the maintenance and development costs: A small-sized source code means less development and maintenance costs. A good refactored code base will always have no duplicate code and this will result in a better software product.

*7.15.2.1 Automation Tools for Refactoring* While doing refactoring, care should always be taken to ensure that the external behavior of these components does not change. Since most software products nowadays are built incrementally, refactoring of existing source code is done very frequently. Most IDEs offer good automated refactoring tools and those tools must be used.

Manual refactoring is never advisable. If refactoring is done manually, then it will consume a lot of time of the software developers. Thus, there will be a significant decline in the productivity of software developers. Manual refactoring is also prone to errors. For example, a method of a class can be referenced at many places in other classes. If this method and the class definition are changed, then those changes have to be made in all the places where that method and the class are referenced. If those changes are done manually, then, at some places, such changes could be missed because of human error. In contrast, when such changes are done with the use of a refactoring tool, this problem will not arise.

### 7.16 Client-Side and Server-Side Scripts

Web programming is a technology that is based on a set of fragmented pieces of technologies. This is because all the layers involved in web technologies reside in very different environments. The web browser runs on the user machine and it is physically disconnected from other layers. The middle layer (in the form of a website) may be running at a faraway location and may have a very different environment from the user machine. The database layer may again be located at some other faraway location on a different operating environment.

There are different types of websites, databases, and web browsers. However, when a user points his or her web browser to a website, it is expected that the user should be able to connect to the website and do his or her work using the features available on the website. How is a browser able to connect with a middle layer (website) that, in turn, is connected to a database when they operate in very different environments? This has been made possible using connection technologies like HTTP, TCP/IP, and CORBA. These technologies allow a client to connect to a server regardless of the physical and logical differences between them. A discussion on these technologies is beyond the scope of this book.

The other challenge in connecting a browser to a website is content delivery. The web browser only understands HTML. HTML can only provide static information from a website. For example, if the user is connected to a website and wants to submit his or her personal information to register on the website, then this is not possible with static content. This kind of functionality is possible only when the website accepts user inputs and, on the basis of the input, provides relevant output. This is possible only through dynamic content. The website should be able to process the user input and provide output using some dynamic content delivery mechanism.

To generate dynamic content, you can use databases. You can connect a website to a database and the database can provide dynamic data for the user interface. A compiled component at the business logic layer can connect with the database and generate dynamic content. Using SQL queries (we will learn more about databases and SQL in Chapter 8), you can manipulate or read the data from the database; thus, the database provides dynamic content.

There is still a problem to be addressed, though. Delivering the dynamic content (that is generated by the database) is still a challenge because the web browsers may be located at faraway places (on the users' computers) and linked through HTTP connection to the website. The question is, how does one send dynamic content (generated by the database) to the web browser (through a website) when the web browser understands HTML alone? For this, the dynamic content coming (to the web browser) from the website needs to be converted into HTML. The server-side scripts help convert the dynamic content into HTML. In fact, there is always some static content (such as page layout design) on a web page. Combining this static content with the converted content (dynamic content converted into HTML) from a website is done using server-side scripts.

Client-side scripts are also useful in web programming. We discussed client-side scripts in Chapter 6. In this chapter, we will see how some client-side scripts implement some functionality such as the user input validation. Then, we will learn about server-side scripts.

#### *7.16.1 Client-Side Scripts*

Client-side scripts are used for user input validation, setting up user preferences, and maintaining user sessions so that the user information is persistent throughout the browsing session of a user.

*7.16.1.1 User Input Validation* When a user provides inputs to textboxes or makes selections using lists, drop-down lists, radio buttons, checkboxes, and so on, it is important to validate such input. For example, suppose a textbox is left empty by the user and the user clicks the Submit button; some vital information that is required for further processing may be missing. Let us look at an example where the user leaves the username field blank and submits the form.

```
Public validate_field() {
  String user_name;
  If user_name = = '' then
  Print ("The username field has been left blank");
  }
```

If the requirement is that some special characters should not be allowed in the username field, then it can also be validated in a similar way. For example, if the username field should not contain the & character, then you can easily validate it.

*7.16.1.2 User Session and User Preferences Management* In Section 7.11.3.3, we saw how the user sessions are maintained in web applications. They are used to make sure that user information is not lost during navigation from one web page to another.

User preferences can be set by providing a menu or a button on the user screen. When the user invokes any of these alternatives, the user is given the option to set the user screen. Some of the popular user preferences include setting the font size and the color of the user screen and changing user screen layout. The user preferences can be stored in a cookie so that when the user visits that website next time, the user preference information can be picked up from the cookie and the user screen can be set automatically. A cookie is a small file that can be installed on the user computer (by the web server). Client-side scripting is used to make this cookie read the information (entered by the user) related to the user preferences. The cookie stores that information and the user preferences are set accordingly.

### *7.16.2 Server-Side Scripting*

Server-side scripts are used to put together all the information coming from the web servers or application servers or any other servers so that the aggregated information can be put on the web pages. Server-side scripts can also be used to completely build the web pages without the need to have any compiled code on the application server. For example, we can have a configuration where the business logic is written in Java files and compiled on the application server. A JavaServer Pages (JSP) file is then created and this file (which is a Java compiled file) is called from within this JSP page. When the web page that contains this JSP script is run on a user web browser, the Java file also runs because it is called from within this JSP file. The other alternative is that you write the entire business logic in the JSP file and do not create any Java compiled code on the application server. We have already discussed web application architecture in Chapter 5.

Let us look at an example of how to use a server-side script. Suppose we have a compiled file called order.class on the application server. This file contains the business logic for order management. It has a method called menu_non_veg() that is used to select a nonvegetarian menu. There is a menu called nonvegetarian combo pack that we need to select in our script so that it will be displayed to the user. Suppose there is also the price for this menu. All the values for the menu and prices are computed in the business logic layer inside the compiled files such as order.class. To display the values computed by the business logic layer, we need to embed the server-side scripts inside the HTML file of the web page, as shown below.

```
<%
Connect to the database;
Get the values from the table for "menu_non_veg" and "price"
fields;
Save these values in the session variables "menu_non_veg" and
"price";
//save the session variable values in variables
String menu = request.getparameter ("menu_non_veg");
String price = request.getparameter ("price");
  out.print(menu);
  out.print(price);
/%>
```

In this piece of code, the <% and /%> tags are used inside the HTML page to instruct the parser that this piece of code is a server-side scripting code. Thus, it needs to be processed by the application server. The request object is a session object that gets and sets the session variables. The getparameter() is the method through which the request object gets the values stored in the session variables.

This code will get values from the fields "price" and "menu_non_veg" from the database table. These values are then stored in session variables. Then, these values are stored in the variables "price" and "menu." These values are then displayed through out.print commands.

### 7.17 Debugging

When you write a piece of source code, you need to check if it is working fine by running a software application that is known as a debugger. Debuggers are included in most IDEs. If your piece of code is not working, then the debugger will provide some error messages. From these error messages, you can find the problem and fix it. You keep fixing the errors until the debugger no longer provides any errors. Once your code is completely fixed, you can proceed to build (compile) your source code. Notice that you can inspect your code manually but the debuggers make the job easier.

Debugging can check many things in your code. First, you can check if your code is syntactically correct. You may have misspelled some variable name or other programming entity (e.g., you could type "els" instead of "else") or you have not declared a variable but used it in your code. The other aspect of debugging is to check for the integration of your code with other systems, to check the control flow of your code, and so on. Most debugging tools can find errors as far as the compiler permits. If the integration points (of your code) with other systems are within the scope of the compiler, then the debugging tools will find out if there are any errors over those integration points as well. However, if the other systems are not within the scope of your compiler, then the debugger cannot determine whether the integration is working well. In most cases, this is a problem because the other systems may be offline or not reachable; thus, checking the integration with those other systems is not possible. In such cases, you cannot check if the integration is working or not.

Even if your code is syntactically correct, it may still have domain-specific (logical or semantic) problems. For instance, your piece of code may perform some computation without errors but the computation logic itself may be wrong. For example, if you type an addition operation for multiplication, then it is a logical error. This may lead to wrong computation results produced by your code. In such cases, the debugging tools cannot help. To check these types of errors, you need to perform domain testing. This area is beyond the scope of debugging, though.

However, you can check for control flow in debugging. For instance, you can check if the loop you have implemented is actually being executed as per the inputs provided. Sometimes, the inputs coming into a loop are above or below the condition values (out of the boundaries) for the loop; therefore, the loop is never executed. Sometimes, even if the loop is executed, it is not executed as per the conditions you have set because either the condition is misstated (e.g., "less than" condition in place of "less than or equal to" condition) or the input values are coming to the loop incorrectly (e.g., because of a mistake, the values are changed before they enter the loop). Using proper debuggers, you can find these issues.

### *7.17.1 Debugging Error Messages*

The descriptive error messages that you receive while running the code will help you fix it. Error messages occur when the compiler finds problems with your code. The more sophisticated your programming language development kit, the better and more descriptive the error message you will get. Thus, it is important for you to have a good programming language development kit. For example, if you install Java Runtime Environment, then you can compile your code but you will not get descriptive error messages even if your code has compiling problems. In contrast, if you install Java Development Kit (JDK) on your machine and compile your program, then you will get descriptive error messages if your program has errors. Development kits such as JDK provide extensive description of the compiling errors, and this feature is a definite plus for software development. You can fix your errors quickly by referring to such error messages; thus, you save a lot of time. On the other hand, if you do not get a good description of the error messages, then it becomes extremely difficult to find the exact location and nature of the problem in your code.

When you are developing a web-based database application, your programming language environment, the application server, and the database server are involved in your application. This is a complex environment. In such a scenario, debugging your application can become extremely difficult because the problem area may belong to your language environment, to the application server, or to the database server. Thus, descriptive error messages become extremely important.

## 7.18 Case Study

We will discuss the business logic of OBAAS that we are developing. We have seen the architecture design and component design of OBAAS in Chapter 5. In Chapter 6, we have also seen the user interface design for OBAAS. In this chapter, we will see a detailed design for business logic implementation of OBAAS and how this detailed design is implemented.

### *7.18.1 Methodology*

The middle tier of OBAAS consists of the business logic implementation and the client-side and server-side scripts. The business logic is completely implemented using a compiled code on the application server. The client-side scripts are used to validate the user inputs. The server-side scripts are used to call the objects that contain the business logic and embed them in the HTML content of the web page.

There are two versions of the OBAAS: OBAAS 1.1 and OBAAS 1.2. OBAAS 1.2 has some additional features such as money transfer and a fee-based cheque book request service.

### *7.18.2 Detailed Design for OBAAS 1.1*

The detailed design consists of a class diagram, an object diagram, a sequence diagram, and a statechart diagram. The detailed design for OBAAS is based on the component and data flow design that we discussed in Chapter 5.

*7.18.2.1 Class Diagram* On the basis of the component diagram, we decided to build four classes. The rationale behind this approach is that all the elements that need to be covered as per the component diagram can be included in these four classes. At the same time, all the related elements are fitted inside each class.

How did we arrive at defining only four classes for OBAAS 1.1 and not five or three or any other number? If you look at the use cases and the component design, you can see that all the elements required to creating our software product belonged to the user or the bank account or bill payment or service request. For example, account balance belongs to the bank account, creating a new account belongs to the user, and so on.

Let us look closely at all the required elements and where they belong.

In Table 7.1, you can see that all the properties and behaviors of each of the four entities that we have identified are the classes that will implement all the functionalities for the OBAAS 1.1 system. Now, we can create our class diagram for OBAAS 1.1, and it is depicted in Figure 7.21.

In the classes we have created, we can see that we have not included database connection management. We will discuss database connection management now as all transactions in OBAAS result in saved or modified records in the database system. From the component design, we can also see that database connection is required on each web page many times. This means that the connection with the

**Table 7.1** Entities and Their Properties and Behaviors in OBAAS 1.1

| Entity | Behaviors | Properties |
|---|---|---|
| User | Authenticate | User ID, user name, user address, user mobile number |
| Account | Update balance, closing balance, create account, close account | Account number, account name, account ID |
| Bill payment | Bill payment request, bill payment response | Paid amount, biller name, biller ID, date |
| Service request | Service request, service response | Service ID, service type, date |

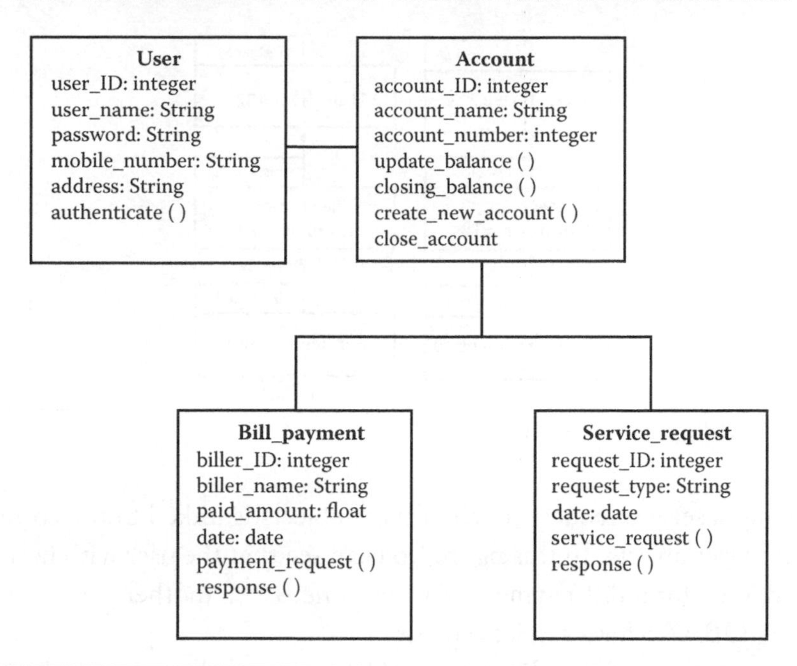


**Figure 7.21** Class diagram for OBAAS 1.1.

database is an activity that is required to be performed many times. You will also realize that each class we have designed in OBAAS will connect to the database many times. Thus, it will be better to pool all the database connection-related activities inside a pool of specialized classes that will deal only with database connection management.

From the requirement specifications (use cases) described in Chapter 4, we can see that for each activity on the user interface, the user is required to provide the username, password, and account number for authentication. This means that each of the activities of checking account balance, bill payment, service request, and account closure are independent of each other. If there is only one response for each of these activities, then the system no longer keeps user information after the completion of each activity. Thus, we do not need to keep the session variables to keep user information after completion of an activity.

As we can see, each activity involves the authentication of a user by checking the database entries for username, password, and account number; the authentication function is used many times. Thus, we need to create a class for the authentication function so that we do not need to write the source code for the authentication function over and over.

We can see that we can reorganize our classes to reuse the source code. However, the methods and class variables that implement our business logic will remain the same.

*7.18.2.2 Object Diagram* The object diagram depicted in Figure 7.22 shows the state of the objects at a time when one object was processing bill payment and another one

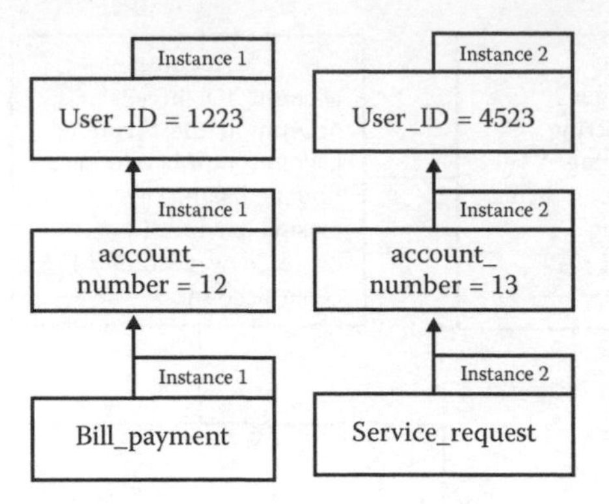


**Figure 7.22** Object diagram for OBAAS 1.1.

was processing a service request. Each of these objects is linked to the corresponding Account and User objects. In this figure, you can see that the user with user_ID 1223 is using OBAAS for a bill payment. At the same time, another user with user_ID 4523 is using OBAAS for a service request.

You can create many object diagrams for OBAAS to fully understand its functionality and design the application. For example, you can create another object diagram where one user is creating a new account and, at the same time, two more users are doing some other transactions in OBAAS.

*7.18.2.3 Sequence Diagram* A sequence diagram for the Bill_payment process for OBAAS is shown in Figure 7.23. The user needs to provide the username, password, amount, and account number into the user interface of OBAAS for bill payment. The user also needs to select the biller from the drop-down list. When the user clicks the Submit button, first, the authenticate() method of the User object is invoked by the

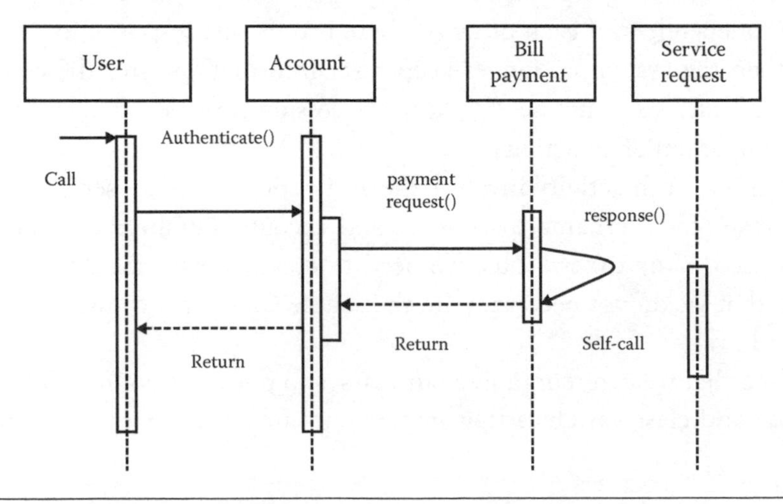


**Figure 7.23** Sequence diagram for OBAAS 1.1.

Account object to verify the user. Later, the payment_request() method is invoked by the Account object. All of these method invocations happen by one mouse click when the user clicks the Submit button. After bill payment, the response() method is self-called by the Bill_payment object.

You can create similar sequence diagrams to fully design OBAAS.

*7.18.2.4 Statechart Diagram* The statechart diagram for OBAAS 1.1 is given in Figure 7.24. Please note that we have used commonly understandable English words in place of actual method names to better understand this diagram. This statechart diagram describes the states through which an object of bill pay type goes through from its initialization to finish. First, this object is created (by a call in the software application). At the initial state, the Bill payment object is idle. When this object is initialized to perform a bill payment transaction, the first state it goes through is when the Bill payment object becomes active and is loaded on the web page. In this state, the list of billers is also loaded in the object. When a bill type (i.e., the biller) is selected by the user, this object moves to the next state, "Select bill." Naturally this state is invoked by a user. Then, the user enters the amount to be paid against the bill. The user then pays the bill by clicking the Submit button on the user screen. At this stage, the Bill payment object is in the "Pay bill" state. The next state into which the Bill payment object will go to is the "Show receipt" stage. This stage in the life of the Bill payment object is in the form of a response page where the Bill payment statement is shown. After the response page is displayed, the Bill payment object is destroyed. This is the complete life cycle of the Bill payment object.

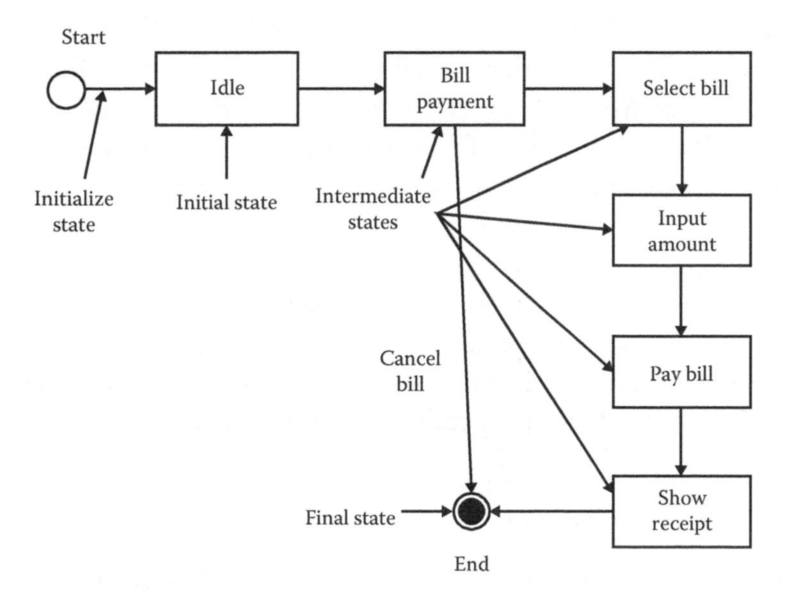


**Figure 7.24** Statechart diagram for OBAAS 1.1.

In Figure 7.24, the mapping of words used with the states of the object with the actual methods and properties is given here:

Bill payment—populate biller list in the drop-down list
Select bill—select biller from drop-down list
Input amount—paid_amount
Pay bill—payment_request()
Show receipt—response()
Cancel bill—cancel button press

You can draw similar statechart diagrams for other objects such as service request and account balance inquiry.

*7.18.3 Detailed Design for OBAAS 1.2* There are two differences between the designs of OBAAS 1.1 and OBAAS 1.2.

- There is a money transfer functionality for OBAAS 1.2 that was not there in OBAAS 1.1. Using this function, users can transfer money from one account to another.
- There is a US$5 fee for each new cheque book request in OBAAS 1.2. However, it is a free service in OBAAS 1.1.

To take care of these changes, detailed design is changed accordingly. Here, we cover the changed class diagram and the changed object diagram. The statechart diagram and the sequence diagram are not provided.

*7.18.3.1 Class Diagram* The class diagram for OBAAS 1.2 (Figure 7.25) includes the money transfer class. All other classes remain the same as in OBAAS 1.1. This new class, "Transfer_Money," is used to take care of the methods and properties that will be needed for money transfer from one bank account to another. Currently, OBAAS only supports money transfer between bank accounts that are kept in the same bank.

*7.18.3.2 Object Diagram* The object diagram depicted in Figure 7.26 shows the state of the objects at a time when one object was processing bill payment, a second one was processing a service request, and a third object was processing money transfer. Each of these objects is linked to the corresponding Account and User objects.

### *7.18.4 Implementation Considerations*

While discussing the class diagram, we mentioned refactoring the classes to make it possible to reuse the source code. This is possible because we can create classes for database connection and user verification. The methods from these classes will be used frequently by other classes.

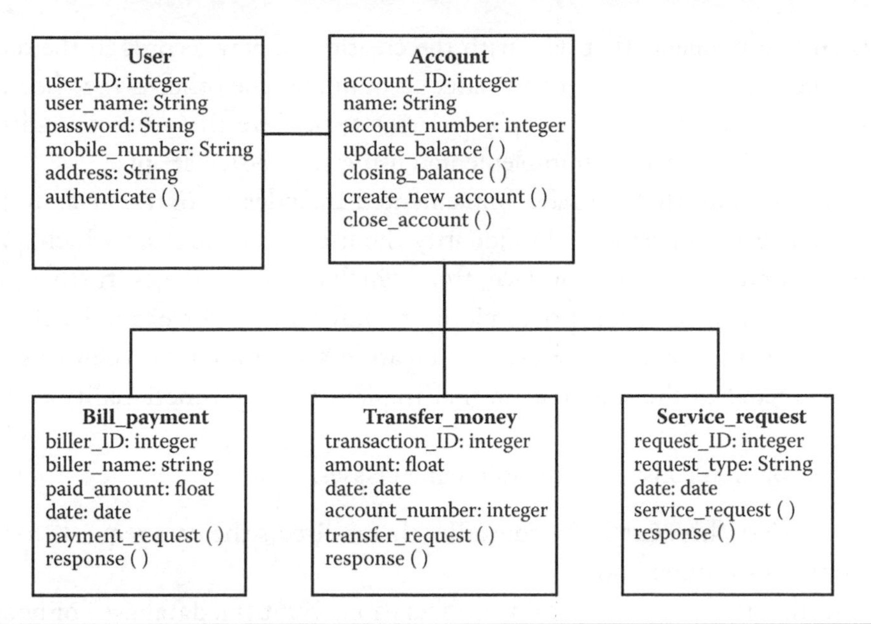


**Figure 7.25** Class diagram for OBAAS 1.2.

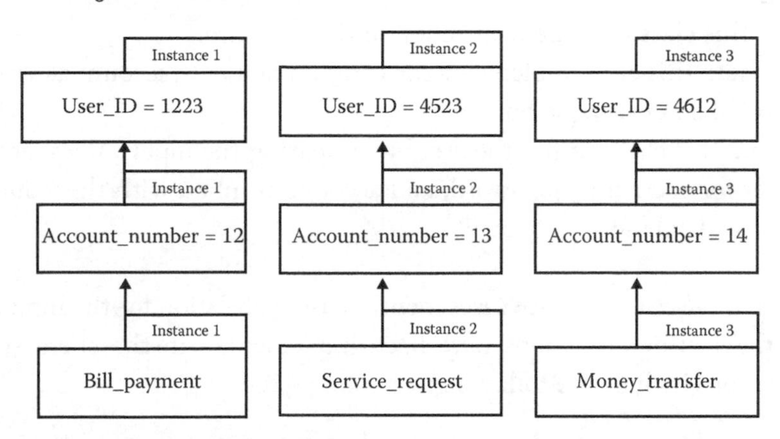


**Figure 7.26** Object diagram for OBAAS 1.2.

The main tasks associated with account balance inquiry, new account creation, bill payment, service request, and account closure can be implemented by using classes. These classes can be compiled and stored at the application server. This alternative is good when a large number of web pages will use these classes. Another alternative is to write the source code for these classes directly on the web pages in the form of client-side scripts. This approach is fine when each class is used only on a few pages; thus, there is not much repetition of source code. We have used this approach to build OBAAS. While classes have been built for reusing the source code for database connections and user verification, the classes that implement all the required functionality for accomplishing the tasks related to account balance inquiry, bill payment, and account closure can be put inside the server-side scripts.

There is another aspect that we have considered. If you look closely at the requirement specifications, you will find that, out of all the functionality in OBAAS, there are only two components that deal with the creation of new records in the database. The first one is the creation of a new bank account and the other is the cheque book request. We decided to create compiled classes to achieve these functionalities. All other functionalities have been implemented using server-side scripts.

The final solution that we have implemented includes a controller as well. This controller has been implemented to identify the user interface from which a request is coming. Based on the user interface, from which a request comes, further processing is redirected to the appropriate piece of source code. For example, if the user request came from the user interface for creating a new account, then this request will be redirected to the source code that implements the functionality to create a new account.

The final solution consists of the following classes:

Create servlet: This class is the controller that redirects the user requests to various pieces of source code.
DBInitializer: This class contains information about the database connection string.
GetCon: This class tests the connection to the database.
RegisterUser: This class implements the creation of a new account as well as the creation of a new cheque book.
VerifyUser: This class verifies the user by comparing the input values, entered by the user for username, password, and account number, with the values existing in the database.

*7.18.4.1 User Validation* The use cases mention user validation for the inputs provided by the user. All user validations have been implemented in the client-side scripts. These validations include the following:

- Checking if the password and reentered password match
- Checking if any field is left blank
- Checking if a string is entered in a field that accepts numeric values alone
- Checking if a numeric input is entered in a field that accepts string values alone

There are also validations to check:

- The amount in a bank account should be sufficient to cover the transfer of money into another bank account. For example, if an account has an opening balance of US$300 and the user who owns this account wants to transfer US$400 to another account, then the validation is there to stop such transactions. The user is provided with a message stating that there is not a sufficient amount in the bank account and thus this transaction cannot be completed.
- For OBAAS 1.2, there is a fee of US$5 for issuance of a new cheque book. This fee is deducted from the bank account of the user. The amount in the bank account of the user should be equal to or more than US$5 to cover the

charge for the issue of the cheque book. Else, the user is provided with a message stating that the amount is not sufficient in the bank account and so this transaction cannot be completed.

All the abovementioned validations cannot be implemented through client-side scripts. This is because these validations need a check in the database to find out the balance in the bank account of the user. For this reason, these validations have been implemented using server-side scripts.

### 7.19 Chapter Summary

The focus of this chapter is the design and coding (programming) aspects of the middle layer. The middle layer contains the business logic of the software product and connects the user interface with the database. In Waterfall methodologies, software design used to be separate from software programming. However, in agile methodologies, both activities are done together.

Software programming involves understanding the requirement as to what computation needs to be done. After understanding the problem, a design can be thought out. Building a large software product (or some features of the product) involves breaking the design into many parts. Software products are inherently complex and breaking down the design into smaller components helps reduce this complexity. Building a few large methods or classes is not good design because understanding such large pieces of code will be extremely difficult. Hence, you need to write many smaller classes and methods that will communicate with each other, thus together forming the required software product features. This type of design involves thinking about what classes need to be made and what methods need to be created to perform the required computation by each class. You also need to find out how these classes will communicate with each other. Communication between the classes or methods is achieved through passing the values (by using parameters) from one class to another. Once the design is clear, the source code can be written to implement the design.

Learning software design and programming involves learning variables, operators, methods, and classes. Variables and methods can have access levels ranging from local to class.

Object-oriented programming is all about creating objects from their classes and then using them as variables. In this sense, classes are just a template for creating objects. Objects have their own data and behaviors. In fact, an object can have a structure in which it can keep data of many different types. This kind of computing promotes reuse. If you can create objects having good structures, then you can use them to get or set many types of data.

Procedural programming languages are older than object-oriented programming languages. However, they still have relevance today because object-oriented programming languages have a limitation that they have difficulty in having serialized access to the components of any objects. Because of this factor, procedural programming languages in the form of interpreted scripts are used to glue the business logic layer to the database layer and user interface layer.

Database programming involves getting user inputs and saving or modifying the data that reside permanently inside a database. Most software products use some kind of database; thus, database programming is one of the most used areas.

Web-based software products are being built in large quantities because the Internet has become very popular and vendors have been creating web-based applications to meet the increasing demand. Web-based software products involve creating user interfaces, creating business logic, and using a database to store permanent data.

When programmers write the code, that code may contain errors such as syntax errors or control flow errors. When these errors are in the code and the programmer tries to save (compile) it, the compiler will raise errors. Finding and fixing these errors is known as debugging. There are good automated tools known as debuggers to perform debugging tasks. These debuggers are provided in IDEs.

Software projects based on agile methodology build software products incrementally. New product features are built on top of the existing product. When the product size becomes large, the existing source code makes it difficult to write new source code for various reasons (e.g., program element names no longer relevant or some source code needs to be moved at some other place because it becomes more relevant there). Programmers make changes to the existing source code to align that source code to a changed design. This activity is known as refactoring.

## QUESTIONS

1. What are some common variable types available in a programming language of your choice? Provide the name of that programming language.
2. What access levels are available for variables in object-oriented programming?
3. Can we declare a method to be public in a class? When should you declare a method to be public?
4. What is object-oriented programming?
5. What is procedural programming?
6. What is a method?
7. What are the state and behavior of an object?
8. Provide some benefits of object-oriented programming.
9. Provide some shortcomings of object-oriented programming.
10. What is a session object? How is it used?
11. What is a class?
12. What is a package?
13. What is refactoring?
14. What is an object?

15. What is a sequence diagram?
16. What is detailed design in software engineering?
17. What is a class diagram?
18. What is a statechart diagram?
19. What is database programming?
20. What is MVC architecture?

## Recommended Reading

Alan Denis (2015), *System Analysis & Design*, Kindle Edition, Wiley, Indianapolis, IN.

Hassan Gomaa (2011), *Software Modeling & Design: UML, Use Cases, Patterns and Software Architectures*, Cambridge University Press, New York.

Bruce Johnson, Walter Woolfolk, Robert Miller, Cindy Johnson (2005), *Flexible Software Design: Systems Development for Changing Requirements*, 1st Edition, Auerbach Publications, Boca Raton, FL

# 8

# Database Design and Construction

**In Chapter 7, we learned**

- **What a software detailed design is**
- **What a software middle layer is**
- **What object-oriented programming is**
- **What procedural programming is**
- **What a class is**
- **What an object is**
- **How to build a web-based software product**
- **What database programming is**
- **What refactoring is**

**In Chapter 8, we will learn**

- **What a database is**
- **What a relational database is**
- **What the elements of a database are**
- **How to make database design**
- **What an entity-relationship diagram is**
- **How to create database table relationships**
- **How to ensure the integrity of a database**

## 8.1 Introduction

Databases are designed to hold a huge amount of data in a manner that these data can easily be accessed and modified when required. The data in a database are permanent in nature unless you delete them: they won't be lost when the power is off or when a software application crashes. They won't be lost if you lose the connection to your website.

In a software application, you can keep some data in the main memory (RAM) during computation but these data are temporary. When you need to make these data permanent, you must store them in a database or a file.

The data you store in a database must conform to some requirements so that they are usable. Some important characteristics of a database include the following:

- Data integrity
- Data consistency
- Data query
- Data manipulation

Databases are very important for most industries that run their businesses using software products. These industries perform their everyday business transactions such as sales orders and purchase orders by storing the information about these transactions in their databases. If any of the transaction data are lost or get corrupted, then these businesses face problems. Therefore, databases are extremely important for these industries. Databases are also used to store data about many other things. Today, social media has become very popular. People use social media to make friends as well as to do business. Storing the personal or business data of the users and keeping them secure and safe are important for the owners of these social media sites.

Databases are an integral part of most software products. Most software products use databases to keep permanent data in a database. You may have come across many websites and may have had a chance to play with some software products. Almost everyone needs the help of search engines for some information. You may also have heard all the hoopla around big data recently. How do these software products work? The power of these software products comes mainly from their databases. Databases store several kinds of data, and with the help of search algorithms, it is possible to search for just about anything.

A database allows searching the items in that database. First, you need to create and store the data in the database. Any data created in the database should not be redundant. The data saved in the database should also be consistent. If wrong and inconsistent data are saved in the database, then the database cannot provide good results when the data are fetched out of the database. Like it is said in the software industry, "garbage in, garbage out": what you store in your database is what you get out of it.

Let us take an example to understand how redundant and inconsistent data can go in a database. When a user wants to create his or her profile on a website, he or she fills in a form with his or her personal information and submits. This information is saved in the database. Later, the user realizes that he or she wants to make a change in his or her profile. The user visits the website and makes changes in his or her profile and submits the form. Because of errors in the logic in saving this information in the database, two records are created for this user: the original record and the modified record. Actually, the original record should have been updated and there should have been only one record for this user in the database. Because of the error in program logic, the database now has redundant data.

Inconsistent data can result from unsatisfactory validation during user inputs. For example, suppose there is a field for entering the sex of the user on the user form we discussed in the previous paragraph. Suppose there is no validation at the user interface or at the database level to check what values a user provides in this field. In this case, it is possible that a user fills in a value like Christian in this field although this field should accept only values of either male or female. When a value like Christian is saved in the database for a field for sex, it is inconsistent data.

The importance of nonredundant (unique) and consistent data cannot be overemphasized. The data stored in a database are used to create many kinds of reports or documents. For example, you apply for a passport; upon getting the passport, you realize that your name or any personal information in the passport is wrong. This wrong information is simply because wrong (inconsistent) information was stored in the database. Similarly, when a manager wants to get a sales report for the month, he or she may do so by querying the database. However, because of the redundant data stored in the database, he or she may get a wrong sales report.

When you design a database, you need to ensure that no garbage is getting into the database. You need to build a good structure in which the data will be stored. If this structure is not good, then you cannot expect to get good results when you fetch data from the database.

A good database design helps in achieving these goals. Software engineers need to know how to create databases and their elements. They also need to know how to use databases to build software products.

## 8.2 Databases and Software Engineering Methodologies

In the Waterfall methodology, the size of the software project teams tends to be large. Project team members possess specialized skills. Each project activity is also large. Thus, database design and construction require software engineers who have specialized skills in those activities. Such software engineers will be solely working on designing and creating the databases. Once they have completed the task for one project, they will proceed to other projects.

In agile methodologies, there are no separate team members who will exclusively take the database design and construction assignments. Database design and creation are mostly done by the same team members who are also involved in software design and programming activities. This is the norm except for very large agile projects. In agile projects, there is another consideration about database activities. If the software product is being developed incrementally, then a complete design of the database is never available to the project team. In any iteration, when the project team gets the list of features to be developed, the project team has to determine the needs for the creation of new tables inside the database. Old tables also need to be examined in light of the new features that need to be developed. New columns may need to be added to the old tables to accommodate the data related to some new features.

## 8.3 Database Types

In general discussions, when people use the term *databases*, they mean both the database engine and the database schemas. However, there is a difference between these two. The database engine can be thought of as a superstructure. In the database engine, you can create many schemas. The database engine is also used for querying and accessing the data in the database schema.

The database schema is the data structure (e.g., database tables in the case of relational databases, which are explained later) and the data that reside in that data structure.

Here, when we use the term *database types*, we are actually talking about the database engine. From the very beginning of the software industry, companies and researchers have tried to create databases that can be used in building commercial and scientific software products. Because of these efforts, many types of databases exist. Some database types include file server based, relational, object oriented, object relational, and NoSQL. Of all these types of databases, relational databases have been the most commonly used. Some of the most popular relational databases include Oracle, Microsoft's SQL Server, and MySQL. Relational databases can hold very large amounts of data (trillions of gigabytes in size).

### *8.3.1 NoSQL Databases*

Web-based software products have an architecture in which the database sits too far away from the users' browsers. One of the main problems faced by the software developers of these web-based products is user input validation. If the users want to validate their input using traditional databases, then the responsiveness of the web-based software products becomes a concern. This is because this type of validation will take an inordinate amount of time to send the user data to a database (which is far away; see the web-based software product architecture discussed in Chapter 5 for a clearer understanding of this aspect) and to get the validation feedback to the user's browser. There was a need for a better way to do it. This need was addressed when XML was invented. XML is a file that can be stored on an operating system and can be queried quickly. This file is generally stored at the web server and can be easily accessed by the user interface (web browser). This provides a quick response if any queries arising from the web browser need to be validated. XML can help in validating the user data quickly without accessing a traditional database that is located at the back end. NoSQL databases are based on this assumption.

Another limitation that the web-based software products face when using traditional databases is that the databases do not allow storing the objects or blobs (large chunks of data) easily. An object can be, for example, a complete sales order with all the fields such as customer information and sales information. Storing a complete object has its benefits. When you want to retrieve a complete sales order record, you do not need to create a large query. You just mention the sales order number and the complete sales order can easily be retrieved using such databases. NoSQL databases can store such objects.

NoSQL databases are essentially a file system. They are not relational in nature. You cannot make a relationship between two NoSQL files. This is their limitation. In contrast, relational databases allow a relationship between two or more database tables, as explained in the latter sections of this chapter.

### *8.3.2 Relational Databases*

Relational databases keep their data in tables. Each table contains many columns. Each column represents a field and each row represents a record. Relational database technology is the most mature among all the database technologies available today. That is why most businesses and government agencies use relational databases to keep their data safely.

Relational databases help you store the data at the most atomic level. If the data are more atomic, then more powerful searches are possible on those data! On the other hand, if you keep your data at an aggregate level, then atomic-level searches are not possible. Any manager knows that he or she cannot make good decisions if he or she does not have accurate data up to the atomic level. This is why it is important that the database system should be able to fetch such data. Similarly, any finance assistant cannot perform his or her task if the payment bills contain faulty data. Relational databases have been successful in performing the tasks of storing and retrieving data.

All the topics related to database design and building in this chapter are related to relational databases. Although an introduction to NoSQL databases is given in Section 8.3.1, the discussion and topics covered henceforth are related to relational databases. NoSQL is mentioned earlier because it is used in some user interface technologies such as AJAX, which is discussed in Chapter 6. We discuss object relational databases in Section 8.8. Object relational databases combine the relational databases with objects. Objects were discussed in Chapter 7.

## 8.4 Database Languages

To search a database, to manipulate a database, to create a database, or to manipulate the data in a database, you need to communicate with the database. This communication is done using database languages.

The database language is divided into two parts: the Data Dictionary Language (DDL) and the Data Manipulation Language (DML). Together, these languages are known as Structured Query Language (SQL). Although each vendor of the databases has its own version of SQL, the differences between those versions are not significant. In most cases, a standard SQL can be used to query any database. In some other cases, slight modification to an existing SQL statement will be good enough for communicating with a different type of database.

*8.4.1 Data Dictionary Language*

There are several database entities such as tables, columns, schemas, triggers, and stored procedures that reside inside a database engine. These database entities need to be created first in order to create or manipulate the data in that database. DDL is used to create these database entities. For example, if you need to create a schema, then the DDL statement is Create schema <schema name>;. Some of the most commonly used DDL statements include the following:

**Create/drop schema:** When a new database needs to be created, the SQL statement is CREATE SCHEMA <Schema_Name>. When you no longer use a schema, you can delete the schema using the Drop schema statement.

**Create/drop table:** Create table is used to create a table while Drop table is used to delete a table from the database. The statement to create a table is as follows:

```
CREATE TABLE <Table_Name> (<Column_Name1> datatype1, <Column_Name2>
datatype2,…) PRIMARY KEY (<Column_Name1>)
```

Here, <Table_Name> is the name of the table. <Column_Name1> datatype1 represents the name of the first column and its data type. The same is true for other columns and their corresponding data types. PRIMARY KEY (<Column_Name1>) represents the fact that <Column_Name1> is the primary key of that table.

**Create/drop sequence:** Create sequence is used to create a sequence while Drop sequence is used to delete a sequence. By using the following SQL statement, we can create a sequence and populate a column of a table with that sequence:

```
CREATE Sequence seq_1
start with 1
increment by 1
maxvalue 999
cycle;
```

A sequence has a start value. In the sample provided earlier, we have given the start value 1 to the sequence. The sequence will have an increment each time it is used. The next value of the sequence is incremented by the "increment by" statement. Here, we have incremented the sequence by 1. Each sequence has a maximum limit up to which the sequence can be used. Here, the maximum value of the sequence is 999. A sequence can be used again once the maximum value is reached. If you want to use a sequence again and again, then you can specify it by declaring a "cycle" statement. If you do not want a sequence to be reused, then you can specify "no cycle." Once a sequence is created, you can use it inside an insert statement by calling a next value in the sequence. For example, consider the following insert statement:

```
INSERT into account value (seq_1.nextval, '00236', 'column2');
```

This insert statement will create a new record in the account table. If the last value of the sequence, seq_1, is 30, then nextval will generate and populate 31 in the column, column2, of the new record created. Some additional details on sequences are provided in Section 8.5.6.

**Create/drop index:** Create index is used to create an index on a database table. Drop index is used to delete an index. The DDL statement to create an index is as follows:

```
CREATE INDEX "index_name" ON "table_name" (column_name);
```

From the above statement, you can see that an index can be created only for one table and not for multiple tables. Thus, each table that you use for searches should have its own index.

### *8.4.2 Data Manipulation Language*

DML is used to create or manipulate the data in a database. Some examples of DML include the following:

**Insert statements:** These are used to create new records (rows) in a table. The insert statement is structured as follows:

```
Insert into <Table_name> <Column_name 1, 2,…> values (x, y,…);
```

The column names in an insert statement are optional. If you need to insert values in all the fields in a record in the table, then you do not need to specify the column names. You only have to ensure that the values being inserted are in the same order as the columns.

You will have to use the column names in an insert statement only when you need to insert values in some specific columns, in which case you will provide the names of the columns and the corresponding values for those columns in your insert statement.

Suppose we have a table Account and it has the columns Account_ID, Account_number, Account_name, Transaction_date, Debit/credit, Amount, and Account_balance. We need to insert the values 101, 00234, ABC Bank, 11/15/2015, Debit, $300, and $4500, respectively. The SQL statement to do this insert will be as follows:

```
Insert into Account (Account_ID, Account_number, Account_name,
Transaction_date, Debit/credit, Amount, Account_balance) values
(101, 00234, "ABC Bank," 11/15/2015, "Debit," 300, 4500).
```

This insert statement will insert a record in the table Account. The "where" clause is not used with insert statements. It is used for deletion or updating of a record in the database.

**Delete statements:** These are used to delete records (rows) from a table. You need to specify which records need to be deleted. If you do not, then all the records in that particular table will be deleted. The delete statement syntax is as follows: Delete from <table_name> where (condition);.

The "condition" will contain the name of a column in that table and a value that may be equal to data in that column. For example, if the table Account has a column Account_number and you want to delete a record where the Account_number field of that record is equal to 00234, then you will have to specify it in the "where" clause of the delete statement like the following:

```
Delete from Account where (Account_number = '00234');
```

**Update statements:** These are used to update the data in any specific field in a column in a table. You will need to specify the column name as well as the existing data in that column to be updated with a value that you specify. The update statement syntax is as follows:

```
UPDATE <table_name>
SET column_name1 = {expr1} [, column_name2 = {expr2}]...
[WHERE where_condition]
```

For example, if the table Account has a column Account_number and you want to update the value of the Account_number field, which is 00234, to 00238, then you will have to specify it in the "where" clause of the update statement as follows:

```
Update Account set Account_number = '00238' where (Account_number =
'00234');
```

DML statements are generally simple but sometimes they are very complex. For example, if an insert statement is based on searching and finding some specific value in a column, then you need to check that specific value in that column. You will be able to do that insert operation later. The "where" clause in a DML statement can cascade too many levels and can become complex. You can learn more about SQL queries from a good database programming book.

### 8.5 Database Entities

A database consists of many entities. There is a top-level entity known as schema under which all other database entities are defined. You need to use DDL to define a new database entity (already covered in Section 8.4.1). You can also delete a database entity, but you cannot update a database entity once it is defined.

### *8.5.1 Schema*

Schema is the topmost-level entity in a database. Schema is a blueprint that describes how the data will be stored and what structures will be available for the storage of data inside the database. A schema is divided into database tables and other database entities such as stored procedures, sequences, primary keys, and secondary keys. Schema can also be considered the database name for a group of tables and other entities that hold the data for a specific company or department. If you want to design a database (e.g., for a company), the first issue you need to address is how you create the tables and columns so that all the information (related to that company) can be kept in that database. In other words, you need to provide a suitable structure to keep your data. This structure is known as schema. The schema consists of the tables and the columns within those tables. For example, if you are building a software product for a bank and its branches, then you will create a schema in a database that will hold all the data related to this bank and its branches. You need to name a schema that will be unique. For example, you can name the schema big_bank (or any other name that is meaningful to you) if you are creating a database for a big bank.

### *8.5.2 Tables*

Tables are the second top-level entity inside a database after the schema. A table may contain many columns. Each column keeps the data at the most atomic level. For example, you can keep the data about the branch names of a bank in one column of a table. A table contains data related to a task or a function. For example, if you are building a database for a bank, then you can have a table for treasury, a table for cash, a table for customers, and so on. If your data that are being kept in many tables are related to each other, then you will need to create relationships among those tables. Tables are related to each other through the relationships established through primary and foreign keys, as explained later in this chapter.

To understand database tables better, let us look at an example. Figure 8.1 shows the table structure for a table named Account_transaction. It has columns named Transaction_number, Account_number, Account_name, Date, Debit/credit, Amount, and Balance. You can create a table similar to this in a database using DDL, described in Section 8.4.1.

| Account_transaction |
| --- |
| Transaction_number<br>Account_number<br>Account_name<br>Date<br>Debit/credit<br>Amount<br>Balance |

**Figure 8.1** Database table structure for the Account_transaction table.

**Table 8.1** Database Table Account_transaction with Records

| Transaction number | Account number | Account name | Date | Debit/credit | Amount | Balance |
|---|---|---|---|---|---|---|
| 101 | 00234 | ABC Inc. | 11/13/2015 | Debit | 300 | 4500 |
| 102 | 00235 | XYZ Inc. | 11/14/2015 | Credit | 500 | 6400 |
| 103 | 00234 | ABC Inc. | 11/16/2015 | Credit | 200 | 4700 |
| 104 | 00234 | ABC Inc. | 11/17/2015 | Credit | 100 | 4800 |
| 105 | 00235 | XYZ Inc. | 11/17/2015 | Debit | 500 | 5900 |
| 106 | 00235 | XYZ Inc. | 11/18/2015 | Credit | 200 | 6100 |

Table 8.1 depicts the database table Account_transaction with records. After creating this table in the database, you need to populate this table. Data can be inserted, deleted, or updated in a table using SQL queries. The SQL language used for these operations is known as DML, and it is already explained in Section 8.4.2.

Each row in Table 8.1 is known as a record in the database. Most of the operations of manipulating the data in a database are done at the record level. For example, the data provided in Table 8.1 are about the transactions belonging to the account holders (companies) of a bank. Whenever a debit or credit transaction happens for a company, which has an account in this bank, an entry will be created in this database table. Suppose the company whose account information is under the account name ABC Inc. has some money deposited in this account on November 14, 2015; a new record will then be created in this table. Notice that in Table 8.1, currently there is no transaction for ABC Inc. for the date November 15, 2015 as this transaction has not been done yet.

### *8.5.3 Primary Keys*

In the relational database model, a table should not have any duplicate rows (records) because it creates problems while retrieving the data. To maintain the uniqueness of records, there should be a column in each table, and this column acts as the primary key. The primary key of a table uniquely identifies each record in that table. The primary key of a table is also used to join that table with other tables in the database. A column with a primary key should have unique data. No data should be repeated in the column that represents the primary key. For example, if you have Sales and Customers tables, then a column sales_ID in the Sales table can be used as a primary key to join the two tables.

Similarly, for an online bookstore database, customer_ID is suitable to be used as the primary key for the Customer table of that database. Alternately, a customer name and address can be used as a primary key. Likewise, book_ID or ISBN can be used as the primary key for the Books table. A primary key is also called a simple key if the primary key is a single column; otherwise, it is called a composite key. A careful decision has to be made as to which column or columns have to be used as the primary key. The primary key has the following properties.

The values in the primary key column must be unique: there should be no duplicate value. For example, for a restaurant database, the username is not a good choice to use as the primary key for the Customer table of that database because there could be two or more customers with the same name. However, the social security number (used in North America) can be used as the primary key because the social security number of a person is always unique and never changes. The primary key must always contain a value and it should not be NULL. Here are some points that can help you choose a primary key for a table:

- The value of the primary key should remain the same. For example, the apartment number of the customers may not be a good choice for a Customer table because that address changes if the customer moves to another location.
- The primary key should be simple. For example, store_ID for a Store table and ISBN or book_ID for a Books table.
- The primary key column should not have a NULL value. A NULL value means there are no data in a field in that column and it creates problems.
- The primary key generally consists of integers but you can also use text. Generally, numeric primary keys are advisable if you want efficiency in your databases.
- The primary key could take any number. As mentioned earlier, factual information such as apartment number is not always a good choice for the primary key of a table. There are two reasons for this: (1) factual information can contain duplicate values and (2) factual information can change. How can you avoid this problem? Most relational databases use factless information to create the primary keys. To create data for primary keys, a technique called auto-increment (for numeric primary key) is used. Using auto-increment, the column that is used as the primary key is populated with factless values. An auto-increment system has been built (i.e., included) in the databases of several vendors. For example, in the Oracle database, there is a facility known as "sequences" that is used for auto-increment. Using a sequence, it is very easy to populate the primary keys. We will learn about sequences in Section 8.5.6.
- Generally, a primary key contains a single column for efficiency purposes, but it can contain more than one column, if necessary.

When you create a primary key, always remember that the data it contains always conform to the integrity rules explained later in this chapter. If you create a wrong primary key, then you will face difficulty in keeping the data in the table.

The database table given in Table 8.1 can be used to create a primary key for that table. Which column (field) looks like the best candidate to be the primary key? Obviously, the column Transaction_number is the best candidate for this job. This is because the Transaction_number column will always have a unique number. Other data traits in this primary key column include arbitrariness, simplicity, and fixed (data in this column will never have to be changed).

A primary key on a table can be created when the table itself is created. Alternatively, if a primary key needs to be created on an existing table, then the table needs to be altered.

*8.5.4 Foreign Keys*

In a database, a foreign key is a column in a table that is used to join that table with one or more other tables in that database. A foreign key on a table ensures that all the data belonging to the foreign key column are always in sync with the data belonging to a primary key column in another table. A pair of primary and foreign keys is the main mechanism through which database tables are joined with each other. It is also possible to join two tables where the foreign key column of one table references a column in another table and that column may not be a primary key (or part of the primary key). However, in such cases, the referential integrity constraints will be difficult to maintain.

Joining of tables is what makes a database relational. A relational database is a strong concept because it allows data to be stored in several tables and yet all these data are available through queries to be used for many purposes. When you are able to store data at the most minute level and fetch them in many combinations, all these stored data provide the real power of information processing. Some of the benefits of this approach include the following:

- Ability to store data at the most minute level. This ensures that most detailed data are captured and saved in the database.
- Ability to join data from many tables. This ensures that meaningful information is provided to the users by joining or screening the data from those tables.

For example, suppose we have two databases that store the same information in different ways. The first database saves information in the most minute level while the second one stores information at an aggregate level. Suppose the information that is to be stored in the database is "object oriented programming" (without quotes). The first database stores this information as object, oriented, programming in three different columns and the other database stores this information as "object oriented programming" (without quotes) in just one column. Now, if a user makes a search for "programming," then the first database will list this entry in the search results. The other database will not be able to find this word because this word is only part of that phrase.

Imagine if it were not possible to join the tables in a database. What consequences could you expect in such a scenario? You would end up storing all the data in just one table. This kind of arrangement would result in data being repeated several times over many records. Similarly, we could also end up storing many items in a single field to avoid repetition of data over many records. We will discuss many of these points in Section 8.6.

Let us look at an example to more clearly understand foreign keys. We have a table Account and we want to have a table Account_transaction (Figure 8.2). The table Account is a master table that keeps information about the account number and account name under which a company keeps its bank account with a bank. The table Account_transaction will keep the information about all the transactions related

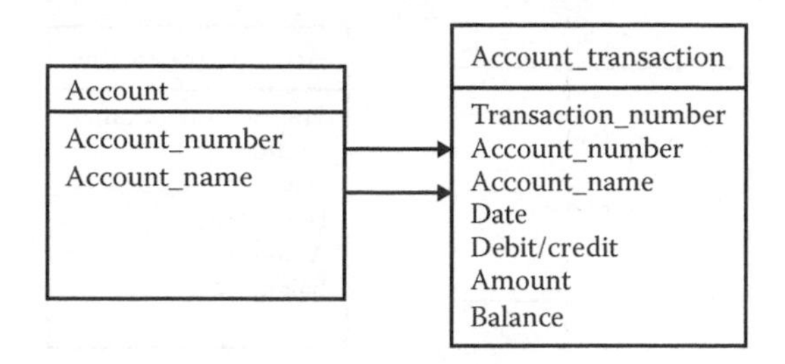


**Figure 8.2** Table join between Account and Account_transaction tables using primary and foreign keys.

to all the bank accounts. Thus, the table Account_transaction is a transaction table. Generally, a master table stores the records of master data. In our case, the master data are the information about the account name and account number under which a company keeps its account with the bank.

How can we link these two tables so that we can link the information stored in the master table to that in the transaction table? It can be done by creating a primary key on a column (or a set of columns) in the master table and then creating a corresponding foreign key on a column (or a set of columns) in the transaction table. Once the primary and foreign keys are created, then the two tables are joined.

We can create a primary key using the combination of Account_number and Account_name columns on the Account table. We can create a foreign key consisting of Account_number and Account_name columns on the Account_transaction table. Assume that we created these primary and foreign keys. Now, when the data start arriving at these tables and a new record is created or an existing record is updated or deleted, because of referential integrity, these two tables will be in sync with each other. Referential integrity constraint implies that a record in the table containing the foreign key column must have a corresponding record in the table containing the primary key column. For example, in our table Account_transaction, if there is a record with data 00234 in the Account_number column, then the table Account must have the data 00234 in the Account_number column of a record. If there is no such record in the Account table, then the table Account_transaction cannot have a record containing this value.

Upon seeing Figure 8.2, you may realize that the join between these two tables is not a good one. This is because the columns Account_number and Account_name are used in both tables. This means that redundant data are being stored. At the same time, there are two issues regarding this primary key. First, there is a chance that the data belonging to either of the columns may be changed in the future. For example, an account name can be changed if the company name is changed. The other problem with this primary key is that it violates a rule that states that a primary key should be placed on as few columns as possible, preferably using a single column. If a primary key consists of more than one column, then it consumes more memory and thus data manipulation or even data retrieval will take more time than when the primary key consists of only one column.

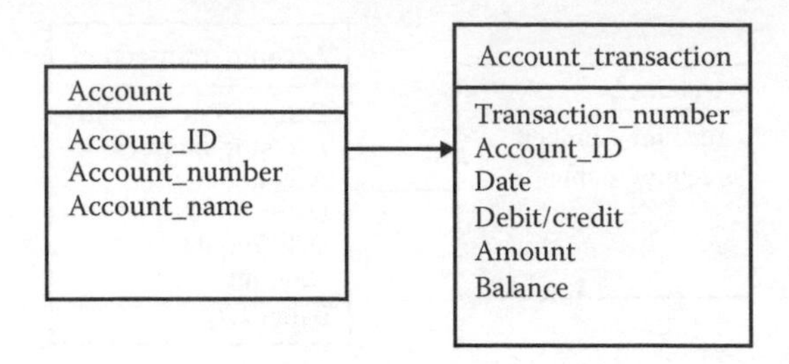


**Figure 8.3** A better join between Account and Account_transaction tables.

For these reasons, it is better to find an alternative. An alternative is to alter the Account table and create a new column that could serve as a better primary key. This way, you can make a better join of those two tables. Let us do it now.

In Figure 8.3, a better join between those two tables is shown. First, the Account table is altered to include the column Account_ID. This column (alone) is now the primary key for this table (Account_number and Account_name together are no longer the primary key for the table Account because a table cannot have more than one primary key). Table Account_transaction is also altered to include a new column, Account_ID (some existing columns from the table Account_transaction are removed, as will be explained later). The column Account_ID has been made a foreign key in the table Account_transaction. Now, these two tables are joined using this new pair of primary and foreign keys.

You can also observe that the two columns, Account_name and Account_number, have been removed from the Account_transaction table. This is because they are no longer needed to be kept in that table. These two columns contain master data and they should be kept only in the Account table. This alteration has helped remove redundant data from the Account_transaction table. Now, the table join between these two tables is looking good.

### *8.5.5 Indexes*

Indexes are used by databases for a faster search of database content. Generally, indexes are created for the primary keys of the tables in the database.

Indexes for databases work the same way as indexes used to read a book. If you may have noticed in books, an index is provided as the last section of the book. If a reader wants to read about specific topics included in the book, then he or she can go to its index section and find the keywords there and which pages of the book contain those keywords. Then, the reader can go to those pages directly and read those topics. Hence, instead of searching the entire book, the reader can use the indexes to find the appropriate pages.

Similarly, if an index is created for a table in a database, when a search is made for a particular value, the searching algorithm will then search the index and will not scan the entire table. This will make that search very fast. Indexes do come with their own cost, though. When an update, deletion, or insertion is made in a table that has an index created, it will take more time to perform these operations than when there is no index. This is because the index has to be rebuilt after each of these operations. Indexes also take additional disc space.

A good strategy for creating indexes is to use them for tables where a lot of searches are made; do not use indexes on tables where update, delete, or insert operations take place more often. For example, we can create indexes for tables in a data store that are not updated frequently but are used for making searches. However, we should avoid creating indexes on transaction tables because they are used more for update, delete, and insert operations.

### *8.5.6 Sequences*

A sequence is used to generate contiguous numbers (1, 2, 3, etc.) to be used to populate (i.e., filling in the data) a column of a table. Sequences are useful to create data for a primary key because the sequences help maintain consistency in the data for the primary key and also help generate unique values for the primary key. This way, the software developers do not need to be concerned about the uniqueness and consistency of the primary key.

Sequences are generally used to populate the primary key columns because the generated data in a sequence are always unique. Data generation for a sequence is automatic. Thus, you do not need to manually populate the data in the primary key column.

The syntax for creating a sequence and other related details was discussed in Section 8.4.1 (under "Create/drop sequence").

### *8.5.7 Stored Procedures*

For fetching or manipulating the data in a database, database programming is used. You can create a similar programming code in the form of stored procedures. Since stored procedures are compiled codes residing inside the database engine, they run extremely fast and can provide search results rapidly when compared to the uncompiled SQL commands that are sent from the application (i.e., when SQL commands are embedded inside the programming language and run from the application server). The downside of stored procedures is that their creation and maintenance need more specialized skills. Again, a programmer not only has to maintain his or her code written on the application server but also has to come time and time again to maintain this separate piece of code saved in a stored procedure. Maintaining the pieces of code at two places is indeed challenging. In addition, stored procedures are poor in terms of exception handling.

Let us look at a stored procedure example to understand how it works. First, look at the following stored procedure syntax:

```
CREATE [OR REPLACE] PROCEDURE proc_name [list of parameters]
IS
Declaration section
BEGIN
Execution section
EXCEPTION
Exception section
END;
```

The stored procedure should have a name (proc_name) under which it should be saved. The list of parameters could be the values that can be passed to this procedure from any other procedure. The declaration section contains the declaration of variables that will be used inside the stored procedure. The execution section contains the statements that will be executed whenever the stored procedure is run. The exception section contains any programming code that will intercept an exception when it occurs during the execution and specify some action that will be taken whenever the exception occurs.

Now, let us look at an example of a stored procedure.

```
CREATE PROCEDURE Account_insert
@Account_ID INT,
@Account_number VARCHAR(6),
@Account_name VARCHAR(15),
AS
BEGIN
INSERT INTO Account
(
Account_ID,
Account_number,
Account_name,
)
VALUES
(
@Account_ID,
@Account_number,
@Account_name,
)
END
```

The procedure named Account_insert given above inserts new records in the Account table. There are three variables declared in the declaration section (@Account_ID,

@Account_number, and @Account_name). When you run this stored procedure, you can fill in the values for all three columns and create a new record.

#### *8.5.8 Triggers*

Triggers are stored procedures that get fired (i.e., triggered) on each database manipulation event. Triggers are generally used to maintain the integrity of the database. For example, in the Sales and Customers tables mentioned earlier, if a record of a new customer is added to the Customers table, then the trigger fires and creates a new record to the Sales table as well. We will need this trigger if these two tables are not joined by any primary key–foreign key combination.

Triggers are not used much because they are sometimes dangerous. If the trigger is not written properly, it could corrupt the database. The reality is that triggers are difficult to handle in many situations; thus, they should be avoided. They should be used only when there is no other alternative. Triggers can be used for some purposes; for example, if a column in a table needs to be populated with the data derived from a formula, then this operation can be linked to a trigger.

## 8.6 Database Design

Database design is important because it involves deciding which data will be kept in the database and what kind of searches will be needed on that data. An important point about databases is that each row (record) of data should be unique. In a table, you can never have two or more rows containing exactly the same data. Another important characteristic is that the data in a primary key column are always unique. In other columns, the data are not required to be unique. In fact, if there is a primary key–foreign key relationship between two database tables, then the data in the foreign key column will not be unique and will be repeating many times.

Database design is done using entity-relationship (ER) diagrams, integrity and referential rules, table joins, and so on. We will learn about all these database design techniques in the following sections.

### *8.6.1 ER Diagram*

Once you are clear on which data need to be kept in the database, the next step is to decide on how to keep those data in the database. For example, suppose you are building an online bookstore and want to keep the data such as book title, author, genre, date of publication, publisher, and price. You may also need to keep the data such as credit card number, credit card expiry date, and credit card holder name for the book buyers. Additional information such as the name of the bookstore, date of sales, and shipment address may also need to be kept in the database.

Let us study how we can build an ER diagram. The relationship between different entities within a database is depicted using an ER diagram. An entity can be a place, person (or organization), event, or object that is part of the information system for which the database is being developed. The main elements of any ER diagram include entity, relationship, and attribute. In the online bookstore scenario, the title of the book, the author of the book, the name of the bookstore, and so on are all entities.

The relationships among the entities follow:

- A book is being sold at the store.
- An author has written a book.
- A book by this author is available at the store.

These statements show the relationships among the entities. A relationship describes the relation between two or more entities. For example, the statement "XYZ has written the book ABC" describes the relationship between the book ABC and the author XYZ. A relationship can be a binary, recursive, or ternary type. A binary relationship can further be a one-to-one, one-to-many, or many-to-many type. A recursive relationship is the relation of an entity with itself.

A recursive relationship is also known as unary relationship because there is only one entity involved here. For example, generally, a professor teaches students. Thus, we have two entities here: professor and students. When there is a relationship between two entities, it is a binary relationship. However, if all the professors are also enrolled in some courses, then a professor is also a student.

In other words, this relationship can be presented as follows:

All professors are students but not all students are professors!

In this case, it will be wrong to present the professors and students as two entities. The reality is that there is only one entity. Thus, the relationship between students and professors is a recursive relationship.

Recursive relationships can be depicted in databases as a self-join. You need to create a join where the table joins itself.

Figure 8.4 depicts the self-join between Student_ID and Professor_ID. This is because each professor is also a student in some course. In Table 8.2, you can see that

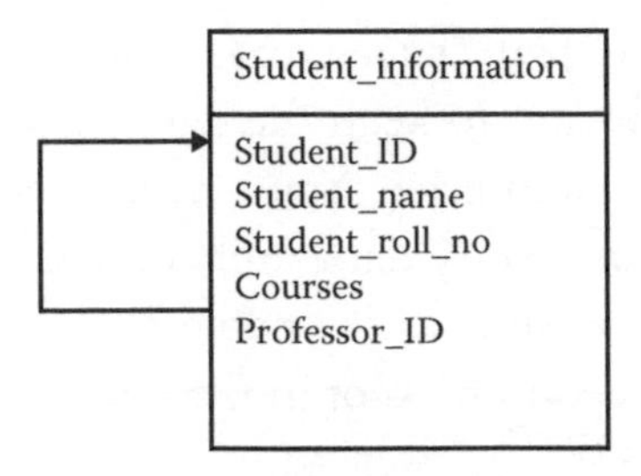


**Figure 8.4** Self-join for a recursive relationship.

**Table 8.2** Database Table Student_information with Data

| Student_ID | Student_name | Student_roll_no | Courses | Professor_ID |
|---|---|---|---|---|
| 101 | Ian | 0011 | Physics | 105 |
| 102 | Greg | 0012 | Chemistry | 106 |
| 103 | Tony | 0013 | Math | 106 |
| 105 | Chris | 0014 | Geology | 106 |
| 106 | Mary | 0015 | Botany | 105 |

a professor named Chris with Professor_ID value 106 is also a student enrolled in a geology course. If you want to know which professors are teaching which students, then you need to do a self-join on the table Student_information.

The DML statement for the self-join will be as follows:

```
select e1.Student_name 'Professor',e2.Student_name 'Student'
from Student_information e1 join Student_information e2
on e1.Student_ID = e2.Professor_ID
```

You can see the result of the self-join in Table 8.3.

We assumed that all professors are also students. In real life, this assumption is not true. A better example of a self-join could be a relationship among the employees and the managers in a company. All managers are employees but not all employees are managers.

An attribute describes a property of an entity. For example, suppose there is a book ABC (which is an entity). There can be many attributes for any entity. For example, the book ABC is priced at $40 and has 700 pages. Here, price and pages are attributes (i.e., properties) of ABC while $40 and 700 are the values of those attributes. It is a good idea to make a unique ID for each entity in the database. This helps in searching as well as in making sure that there are no duplicate data for the same entity in the database. The ID along with some other essential attribute can be considered the key attribute of an entity. Key attributes can be used for creating primary keys.

In ER diagrams, a rectangle represents an entity, a diamond represents a relationship, and an oval represents an attribute.

Figure 8.5 shows an ER diagram for author and book entities. These entities are related by "writes" relationship. An author and a book can have a one-to-one and even a many-to-many relationship. A book can have many authors and an author can write many books. Both kinds of relationships can exist.

**Table 8.3** Result of "Select" Statement after Self-Join on a Table

| Professor | Student |
|---|---|
| Chris | Ian |
| Chris | Mary |
| Mary | Chris |
| Mary | Greg |
| Mary | Tony |

**Figure 8.5** ER diagram for author and book entities.

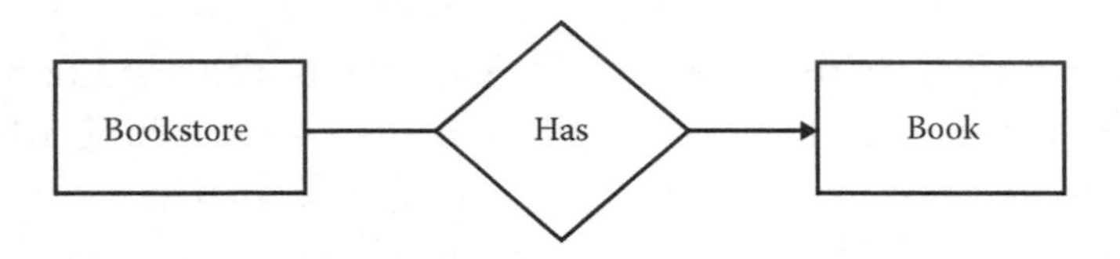


**Figure 8.6** ER diagram for bookstore and book entities.

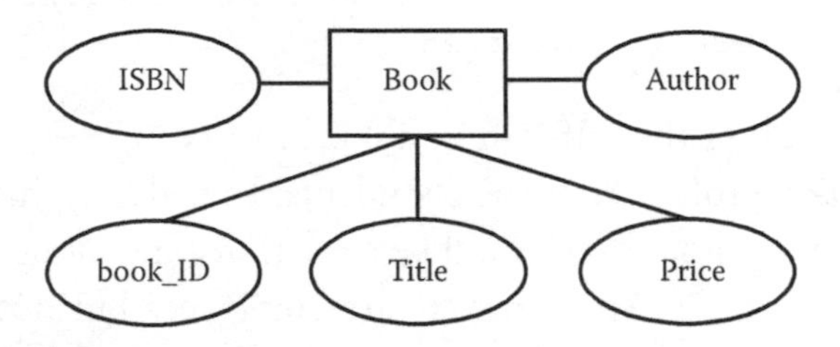


**Figure 8.7** Attributes of a book.

Figure 8.6 depicts an ER diagram for the bookstore and the book. Again, a bookstore and a book can have a many-to-many relationship. A bookstore can have many books and a book can be at many bookstores.

Figure 8.7 depicts some attributes of a book. A book can have several attributes like ISBN, book_ID, title, price, and author. The book_ID and ISBN can be key attributes here.

You can build the entire schema for a database using ER diagrams. Here, you have seen the attributes of a book, the relationship between a book and an author, and the relationship of a book with a bookstore. You can build similar attributes for bookstore, author, and so on. You can also join the entities with each other when they have dependency through a primary and foreign key combination. This way, you can build the entire schema for the online bookstore.

### *8.6.2 Normalization*

When we create tables and columns in the database schema, we end up having one or more columns with duplicate data in that (those) column(s). Normalization is the process of eliminating the duplication of data.

A novice person may ask, why do you need many tables in the database and why can you not put the entire data in one big table? Good questions. Let us examine them.

Suppose you want to create a database for a bank. You want to keep data about the customers (i.e., the customer address and customer name) and the customer accounts (transaction date, account ID, debit/credit, and balance). If we create just one table to keep all these data, then it looks similar to Table 8.1. We also have Table 8.4 showing the data from a table join. Although these two tables show very similar data, they are actually very different. We will learn about the difference between these two in this section.

For good database design, duplication of data is not good. To make sure that the entry of data into the database happens at only one place, you may need to create many tables instead of just one or a few tables. This is done using normalization. We have already seen in ER diagrams that there can be a binary relationship between entities. This binary relationship between the entities can be taken further to design the tables for those entities. When creating the tables, we can use normalization to divide the tables and make sure that each table contains only one entry of any data. Normalization is done using three or more steps. After the first step, the first normal form of the data can be achieved. It can be further refined in the succeeding steps to get the second and third normal forms. Let us see more details on normalization.

**First normal form:** A table is in the first normal form if every field in every row of that table contains a single atomic value but not a list of values. This is also called the "atomic" property.

Table 8.5 presents a database table that is not in first normal form. This is because the data in the columns Tr_Date and Debit/Credit have more than one value. This happens because, on the same date, more than one debit/credit transaction happened and the columns have to accommodate two entries.

**Second normal form:** A table is in the second normal form if some of the non–key columns are dependent on the primary key. Still some other non–key columns may

**Table 8.4** Customer Information and Transaction Table

| Tr_ID | Cust_name | Cust_address | Tr_Date | Account_ID | Debit/Credit | Balance |
|---|---|---|---|---|---|---|
| 1 | Gary | 1 Park Street | 03/15/2014 | 23456 | 500 | 2340 |
| 2 | Andy | 2 Park Street | 03/15/2014 | 34567 | 200 | 4500 |
| 3 | Andy | 2 Park Street | 03/17/2014 | 34567 | 1000 | 5500 |
| 4 | Gary | 1 Park Street | 03/17/2014 | 23456 | -2000 | 340 |
| 5 | Casper | 3 Park Street | 03/17/2014 | 34568 | -400 | 3800 |

**Table 8.5** Database Table That Is Not in the First Normal Form

| Tr_ID | Cust_name | Tr_Date | Account_ID | Debit/Credit | Balance |
|---|---|---|---|---|---|
| 1 | Gary | 03/15/2014<br>03/15/2014 | 23456 | 500<br>500 | 2340 |
| 2 | Andy | 03/15/2014<br>03/15/2014 | 34567 | 200<br>300 | 4500 |
| 3 | Andy | 03/17/2014 | 34567 | 1000 | 5500 |
| 4 | Gary | 03/17/2014 | 23456 | -2000 | 340 |
| 5 | Casper | 03/17/2014 | 34568 | -400 | 3800 |

not entirely be dependent on the primary key. The table should also be in the first normal form. If there are multiple columns in the primary key, then every non–key column shall depend on that entire set of columns (that made up the primary key) and not on any part of it. If a table is in the first normal form and contains a single column as the primary key of that table, then that table is automatically in the second normal form. In the second normal form, some columns may have data repeated (since they do not depend on the primary key).

In Table 8.1, the data are in the second normal form. You can see that the data in the columns Account_name and Account_number are repeated. However, no column has more than one value in any field; therefore, this table is already in the first normal form. In addition, a single column (Transaction number) represents the primary key. Thus, this table is in the second normal form.

**Third normal form:** A table is in the third normal form if the non–key columns are not dependent on each other and the table is already in the second normal form. That is, the non–key columns of the table are dependent on the primary key alone. In Table 8.1, since the Account_name and Account_number columns have repeating data, this implies that these two columns need to be moved to another table if you want this table to be in the third normal form. What we can do is create a new table to move these columns. We have done so in Figure 8.3. Figure 8.3 depicts a primary key–foreign key join between Account and Account_transaction tables. We have created a new column, Account_ID. This column is the primary key in table Account and is a foreign key in table Account_transaction. Now, the table Account containing the primary key will be in the third normal form. Similarly, the table Account_transaction will also be in the third normal form. Both tables will have no repeating data.

Table 8.4 shows the data (records) for the table join depicted in Figure 8.3. We had already mentioned at the beginning of this section that the data (records) shown in Tables 8.1 and 8.4 look similar but there is a big difference. The difference is that in the case of Table 8.1, all the data (records) are stored in just one table. Hence, duplicate data are entered in the columns Account_number and Account_name. In Table 8.4, although the data (records) for the columns Cust_name and Cust_address seem to be duplicated, the data in these columns are actually stored in the table Account. The other data being shown in Table 8.4 come from the Account_transaction table. These two tables (Account and Account_transaction) have a join as shown in Figure 8.3, and the combined data from these two tables (Account and Account_transaction) are shown in Table 8.4. When we show the data (that are coming from two or more tables) at one place through some table joins, this entity is known as a view. A view in a database works like a mirror. A view never has its own real data but the data actually come from the tables on which the view is based on.

There are other normal forms, namely, Boyce–Codd normal form, fourth normal form, and fifth normal form, but discussing them is beyond the scope of this book. You may see a good database design book for details on these forms.

*8.6.3 Relationship among Tables*

If a database contains unrelated tables, then it does not provide any additional help compared to a regular spreadsheet. However, if the tables are related, then you can create powerful queries out of those tables to retrieve useful data. While designing a database, you should first identify the relationships between the tables. Different relationships that may exist between the tables include the following:

- One-to-many
- Many-to-many
- One-to-one

*8.6.3.1 One-to-Many* In a bookstore database, a store can have zero or more books, while a book is available at that store only (if that store has no other branches). In a company database, one chairman has many employees under him or her, while all those employees have only one chairman. These are some examples of one-to-many relationships.

It is not possible to represent a one-to-many relationship using a single table. For example, in the bookstore database, we may begin with a table called Books, which may store the information such as ISBN, title, author, and genre for each book available in that bookstore. To store the information about the bookstore, we could create a Store table with columns book1, book2, book3, and so on. However, the immediate issue is: How many columns are needed for that table? A one-to-many relationship needs to be created between the tables to address this issue. Figure 8.8 depicts a one-to-many relationship between the tables. To create a one-to-many relationship, we need two tables: a Store table to store the information about the store with Store_ID as the primary key and a Books table to store the information about the books with book_ID as its primary key. Then, a one-to-many relationship between these two tables can be created by storing the primary key (Store_ID) of the Store table (the parent table or the "one" end of that one-to-many relation) in the Books table (the child table or the "many" end of that one-to-many relation).

The Store_ID column in the child table, Books, is known as the foreign key. A foreign key of a child table is related to the primary key of the parent table. The parent

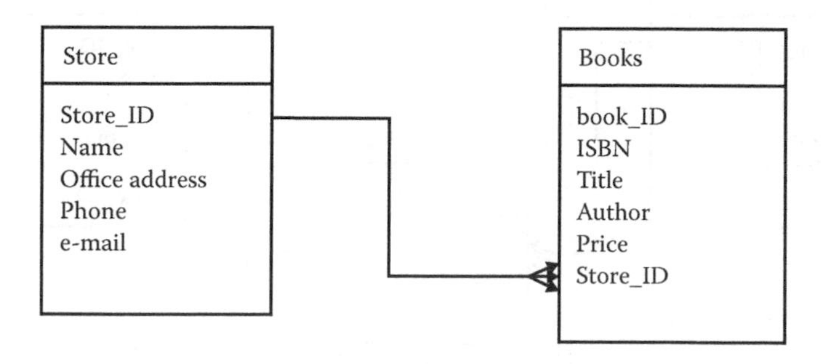


**Figure 8.8** One-to-many relationship among tables.

table is referenced by the child table using this primary key. For example, in Figure 8.8, Store_ID is a foreign key of the child table (Books) but Store_ID is a primary key of the parent table (Store). It is possible to have zero or more rows in the child table for each row in the parent table. In addition, there is one and only one row in the parent table for each row in the child table.

*8.6.3.2 Many-to-Many* In a many-to-many relationship, one or more rows in a table are related to zero or more rows in another table. If you consider a bookstore database, each book has one or more persons as the authors but each person may write zero or more books. Let us illustrate this with our bookstore database. We begin with two tables: Author and Books. The Author table contains information about the author (such as author_ID, name, address, and e-mail) with author_ID as its primary key. The Books table contains book_ID, ISBN, and so on, as shown in Figure 8.9. To create a many-to-many relationship between two tables, a new table (i.e., a third table) is needed. This table is known as a junction table. We created a Book Details table as the junction table in Figure 8.9. By using the junction table, the many-to-many relationship can be implemented as two one-to-many relationships. In Figure 8.9, for each row in the Author table, there could be zero, one, or more rows in the Book Details table. For every row in the Book Details table, there is one and only one row in the Author table. For each row in the Books table, there could be zero, one, or more rows in the Book Details table. For every row in the Book Details table, there is one and only one row in the Books table.

*8.6.3.3 One-to-One* In a one-to-one relationship, each row in a table is linked to one (and only one) row in another table. The number of rows in these two tables should be exactly the same. One-to-one relationships are rare. This is because when you have transactions, each transaction is going to use the master data from the master tables. Each master table's primary key is going to be used many times in each transaction table. Master and transaction tables are explained in Section 8.6.5.

In some databases, there is a limit on the maximum number of columns a table can have, in which case you can split a table into two different tables and then

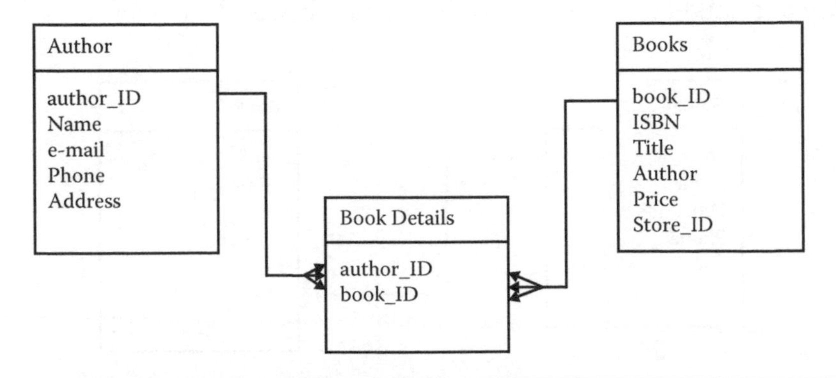


**Figure 8.9** Many-to-many relationship among tables.

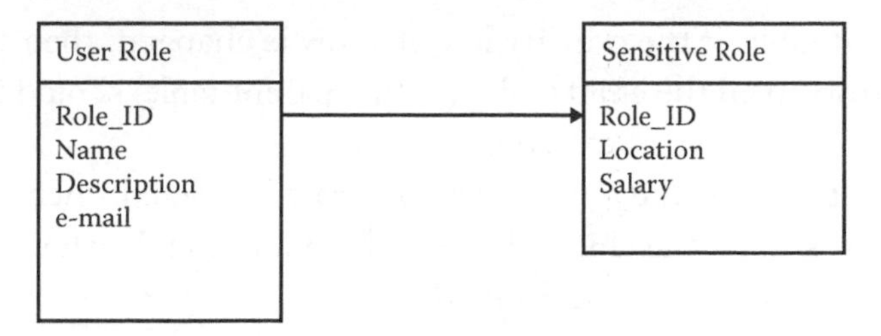


**Figure 8.10** One-to-one relationship between two tables.

maintain a one-to-one relationship between those split tables. If you have some sensitive data, then you can use one of those tables for storing those sensitive data and keep the remaining data in the other table. In Figure 8.10, you can see two tables, User Role and Sensitive Role. You can keep sensitive user roles in the Sensitive Role table and the rest of the user roles in the User Role table. Similarly, you can separate the rarely used columns of a table into another table to improve the performance of the database.

### *8.6.4 Integrity Rules*

Integrity rules pertain to the data contained inside the tables. You need to define some rules to enforce them when some data are written to the tables in a database. If there is an attempt to feed some wrong data into any of the tables, then the integrity rules will prevent this from happening. There are two integrity rules: entity integrity rule and referential integrity rule. Both rules are automatically enforced when you define the primary and foreign key tables.

*8.6.4.1 Entity Integrity Rule* There should not be a NULL value in the primary key. Otherwise, the primary key cannot uniquely identify the row in which the NULL value is. In the case of a composite key (that contains several columns), there should not be a NULL value in any of those columns.

*8.6.4.2 Referential Integrity Rule* Each foreign key value should match a primary key value in the parent table. The referential integrity rule results in the following behavior:

1. It is possible to insert a row with a foreign key in the child table only if that value (i.e., that row) exists in the parent table. As an example, let us consider the Account and Account_transaction tables discussed earlier. In the Account table (i.e., the table with the primary key), if a record (i.e., data) for a customer named "Gary" exists, only then will it be possible to have the transaction data related to Gary in the Account_transaction table (see Figure 8.3). The parent table is the one that contains the primary key and the child table is the one that contains the foreign key.

2. In the parent table, if the primary key of a row is changed, then all the corresponding rows in all the child tables (of that parent table) should be modified. This behavior is known as cascading update.
3. In the parent table, if the primary key of a row is deleted, then all the corresponding rows in all the child tables (of that parent table) should be deleted. This behavior is known as cascading delete.

Referential integrity is important because if some changes are made in the parent table and these changes are not reflected in the child table, then it can lead to many integrity problems. For example, suppose we have an Account table that stores account information about the bank customers and an Account_transaction table that stores account transaction information of all the transactions done for all the accounts maintained with the bank. If there is no referential integrity relation between these two tables, then it can happen that when a customer account is closed with the bank, his or her record is deleted from the Account table but his or her account transaction information is still kept in the Account_transaction table.

The table in Figure 8.3 can be used to see how referential integrity works. The Account_transaction table has a foreign key on the column Account_ID and the corresponding primary key is put on the column Account_ID of the Account table. If any change has to be made to any data in the Account table, then the corresponding changes will also occur in the Account_transaction table because of the referential integrity. These changes will be done automatically but you can control whether you want a manual intervention when it occurs. Manual intervention can be introduced by either using the stored procedures or controlling the change in the corresponding data programmatically from the business logic layer.

We can control how referential integrity works on the child table by using the DDL directly when we create the table.

### *8.6.5 Master and Transaction Data*

If you are building a software product that is transaction oriented, then a lot of transaction data will be generated and added on a daily basis to the transaction tables of the database of that software product. If you are not careful about the database design, then thousands of unnecessary data entries will be stored in the database.

In transaction-oriented databases, tables can be divided into two categories. One category of tables will store the master data and the other will store the transaction data. Now, what are master data and transaction data? Transaction data are the data that are created as a result of some transactions. For example, sales data are an example of transaction data. Every day, hundreds or thousands of sales could be made by a shopping store. The data that are generated by these sales transactions are transaction data. In contrast, master data belong to the defining entities. For example, shopping store details such as its address, nature of business operations, and variety of goods available can be considered master data. Master data are almost permanent and do not change very often in contrast to transaction data. Master data are always created only once. For example, you create the store information only once. Thus, master data are very different from transaction data.

We can see that the information about master data and transaction data is useful for designing our databases. Despite the rules of normalization, if you stick to this fundamental rule of understanding and dividing your data into these two categories (master data and transaction data), you can create your tables as either master or transaction tables. Each master table should be based on an entity that has data related to that entity. Later, you can plan the transaction data and create tables so that data redundancy in those tables can be avoided. Usage of appropriate primary and foreign keys will definitely help here. For example, provide a primary key in the master tables and the corresponding foreign key in the transaction tables to make sure that no master data are repeated in the transaction tables. Again, create a separate transaction table for each major transaction data to ensure that the transaction data are not repeated as well.

A database for a bank can have one or more tables for customers, branches, products, services, and so on. All these tables contain the master data. Then, we can have a transaction table that can contain transaction data such as sales receipt, sales date, and payment amount, to name a few. For example, we have Table 8.6, which is a master table because it contains master data such as customer name, customer address, and customer ID. The actual field names used in this table are Cust_ID, Cust_name, and Cust_address.

We also have a transaction table, Table 8.7, which has transaction data such as transaction ID and customer name. The actual field names used in this table are Tr_ID, Cust_ID, Tr_Date, Account_ID, Debit/Credit, and balance.

You can create master and transaction tables in such a way that they can be easily joined. For example, Tables 8.6 and 8.7 can be easily joined. The Cust_ID column can be set as the primary key in the master table (Table 8.6), and the Cust_ID column given in Table 8.7 (transaction table) can be the foreign key in that table.

**Table 8.6** Customer Account Master Table with Primary Key

| Cust_ID | Cust_name | Cust_address |
|---|---|---|
| 1 | Gary | 1 Park Street |
| 2 | Andy | 2 Park Street |
| 3 | Casper | 3 Park Street |

**Table 8.7** Account Transaction Table after Removing Master Data

| Tr_ID | Cust_ID | Tr_Date | Account_ID | Debit/Credit | Balance |
|---|---|---|---|---|---|
| 1 | 1 | 03/15/2014 | 23456 | 500 | 2340 |
| 2 | 2 | 03/15/2014 | 34567 | 200 | 4500 |
| 3 | 2 | 03/17/2014 | 34567 | 1000 | 5500 |
| 4 | 1 | 03/17/2014 | 23456 | -2000 | 340 |
| 5 | 3 | 03/17/2014 | 34568 | -400 | 3800 |

Once we understand master data and transaction data, we can create separate tables for these types of data and then join those tables (using a table join) so that the data between them are linked.

You can observe how ER diagrams, type of data (master or transaction), and normalization rules can be implemented to make your data in the database atomic and avoid duplicate data anywhere.

### 8.7 Database Management Systems

The term *Database Management Systems* or *DBMS* was traditionally coined to mean the computer systems that were using a database for doing tasks such as payroll, finance, and accounting. Electronic Data Processing (EDP) is also used to mean the same. All of these were early systems that were commercially used by companies and government agencies for these purposes. Since then, many other types of DBMS have come along. In all these software products, the role of databases is extremely important; they have become an integral part of most commercial and governmental computer systems. We have products like Enterprise Resource Planning applications that use databases.

A DBMS can be divided into many layers. The most common are the user interface, business logic layer, and database layer. We have already learned how to create the user interface and business logic layers. Now that we have learned about databases, specifically relational databases, in this chapter (Section 8.8 will give details on object-oriented databases), we can create the database layer as well. This will complete the discussion on developing a DBMS.

### 8.8 Object Relational Databases

In object-oriented programming, we create classes to create objects. We discussed classes and objects in Chapter 7. For any software product that will use databases, it makes sense to create classes that create objects having exactly the same structure as the database schema. If this happens, the project team can save time and effort by making just one design instead of creating two separate designs, one for the business logic and another for the database schema. The other factor is that the same structure for the objects and the database schema will result in faster business logic development. It will also enhance the clarity of the source code and maintainability of the system.

To understand this concept, let us create a class and a corresponding database entity. In Figure 8.11, you can see that we have a User role class and a User role table, both having the same structure. Now, if we create a class diagram and design the software product, then we can again use the same class diagram to create the database schema. This is because the classes and their relationships will be the same as the database tables and their relationships. Hence, you need to design once and use it for both purposes. This is the essence of object relational databases.

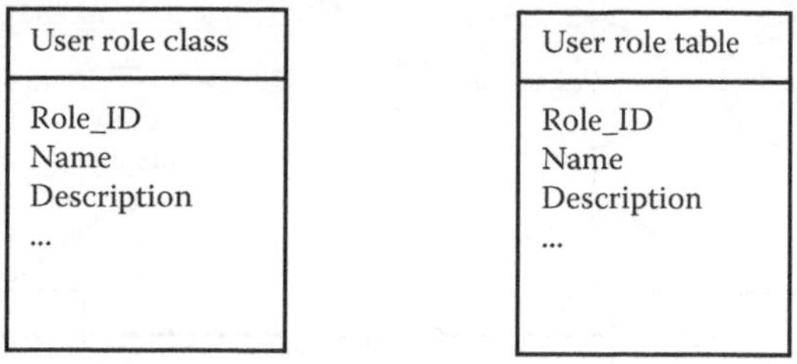

| User role class |
| --- |
| Role_ID<br>Name<br>Description<br>... |

| User role table |
| --- |
| Role_ID<br>Name<br>Description<br>... |

**Figure 8.11** Class and table structures with identical data.

Object relational mapping between objects and database schemas can be set from either the business logic layer to the database or vice versa.

A limitation of object relational database technology is that it works only if the software design consists of direct mapping of middle-tier components (classes and objects) and database entities. Software architectures such as the Model–View–Controller do not have direct mapping between middle-tier components and database entities. Thus, the object relational database concept does not work in these cases.

This object relational database technology is still in its infancy and is being developed by major software vendors such as Microsoft Corporation and Oracle Corporation. Once it is fully developed, you can create your classes first and then the database schema can be created automatically from those classes. Hence, you do not need to design and create your ER diagram and your schema in this case.

## 8.9 Case Study

For the online bank account access system OBAAS, we have to create a database schema. As we can see from the requirement specifications, there are customers (users) who have bank accounts. The customers are provided with a new system, through which they will be able to access their own bank accounts online by means of an Internet connection.

We have already provided the requirement specifications, the software architecture, the user interface design, and the business logic design in previous chapters. Here, we provide the database schema for this system.

As envisaged in the software requirement document, there are two versions of OBAAS. We decided to create just one data model for both systems. The database schema will serve both designs of OBAAS.

### *8.9.1 ER Diagram*

From the software requirement specifications (SRS), we gather that we need to store permanent data for the entities "user" and "account." We also need to store data for the transactions performed by the user.

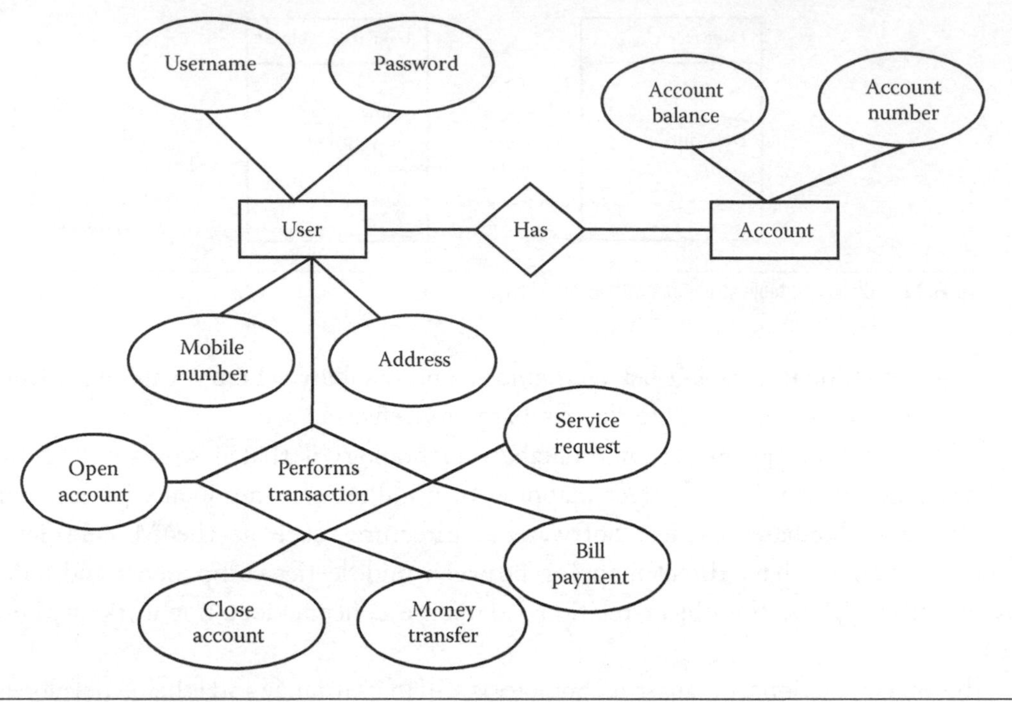


**Figure 8.12** ER diagram for OBAAS.

On the basis of the requirement specifications, the following attributes are observed:

**User:** Each user of OBAAS has the following attributes: username, password, mobile number, and address.

**Account:** Each bank account in OBAAS has the following attributes: account number and account balance.

**Transactions:** Each transaction performed by the user must have a transaction number. Transactions performed by a user could be a money transfer, bill payment, or service request.

On the basis of the data requirements, a data model has been made and an ER diagram has been drawn. Figure 8.12 shows the ER diagram for OBAAS.

### *8.9.2 Schema*

On the basis of the ER diagram, the schema and database entities have been created. Although there are many types of transactions done in OBAAS, some of them are similar in nature. For example, bill payment and transfer fund transactions are very similar. In both transactions, fund transfer happens from one account into another account (in the case of bill payment, fund transfer happens from the user account to the account of the biller).

Transactions such as close account, transfer fund, bill payment, and balance inquiry exclusively deal with the account entity. However, service request transaction deals with a service entity. Thus, we can see that, for our purpose, only two tables are needed: one table deals with the user and account entities and the other keeps information about the service requests.

**Account and user:** For the account and user entities, a NEWACCOUNT table has been created.

```
CREATE TABLE "NEWACCOUNT"
( "ACCOUNTNO" NUMBER,
  "USERNAME" VARCHAR2(40),
  "PASSWORD" VARCHAR2(40),
  "REPASSWORD" VARCHAR2(40),
  "AMOUNT" NUMBER,
  "ADDRESS" VARCHAR2(400),
  "PHONE" NUMBER,
  "CATEGORY" VARCHAR2(1),
  CONSTRAINT "NEWACCOUNT_PK" PRIMARY KEY ("ACCOUNTNO") ENABLE
);
```

**Service request:** For taking care of service requests, a table named CHEQUEBOOK has been created.

```
CREATE TABLE "CHEQUEBOOK"
( "ACCOUNTNO" NUMBER,
  "SERVICENO" NUMBER,
  "SERVICETYPE" VARCHAR2(1),
  CONSTRAINT "CHEQUEBOOK_FK" FOREIGN KEY (ACCOUNTNO) REFERENCES
NEWACCOUNT(ACCOUNTNO) ON DELETE CASCADE
);
```

We also need a sequence to populate the data in the primary key field ACCOUNTNO in the NEWACCOUNT table.

```
CREATE SEQUENCE BANK_SEQ MINVALUE 1 MAXVALUE 999999999999 INCREMENT
BY 1 START WITH 1 NOCACHE NOORDER NOCYCLE;
```

**Explanation:** The table NEWACCOUNT is the main table in the OBAAS database and it holds most of the data used in OBAAS. It holds information for username, password, phone number, address, and account number.

In OBAAS, at present, we are creating a service number whenever a cheque book is issued. We are using the SERVICENO column of the CHEQUEBOOK table to populate the service numbers. For these services, we have currently created the table CHEQUEBOOK. It is named so because currently there is only one service: cheque book issue. In the future, when other services are introduced, this table name will be changed accordingly (as part of refactoring). In the future, there could be many services introduced by the bank. Hence, we have created a service-type field (i.e., a column) in this table to represent different types of services.

The CHEQUEBOOK table is a transaction table. The ACCOUNTNO field in this table is actually a foreign key and the corresponding primary key is the ACCOUNTNO column of the NEWACCOUNT table. We have defined a primary key–foreign key relationship between these two tables. In the foreign key definition, we have also added the DELETE CASCADE clause. This is because when we delete an account from the system (to close that account), all the child records related to that account (in the CHEQUEBOOK table) should also be deleted. The DELETE CASCADE clause exactly does this task. For such a scenario, a foreign key is not needed.

As per the SRS, there are no reports to be shown for the transactions. For example, the SRS does not mention the transaction history reports for the customers. Hence, we are not generating any transaction numbers for the transactions, including the money transfer and bill payments transactions (when they happen). Thus, in the database, we have not provided any columns to keep these transaction numbers.

#### *8.9.3 Database Transactions*

Database transactions such as insert, update, and delete have been used in OBAAS when users interact with them. For example, when the user wants to create a new account in OBAAS, an insert operation takes place in the NEWACCOUNT table so that a new record can be created in this table for the user. Similarly, when the user wants to pay a bill, the NEWACCOUNT table is updated. The biller's account is credited and the user's account is debited by the amount of the bill that needs to be paid.

We have used select, insert, update, and delete transactions using the DML language. We have also used create and delete transactions using the DDL language.

#### *8.9.4 Future Enhancements*

Currently, we need to populate two fields on the user interface: "service type" and "list of billers." We are populating them using hard-coded values. In the case of "list of billers," we are manually creating a record in the NEWACCOUNT table for billers with a value in the CATEGORY field as B. For customer accounts, the value for the CATEGORY field is U.

For populating the values in the service request–type field on the user interface, the hard-coded value of the cheque book request is set.

In the future, tables can be created in the database to populate the values for the service request types. Similarly, the user interface can be created for creating records for billers so that these records do not need to be created manually.

### 8.10 Chapter Summary

Databases are an important component of software products. Databases are used to store permanent data that are used by a software product. A database forms the back end of an *n*-tier architecture of any software product. Although many types of databases have been created, relational databases are the most successful and widely used.

You need to create a database schema before you can create the database entities. Database design consists of creating the ER diagrams. An ER diagram depicts the entities and their relationships. Once the ER diagram is designed, you can create entities such as tables, indexes, and sequences using DDL. You also will have to create relationships using primary and foreign keys.

To ensure that there are no redundant and inconsistent data being generated in the database, you need to follow some rules. These are known as normalization rules. The data stored in a database can be in the first normal form, second normal form, or third normal form. The first normal form depends on whether a list of values is getting populated in one field of a column of a table. The second normal form depends on whether all the data being populated in all the columns are already in the first normal form and each non–key column is dependent on the primary key. The third normal form depends on whether all the data being populated in all the columns are already in the second normal form and the non–key columns are independent of each other.

A database can be queried or updated using DML. Some of the DML statements include insert statements, delete statements, and update statements.

## QUESTIONS

1. What is a database?
2. Why are databases important?
3. What factors are responsible for making a database powerful?
4. What is a NoSQL database?
5. What is a relational database?
6. What is referential integrity?
7. How can you implement referential integrity?
8. Elaborate the problems if referential integrity is not present between some related tables.
9. What is a primary key?
10. What is a foreign key?
11. What is a sequence in databases? How can you get the values for a sequence in the databases?
12. What is an index in databases? Why is an index important?
13. Describe cases where indexes can create problems in databases.
14. What is DDL?
15. What is DML?
16. What is a stored procedure?
17. What are master data? Provide an example.
18. What are transaction data? Provide an example.
19. Why are master and transaction data important for database design?

## Recommended Reading

C. J. Date (2012), *Database Design & Relational Theory*, 1st Edition, O'Reilly, Sebastopol, CA.

Jan Harrington (2002), *Relational Database Design Clearly Explained*, 2nd Edition, Academic Press, San Diego, CA.

Sam Lightstone, Toby Teorey, Tom Nadeau, Morgan Kaufman (2007), *Physical Database Design*, 1st Edition, Burlington, MA.

Gavin Powell (2005), *Beginning Database Design, 1st Edition*, John & Wiley Sons, New Jersey.

# 9

# SOFTWARE TESTING (VERIFICATION AND VALIDATION)

**In Chapter 8, we learned**

- **What a database is**
- **What a relational database is**
- **What the elements of a database are**
- **How to make database design**
- **What an entity-relationship diagram is**
- **How to create database table relationships**
- **How to ensure the integrity of a database**

**In Chapter 9, we will learn**

- **What software testing is**
- **What software verification and validation is**
- **What the levels of testing are**
- **What unit testing and integration testing are**
- **What system testing and user acceptance testing are**
- **What functional testing and nonfunctional testing are**
- **What alpha testing and beta testing are**
- **How testing can be done in maintenance and production**
- **What the steps involved in testing are**

## 9.1 Introduction

Software testing is important because you need to ensure that the software product you are building is defect-free. A defect-ridden software product will result in either the users not being able to use the product or the users not being able to use it productively.

A defect-free software product is important. To make a software product defect-free, it should be tested thoroughly. All the defects that are found (through testing) should be removed. In software testing, the software product is checked and evaluated to see whether the software product is giving outputs correctly. If these outputs do not match the expected result, then it is a software defect. This software defect should be fixed so that the software product becomes defect-free.

Although a 100% defect-free software product is desired, there are some practical limitations in achieving this goal. Software testing is mostly a manual and effort-sensitive activity. There are time limits within which a software product must be released for use. Since software testing requires effort, there is a cost factor involved. Thus, there are always cost limits within which testing activity needs to be completed. Because of these practical limitations, there is always a compromise between budget, time, and the quality of testing. Thus, the project stakeholders try to maintain a balance between these parameters. A reasonably good quality software product then becomes a priority in any software development project.

During testing, a software product is tested to make sure that it conforms to its functional and nonfunctional requirements. Functional requirements are related to the features of the software product. Nonfunctional aspects of a software product include safety, security, performance, and portability.

Testing activities include test case writing, test case execution, defect logging, defect tracking, defect fixing verification, and defect closing. Additional testing activities include test bed preparation and test case automation.

Testing activities need to be managed. Testing management activities include test planning and monitoring. Testing management also includes test result reporting.

**Usage notes:** Many activities related to software testing overlap with each other; therefore, it is difficult to categorize all these activities. Because of this difficulty, this chapter is divided into different sections to ensure that the overlapping areas are covered without creating any confusion. The following is an outline on different testing types and the sections in which they are discussed.

- Verification and validation: Verification includes code reviews, design reviews, and requirement specification reviews. Verification is discussed in Section 9.6.
- Testing levels: Validation is the same as testing levels; therefore, it will not be duplicated in the verification and validation section. In testing levels, we will discuss unit testing, integration testing, system testing, and user acceptance testing. Testing levels is briefly explained in Section 9.5. Unit testing, integration testing, system testing, and user acceptance testing are explained in Sections 9.7 through 9.10, respectively.
- Other important tests: Here, we discuss other testing types, such as sanity testing and smoke testing, and special testing types, such as alpha testing and beta testing. Such tests are discussed in Section 9.11.
- Testing techniques: Here, we discuss boundary value analysis, equivalence partitioning, decision table testing, and so on. These are further examined in Section 9.12.
- Testing life cycle: Here, we discuss test case design, test execution, defect life cycle, and so on. Test case design is tackled in Section 9.12. Test preparation is examined in Section 9.13. Test execution and defect life cycle are discussed in Section 9.14.

### 9.2 Software Testing and Software Engineering Methodologies

In the Waterfall model, software testing has been traditionally relegated to carry out only integration and system testing (explained later in this chapter). This is because only after the source code has been written and the developers have handed over the software product to software testers can software testing be done. At this stage, software testers can only perform integration and system testing.

When the benefits of software testing were realized, software testing was introduced and conducted throughout the life cycle of software development. Because of this, software design specifications and source codes were verified when they were ready for inspection. Later, complete software testing, including unit testing, integration testing, and system testing, was carried out. These kinds of testing ensured that no defects could pass right from the stage when software design gets started to the rest of the stages of software development. Because software testing is only about testing running software (i.e., a piece of source code that can be executed), new terminology was used for testing the items such as design specifications in which there is no running software. The entire gamut of activities became known as verification and validation.

In verification, the source code is not executed. In addition, testing the items such as design specifications is also part of verification. In validation, the source code is executed. Verification and validation are also known as static and dynamic testing, respectively. Verification and validation are explained later in this chapter.

On agile projects, testing activities are performed differently from Waterfall model projects. For example, in eXtreme Programming (XP), which is based on agile methodology, test first development (test-driven development) is carried out, as explained in Chapter 2.

In agile and spiral software development methodologies, software products are developed using incremental integration so that new features are added to the existing software products continuously. This process poses a grave risk because the existing functionalities of the product may be broken in the new versions. This happens because when new features are added to the existing software product, many times, the existing source code will be changed to accommodate the new features. In fact, sometimes a lot of existing source codes are rewritten to create a new version because the old source codes were not scalable to allow the addition of new features. Similarly, any old source code can be refactored to make it scalable for the addition of new features. This results in changes to the old (i.e., existing) source code of the product. In these scenarios, once the new features are added, many old features may not work at all or may work very differently from what they are supposed to do. It is important that the old features continue to function in the same way for the existing customers of the software product. Because of this, during software testing, old features are also tested when the new software product is released or when new features are added to the existing product. This kind of testing is known as regression testing.

There are more differences between software testing in agile projects and Waterfall model projects. As explained in Chapter 2, pair programming is used in XP. Pair programming ensures that while one software developer writes the source code, the other developer checks the source code for errors in implementing business logic. This ensures that the software product is defect-free. Since continuous integration of the source code is carried out in most agile projects, smoke testing is also carried out on a regular basis to ensure that integration errors can be found and fixed immediately (smoke testing is explained later in this chapter). In essence, software testing plays a central role in agile projects.

### 9.3 Introduction to Different Types of Software Testing

Different software testing types are needed to test the functionality of a piece of software (or any other software artifact). Software systems are prone to hacking and data theft because, generally, they are connected to the Internet. Therefore, a software system must have foolproof security to prevent hacking and data theft. Similarly, a software product should be dependable enough so that the user can reliably use it without facing problems such as crashing, receiving wrong output, and monetary losses owing to wrong calculations related to financial activities.

The fundamental question in software testing is how testing is performed and on what artifact. As discussed in Chapter 2, during software development and maintenance, we have several artifacts such as requirement specifications, software design, software source code, user documentation, and the software in the production environment. Testing of all these artifacts must be carried out to ensure consistent quality for the finished software products.

Let us get acquainted with some major categories of tests. Figure 9.1 presents several types of tests, their levels, hierarchies, and some of their relationships to each other. It also shows the levels at which these tests are performed and the verification and validation activities. We will learn about testing activities in latter sections of this chapter.

### 9.4 Introduction to Verification and Validation

Traditionally, software testing has been restricted to testing a software product dynamically. That is, while the binary code of the software product is running, some input values are provided (either through the user interface or through some integration points of that product) and the output values are checked to see if the product is working as per the specifications. This is known as dynamic testing. However, now testing is done for all the artifacts (not just for running binary code) that are generated during the entire software development process. Dynamic testing of some artifacts such as requirement specifications is not possible because they cannot run. You can test them only in their static condition.

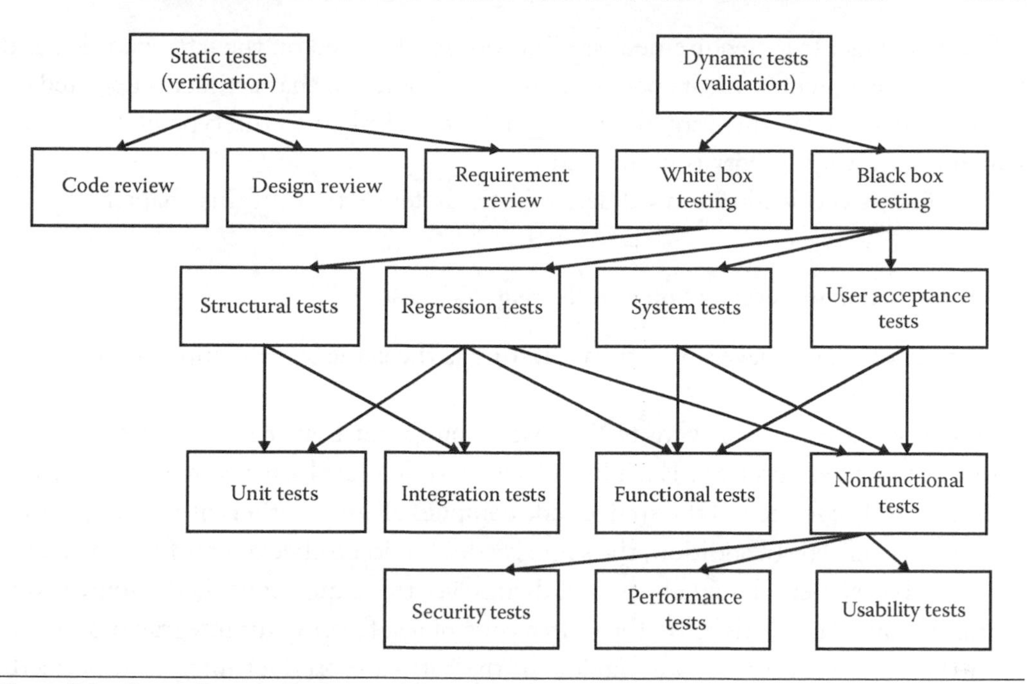


**Figure 9.1** Some software testing types and their relationships.

Since software "testing" is not the appropriate term for all these dynamic testing and static testing activities, a new term has been coined by Barry Boehm. It is known as verification and validation. All the tests that are carried out under static conditions are known as verification, and all the tests that are carried out under dynamic conditions are known as validation. Some people also refer to these terms using statements like "Are we building the right product?" and "Are we building the product right?" When we conduct a test to address the first question, it is verification, and when we conduct a test for the second question, we are doing validation. The complete verification and validation processes are depicted in Figure 9.2. In this figure, in each box, a

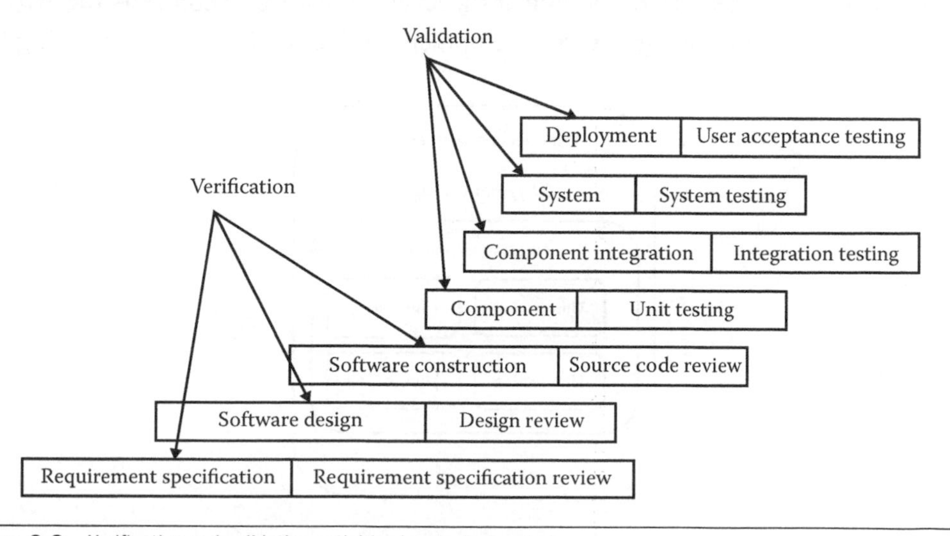


**Figure 9.2** Verification and validation activities in a typical project.

software artifact (e.g., requirement specification) is depicted on the left-hand side and the testing (e.g., requirement specification review) done on that artifact is depicted on the right-hand side. This way, the testing activity is linked to the type of artifact on which that testing activity is performed.

We discuss verification and validation in the latter sections of this chapter.

## 9.5 Introduction to Levels of Software Testing (Validation)

Performing different levels of software testing is the same as validation, as envisaged by Barry Boehm.

When we think of software testing, we often forget that it is the most profound activity in software projects. Each line of source code must be tested to determine if it is syntactically correct and the source code compiles cleanly. Otherwise, a compilation error will be raised. In addition, the source code should be checked for functional correctness. Developers should always ask themselves these questions: Is the source code doing the calculations right? Is the source code of a software unit integrated perfectly with the other units? Is a code module of the software product integrated perfectly with the other modules? Is the entire system working as per the requirement specifications? Here, we see that the software needs to be tested at various levels, including when the source code is written.

As shown in Figure 9.3, software testing can be done at many levels. Different types of tests are conducted at different levels. Unit testing, integration testing, system testing, and user acceptance testing are all part of "levels of software testing" (validation) and they are discussed later in this chapter.

### *9.5.1 V Model*

The V model is another way of looking at the levels of testing. It connects the artifacts that are developed during a software development project with the level of testing that

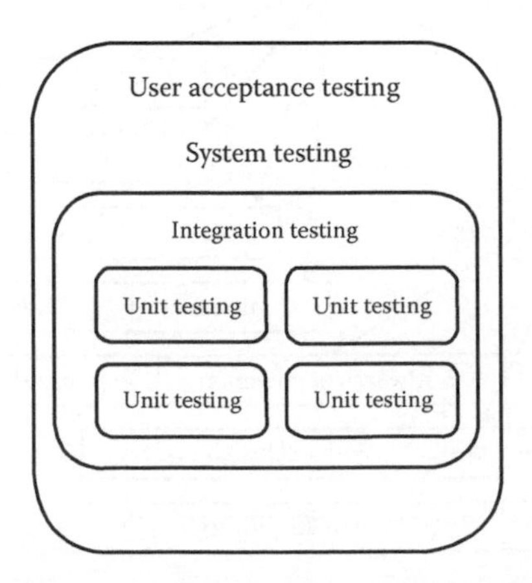


**Figure 9.3** Levels of software testing.

is performed on those artifacts. The V model is used to capture the essence of system testing and user acceptance testing. Different levels and types of testing that are carried out during these testing activities are best depicted by the V model developed by Barry Boehm.

The V model relates the major phases in the software development life cycle with the validation that needs to be done on the artifacts that are developed in those major phases. Unfortunately, the V model does not cover the verification part of testing.

The V model is not based on time-phased life cycle activities. Instead, it only relates the validation activities to be done on a phase after that phase is completed. Figure 9.4 depicts the V model. This figure indicates that user acceptance testing (the last activity in the software development life cycle) is done based on the requirement specifications because user acceptance testing and requirement specification are at the same level in the figure. As shown in the figure, system testing is done based on the software design because these two are also at the same level. Similarly, the other activities that are at the same level are associated with each other.

If we observe closely, we can see that the validation activities maintain a reverse chronological relation with the phases of the software development life cycle. In Figure 9.4, you can see two long arrows, one of which is pointing down and the other pointing up. These arrows indicate the chronological order in which the phases and validation activities are carried out in the software development life cycle. Even here, this chronological order is not followed strictly for unit and integration testing because, in this figure, unit testing occurs after integration testing. Still, the V model

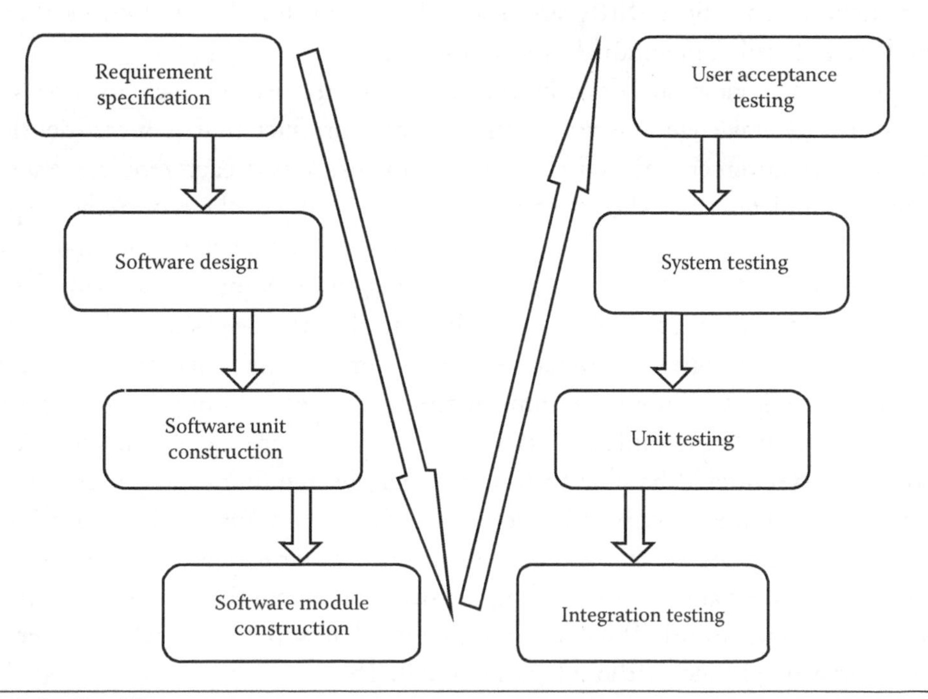


**Figure 9.4** V model.

is important because it establishes a relationship between validation activities and phases in the software development life cycle.

## 9.6 Verification

Verification is also known as static testing because an artifact is tested in its static condition but not in the running condition. Now, we will discuss different kinds of tests available in static testing.

Verification activities are mostly formal processes because they are performed (on different artifacts) by people who are different from those who developed these artifacts. Verification activities include requirement reviews, design reviews, code walkthroughs, code inspections, and code reviews.

The people involved in verification processes include project managers, peers (e.g., software developers from other teams), software developers, designers, and business analysts. The person who developed the artifact will present it to other people for verification. For example, when a software design needs to be verified, the software designer will present it to the other members of the team. Verification activities involve either formal or informal meetings.

### *9.6.1 Requirement Reviews*

Requirements are important; thus, they need to be verified thoroughly as to what the software product, once it is built, should do and not do. When all the software requirement specifications (SRS) are ready, the testing team should check them for testability, ambiguity, maintainability, and so on.

Requirement reviews are done by the testing team. Testing team members, who have business knowledge relevant to the software product that is being developed, perform the requirement reviews. Their task is to ensure that each requirement specification is complete. There should be no ambiguity in any of the requirement specifications. For example, if a requirement specification states that the performance of the software product should be adequate, then this is an ambiguous statement. Instead, the requirement specification should clearly state that the user should be able to get a response from the software product within 5 s (or a similar concrete measurement) after the user clicks a Submit button on the user screen. Similarly, a requirement specification should be testable so that when the software product is ready to be tested, a software tester should be able to test it as per the requirement specification. For example, a requirement specification states that the user should be able to log into the software product using a username and a password. However, the requirement specification does not state anything regarding the situation where the user does not provide a username or a password. Thus, it is not known what will happen when a user operates the software product without logging into it. This condition cannot be tested. The requirement specification must state all the preconditions and all the states in which the software product can be used.

Maintainability of a requirement specification should also be considered when the SRS document is made. For example, a requirement specification should indicate the software features that may be required in the future. This will help the development team create a software design that will allow future extension if required.

### *9.6.2 Design Reviews*

Once you have a good understanding of the SRS, the software design should be checked to see if it corresponds to those requirements and that the software product is capable as per the requirements. The software design should also be reviewed for the feasibility of constructing the software code (i.e., source code) and to see if the design provides a good structure. A software structure is difficult to change once it is made. Changing the software structure is time-consuming and costly after it has been developed. Hence, a good software structure should be made. For example, designing large classes with many methods and class members results in bad software design. This kind of software structure should be avoided. A good software structure should also have openness so that extensions can be made in the future if required. Software testers can find out if all these aspects about software design are in the software design.

Care should be taken to determine that the software design is not adding any unwanted feature to the software product and that the product is made for the required functionality only. At the same time, it should also be checked to ensure that all the requirements are translated correctly into the software design and no software requirement has been designed incorrectly. In other words, the design review should make sure that there is a one-to-one mapping between the software requirements and the software design.

### *9.6.3 Code Walkthroughs*

After a piece of software code is finished, it should be checked formally by the project team to ensure that it is serving its purpose as per the software design. The source code should also be checked for maintainability. If the source code is well structured, simple, well documented, and scalable, then it can be easily maintained. Code walkthroughs should also be used for checking the logic in the source code for its soundness.

Code walkthroughs are done in a team meeting where the software engineer who has written the source code shows and explains the code to the reviewers (other software engineers and the project manager). This is an informal meeting and the code walkthroughs are done every time a unit of source code is completed by a software engineer.

### *9.6.4 Code Inspection/Code Review*

Code inspection is an informal method of checking the source code for errors. Code inspection can be done manually or by using some tools such as debuggers. Some of the errors that can be trapped inside the source code include dead codes, unused variables, logic errors, and assignment of wrong values to the variables. Code inspections can be done informally by the developer himself or herself, along with some other developers, to check logic and scalability. Inspections also check whether the code reflects the software design well. When we refer to the term *code review*, it generally means that it is carried out formally by people other than the developer.

Code inspection appears to be similar to debugging, but it is different. Debugging involves fixing the source code for syntactic errors so that the source code compiles cleanly, but code inspection involves finding the logic errors as well as the source code that never runs (dead code).

## 9.7 Unit Testing

Unit testing is part of validation. It is done by the software developers or software testers. Unit testing is considered part of the software development life cycle; thus, it is not included in the software testing life cycle. It is also difficult to include unit testing in the testing life cycle because it is difficult to separate unit testing from software development.

When the developers write their code, they are very confident that the code is correct. In reality, the source code they write may have some defects. The reasons are many. Let us look at some of them.

- **Design clarity issues:** Even if the software design is good and contains no perceived defects, the developer may interpret the design differently from what it is intended to do.
- **Incomplete logic:** The design logic should be captured completely in the source code. Sometimes, it so happens that some parts of the design are not captured in the source code.
- **Unintended errors:** A developer may be thinking right when he interprets the design correctly, but when he actually writes the code, he does it wrong because of his oversight or carelessness.

Once the developers begin the process of developing a source code, they usually start checking it by debugging. Once the code is debugged completely, it may compile without problems. At this stage, the code is free from compilation errors only. What about logic errors, calculation errors, and so on? The code may not be error-free in such cases. In fact, these errors are defects that need to be removed through unit testing. How do the developers perform their unit testing?

In procedural programming languages, a unit of source code is a program containing one or more functions or procedures. In today's world of object-oriented programming, a unit generally may correspond to a class. When a developer writes a class, he or she checks for errors as per the software design. When all the classes are checked and verified at the unit level, all the internal errors such as logic errors and calculation errors are checked and rectified if any such errors are found. In some cases, when a class is too big, the class can be broken down into many parts to create many unit tests for each of those parts. Generally, the methods insidc a class are tested because they contain the business logic. If a class is too big, then it contains many methods; thus, you need to separate these methods and create unit tests for each set of methods.

Although the name of the testing does not imply it, unit testing is all about testing the business logic of your source code (Figure 9.5). While you debug, you are checking if your code has any compilation problems. If your code has no compilation problems, then the next step is to check your business logic. If your business logic in the class is correct, then you do not have to worry about business logic problems afterwards. All business logic problems must be caught and repaired at the unit testing level. They should not go unchecked into a later stage in the software development life cycle. This is very important. If you are able to implement your business logic correctly and it is unit tested thoroughly, then your source code will never have any business logic defects that can go undetected downstream. Hence, unit testing is very important!

Tcst your code for both negative and positive results. The negative results will make sure that your unit of source code works by trapping the error conditions. If you have not provided a mechanism in your source code to trap all these user- and system-generated errors, then your code will fail (the system may crash or throw exceptions). Thus, negative testing is directly related to handling exception errors.

To ensure that there are no defects in the software product, testing should start even before the developers start writing their source code. In object-oriented programming, when a developer writes a class, a test case should be written to test that class and its methods. This level of testing is known as unit testing because it is done at a unit (class) of source code. Only when all the unit tests are done and a piece of source code has successfully passed such tests should a developer integrate that piece of source code with the software build where all the fresh source code is integrated (software builds are explained in Chapter 12).

A unit test should check the business logic that is implemented in the methods inside the classes. A method does some computation and provides some output based on the inputs it receives either from the user or from any other software program

Unit testing = testing business logic

**Figure 9.5** Unit testing is all about testing the business logic.

unit. In unit tests, input values from boundary conditions should be provided and output values should be checked to verify if the method is performing the correct computation.

Suppose a method is created to compute the sum of two integers and these integers should have values within the range of 1 to 100. Then, in unit testing, you should create test cases that will provide inputs in the range of less than 1 (e.g., –3) to more than 100 (e.g., 105). These two test cases will give negative results (the tests will fail) when they are run against the method that you are testing. Thus, these test cases are known as negative test cases. You should also create a test case that will provide values within the range (e.g., 48). This test case should pass because it will meet the criteria of the business logic. This test case is known as a positive test case. When you run this test case and if this test case passes, then the business logic implemented in your method is fine for this test case. If this test case fails, then your business logic implementation is wrong and you need to fix it.

You can write your test case manually. However, writing and maintaining test cases manually is a laborious task. Nowadays, tools are available to help create, maintain, and automatically run test cases. When you maintain your source code (e.g., by refactoring), you can use the suite of test cases (that were written earlier) to perform regression testing (we will learn about regression testing later). Unit test cases (saved in a repository) should be run each time there is some change in the design. Since these unit test cases are executable pieces of code, they can be run automatically.

Let us understand how unit testing works by considering an example.

Suppose we have a business scenario where we need to add two integers. As mentioned earlier, the range under which this addition is done should be 1 to 100. If any of the integers is either less than 1 or more than 100, then the system should throw an exception error. Hence, we can have a class like below:

```
class addition {
        public boolean adding (integer one, integer two){
        if one > = 1 and one < = 100 and two > = 1 and two < = 100 {
        integer x = one + two;
        print (x);
        return true;
        }
        else
        return false;
        }
}
```

To test this class, we need to create a test class. The test class will have a test method. Generally, a test method will test one target method. However, if the target method is too big, then you can end up creating many test methods to test several parts of that target method.

```
class testAddition {
      public method void testAdding () {
      boolean x = addition.adding(-3, 5);
      }
}
```

The above method is a negative test case because it is checking the values outside the boundary of the acceptable limit (because -3 is less than 1). Therefore, that test case should fail. The simple addition operation we are doing here can be tested with just three test cases. The other test case can test with a right answer to the addition operation by considering 4 and 78. In the final test, you can test the method with values beyond the test range (e.g., 7 and 109). When we have to test the business logic where the correct answer could be within a range, we have to create at least three test cases. One test case will check the values on the lower end of the range and another will do so on the higher end of the range. Both of these will be negative test cases and they should fail. The positive test case will check with values within the range and should pass.

This way of constructing test cases to test a software unit (e.g., a class) is known as boundary value testing. Partitioning the test cases into different types of test cases (e.g., positive and negative test cases) is known as equivalence partitioning. Analyzing the scenarios for designing test cases in this manner is known as boundary value analysis. Equivalence partition and boundary value analysis are further explained in Sections 9.12.2.2 and 9.12.2.3, respectively. We have depicted how to find the minimum number of required test cases in a simple scenario. For complex scenarios, there are formulas to determine this minimum number. Further discussion on these techniques is beyond the scope of this book. You can read a good book on software testing to find out more about these techniques.

The earlier example is about testing a class that is correctly implemented. What is given in the design document is to perform the addition of two numbers if both are within the range or throw an exception otherwise.

Now, let us see a case where the software developers made a mistake in implementing what the software design prescribed. Suppose the design document stated that a calculation is required for a fixed deposit (known as a certificate of deposit in North America) in a bank account at an interest rate of 8.5%. A software developer wrote the following piece of source code:

```
public class interest_compute {
      public integer interest_compute (float interest) {
      integer principal = 293;
      integer accrued_amount;
      accrued_amount = principal + principal x interest/100;
      return accrued_amount;
      }
}
```

If you call this method in your test class and run it, it will not give you compilation errors. However, suppose you give it an input of 8.5; it will not give any compilation errors but it will give you wrong test results. Why? This is because the method returns only an integer but not a float; therefore, all decimal place values will be truncated. This clearly shows that the software developer either made a mistake in understanding the design or did it unintentionally. If you are testing this piece of code, then you should be able to find these kinds of errors so that the software developer can correct his or her code. The correct solution should be that the variable accrued_amount is declared as a float type but not an integer.

Here, we have provided some simple examples to learn unit testing. However, on real projects, there will be many distracting things that may entangle a class that is being tested. In those cases, you may need to take the class out from the entangling environment by porting (copying) it and putting it inside a testing tool environment using some unit testing tools. A technique known as mocking is used to provide an environment that will bring the class out for testing. The class will be imported in the testing tool and some simulated environment will be created for testing. For example, if the class being tested interacts with a database in one of its methods, then a database connection will be simulated so that the class can be easily tested.

In cases where a method is calling other methods, testing becomes more complex. In that case, you may need to find out what those methods are doing and, on the basis of the computation taking place inside those methods, you can create your test cases.

However, as mentioned previously, creating and maintaining unit tests manually is a time-consuming exercise. Unit testing should be done using some tools that are available.

### *9.7.1 Database Unit Testing*

So far, we have seen how business logic is tested and how we can ensure that it works. However, how will you test a business logic that is doing some manipulation in a connected database? Doing so entails many challenges:

- A database is a separate entity and is available only through a connection. If this connection is broken or the database is not running, then you cannot perform database testing. This means that, when you are performing unit testing, first, you need to do integration testing. This integration testing is in the form of testing the database connection first and ensuring that this connection works.
- The data in a database are permanent. If you manipulate some data in the database, then the changes that occur in the database are permanent. This is contrary to the notion of unit testing where the state of the data inside a data structure does not change after you perform your unit testing. In the case of non-databases (e.g., variables used in writing business logic to

hold some data temporarily), the data inside a data structure are temporary because they are stored in the primary memory. Once the program terminates, the value assigned to a variable also gets lost because the value of the variable is reset. However, if you update the value of the data in a database, then this change of value is permanent and you cannot replicate the testing. This can be best understood by a delete operation example for a database. Suppose you have a customer "Alice" in a customer database and you deleted the row containing the record having the customer name Alice during your unit testing. Now, if you want to run the same test script (a test script is an executable piece of source code that, when invoked, will perform some test on the software program under test) again, your unit test will fail. Why? This is because, in your test script, you mentioned this customer record with customer name Alice to be deleted from the database but this record was already deleted in the previous test. Hence, this record no longer exists in the database; thus, if you run your test script, it will result in an error.

Now that you realize the challenges of unit testing a database-connected software program, let us find ways to tackle them.

Suppose we have the following class with a method that accesses a database and does some data manipulation in that database.

```
class data_access {
      public delete_row () {
      string y = 'Alice';
      integer x;
      search in the database table and find the record number where
customer name is Alice and assign it to x;
      if x = record number of Alice then
      delete row from customer table where customer name = y;
      }
}
```

There are two ways to test a class like this. One way is that you do not connect to a database at all and just mock the database connection. There are tools available that can mock a database connection and provide you with a test result. However, this approach does not actually test a database connection and verify if the business logic actually works against a real database. What if your business logic is correct but the database is not implemented correctly? It can happen that the column and table names you provide in your business logic may be different from what is actually used in the database.

This discussion also leads to one interesting aspect. Not only are you unit testing your business logic, but also you are testing the database structure. Thus, this kind of unit testing can be called an extended unit testing.

Testing the business logic against a real database is important because there is no other test available to actually find out whether the business logic works against the connected database.

So, let us find a framework that can allow us to test our classes and methods that manipulate a database.

- The first thing is to test the connection with the database. If it works, only then should you perform unit testing. If not, fix the connection first. Thus, a database connection test should be part of the unit testing.
- Unit testing should create its own test data. Creation of test data is important because you cannot manipulate some existing data.
- Unit testing should delete the test data from the database after completion of testing. This will ensure that the database is not cluttered with useless data after testing.

### 9.8 Integration Testing

Integration testing is part of validation. It is done by the software developers or software testers. Integration testing is considered as part of the software development life cycle; thus, it is not included in the software testing life cycle. It is also difficult to include this testing in the testing life cycle because of the difficulty in separating it from software development.

Once the units of the source code are complete and they are unit tested, they need to be integrated either with the main build of the application or with a branch build where the related units will be integrated. Several issues may arise when one tries to integrate many units of the source code with each other. The interfaces may not be compatible with each other; thus, the components may not be integrated. Integration of different units can happen only when the interfaces are compatible. Interfaces can be incompatible for many reasons. Nowadays, a large number of applications are developed by geographically distributed teams. Each team develops its own piece of application in isolation to the other teams. This leads to some variations in the way the components of the application are implemented at different locations. Similarly, each developer, even inside the same team, implements the units of source code in his or her own way. This, again, causes differences in the implementation of different components. By testing each component for integration, before these components are actually integrated, we can make sure that they will in fact be integrated without problems.

Integration testing = testing interfaces

**Figure 9.6** Integration testing is all about testing the interfaces of the class.

Unit testing ensures that the business logic is implemented correctly in each class. Unit-tested classes now need to be integrated with other classes that are present in the software build. Integration testing is done to ensure that your unit-tested classes integrate properly with other classes. Integration of classes happens through their interfaces. A class may have methods that are called inside a method contained inside another class. If the parameters of the called method do not match the method that called it, then there will be integration issues between these two classes. Using integration testing, it is possible to determine whether the interfaces of two classes are compatible with each other. Thus, integration testing is all about testing the interfaces. Figure 9.6 depicts this reality.

For example, if you look at the class interest_compute in Section 9.7, you can find that it has the interest_compute method. This method takes a float parameter and returns an integer value. During unit testing, we found that this was giving wrong results. Hence, it was corrected by changing the variable type of the computed result from integer to float. After correction, it takes a float parameter and returns a float value as well. Now, you can create another method where you call this class. Here is the called class:

```
public class show_accrued_amount {
      float int_rate = 8.5;
      integer x = interest_compute. interest_compute(int_rate);
      print (x);
      }
}
```

If you test this class, it will give you wrong results. This is because you called the interest_compute method from the interest_compute class with a wrong data type (i.e., integer *x*). Thus, the results will be wrong because the result (fractional part) is truncated.

There could be many ways in which calling a method from a different class will go wrong. In integration testing, you need to test all possible scenarios.

In the previous example, although the business logic is wrongly implemented during integration (method call), the class will not create any compilation error. However, in many other cases, where integration is not done correctly, there will be compilation errors. For example, if we pass a string data–type parameter to the interest_compute method, then there will be a compilation error. In this case, after integration, the software build will be broken. In most integration testing, if the integration is not proper, then a compilation error can occur.

There are three main methods of integration, namely, big bang, top-down, and bottom-up. As the name indicates, the big bang approach tries to integrate all the modules or components of the system in one step. This way, a lot of time can be saved. The main drawback with this approach is that, even after conducting the unit testing of all the components that need to be integrated, there could be some minor differences in the interfaces, and these differences may prevent the integration from happening. It is also difficult to find the real cause of integration failure at such an aggregate level. This leads to numerous problems in performing integration and, in turn, in conducting integration testing.

Top-down integration testing is used when the top-level modules of the system are integrated and tested first. Later, the internal parts of these modules are integrated with each other and subsequently tested. This kind of testing is employed when there are many branches and deep hierarchies inside those top modules. When integration testing is done, each branch and hierarchy is checked for integration errors.

Bottom-up integration is the most commonly used integration methodology (Figure 9.7). Development starts by writing the source code for the most bottom-level units. When they are ready, they are integrated with each other and also tested for any integration errors. When this level of integration is completed, the higher-level integration is done. The bottom-up integration method is used when the continuous integration method (explained later in this chapter) is used for software development. Integration can be done in many ways depending on the frequency with which the software code is checked in the main build.

The frequency of integration of the source code and subsequent testing can vary widely (Figure 9.8). At one extreme, integration is done very infrequently. Sometimes, it could be months since the last integration of new (fresh) source code with the main build. In some projects, integration is done nightly.

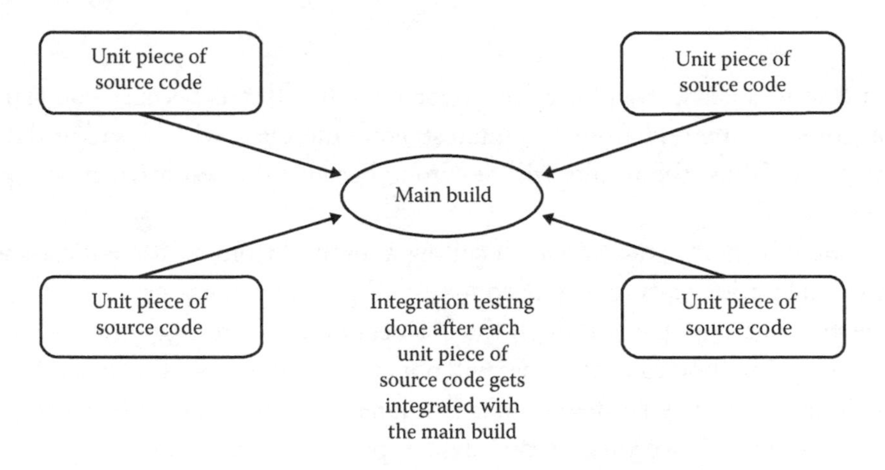


**Figure 9.7** Bottom-up and continuous integration testing.

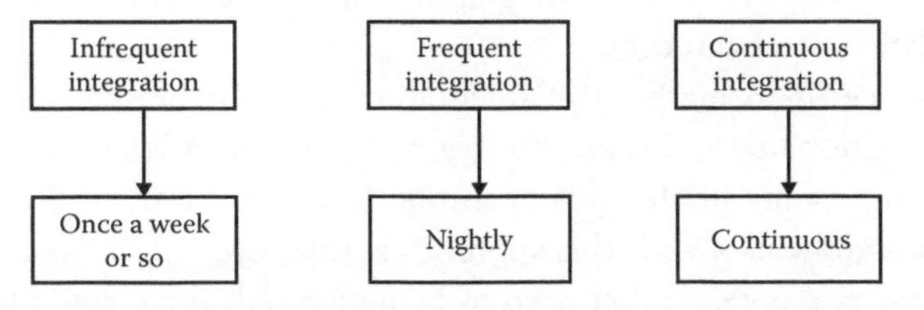


**Figure 9.8** Frequency of integration and integration testing.

At the far end of this continuum, integration is done continuously in real time (which is the most common method of integration). In continuous integration, whenever a developer finishes the unit testing of his or her code, he or she immediately checks in (i.e., submits) his or her code to the main build. Integration testing is done immediately to check whether the (existing) build is intact or whether it has been broken by the addition of that fresh source code. If the build has been broken, then it is the responsibility of the developer, whose code has broken the build, to fix his or her code and check it in again and perform integration testing again. He or she has to repeat this process until his or her code succeeds in making a successful build.

As you can imagine, all the scenarios described above belong to both bottom-up and top-down integrations. In real life, it can be a mix of both types of integration. For example, generally, a developer creates a class and integrates it with the software build. This is the bottom-up way of designing a software product. However, in many instances, a developer may need to create a class interface first and then create the concrete classes from this class interface. This is an example of top-down integration. We have learned about software components and class interfaces in Chapters 5 and 7. See both chapters to better understand these subjects.

There are many tools (from an open source to proprietary vendors) available in the market that can do integration testing automatically. Whenever fresh code is checked in the main build, the tool runs to check the integration. If the integration fails, then the concerned people are informed automatically by e-mail.

Strictly speaking, integration testing is not done explicitly. When you integrate your code with the main build of the software product and find that there are compile issues, only then will you know that there are integration issues with your newly built code. Integration of classes with each other is mostly about method calls. If a class that receives a method call does not provide a proper interface, then you will have integration issues and your code will not compile with the main build. The problem could be attributed to a wrong parameter data type, the wrong size of data, and so on. Hence, such types of integration issues are essentially attributed to interface problems.

In some cases, where a class has many method calls, it becomes difficult to determine the integration issues if they arise. In those cases, the class can be broken down into smaller classes. Each class can then have one method to test; thus, testing becomes easier.

Integration testing is purely technical in nature. It does not test the business logic. Business logic is always tested during unit testing.

Test-driven development (i.e., test first development) is a great environment to ensure that your software product does not suffer from business logic problems because all the business logic issues are trapped at the most minute level (i.e., class level). See Chapter 2 for details on test-driven development. Continuous integration and continuous integration testing, on the other hand, ensure that you never have to worry about integration issues. This is because integration issues in a continuous integration environment are handled at the most minute level (i.e., during the integration of a class). If integration is done at a higher level, then there will be many constituents (i.e., many classes) involved during the integration. There could be many errors at one time when the integration happens at a higher level. In that case, going through and fixing all those errors will be a problem and will take a lot of time.

## 9.9 System Testing

System testing is part of validation. It is also known as black box testing because the source code is not used or required for this testing. This testing is done by running the executable code of the software product. It is done by the software testers.

System testing is actually a group of various testing types done together to check various aspects of the software product. For example, functional testing is carried out to check the business logic aspects of the software product. On the other hand, nonfunctional testing is done to check aspects such as performance, security, and usability of the software product.

During system testing, we check if the software works as per the requirements and design specifications. We do system tests on the application (system) alone if it does not need to be integrated with other software or hardware systems. However, if some integration with external system(s) is needed, then we carry out system testing after such integration with these systems, in which case there could also be separate integration specification documents for such external systems. If this is the case, then these documents also become the basis for our testing.

At the system testing level, it is impossible to perform any testing at the source code level. This is because, at this level, the entire source code is already integrated. Therefore, working with that integrated source code to do the testing will be cumbersome and difficult. At the same time, the goal of testing at the system level is to test the software product to ensure that it works as per the software design and requirement specifications. Hence, at this level, testing the source code becomes irrelevant. All the testing is carried out at the executable binary code (machine code) level. Testing is done by running the system using the user interface commands available at different screens (i.e., user screens). All the testing at this level treats the system as one big black box. This is why all the testing conducted at the system level is also known as black box testing.

At the system testing level, a lot of tests are performed. Some of the important tests include graphical user interface testing, performance testing, system integration testing, security testing, usability testing, error handling testing, compatibility testing, smoke testing, sanity testing, regression testing, and reliability testing. These tests can be divided among many testing types. All of these tests can be broadly categorized as functional and nonfunctional tests.

### *9.9.1 Functional Testing*

Functional testing is part of system testing, user acceptance testing, and regression testing (Figure 9.1). Through functional testing, the features of the software product are checked against the functional specifications mentioned in the requirement specifications by running the software product that is under test. Functional testing is always done by running the executable binary code (compiled code) of the software product. It is never performed at the source code level. This is mainly because testing is done to validate the software design or the requirement specifications and not to validate the software construction aspects.

All functional testing is carried out by testing engineers.

### *9.9.2 Nonfunctional Tests*

We provided some details on nonfunctional requirements in Chapter 4. Nonfunctional requirements should also be tested for a software product. These special tests are known as nonfunctional tests. Among the nonfunctional tests, the tests that seek to validate security measures are known as security tests. Likewise, we have usability, localization, performance, internationalization, and installation tests.

Security tests test security aspects such as unauthorized access and planting of unauthorized software programs. Security tests include testing the login functions and authorized/unauthorized access areas, and testing for planting of unauthorized software programs.

Usability testing involves finding the screen flows and workflows for completing a transaction, the number of steps or clicks needed to perform a transaction, aesthetics aspects of user screens, and so on. If a user has to perform a large number of unnecessary steps to complete a transaction, then it cannot be considered a good software design, and the design needs to be corrected. If the user screens are cluttered with useless information, then, again, it is not a good design and needs to be corrected.

Many software products are used by the people who are located at different geographical locations. These people may be using different languages and conventions. For example, in some parts of Asia and Europe, the date format used is different from that used in North America. Similarly, in some parts of Europe, a comma (,) is used for decimal places but a period (.) is used for the same purpose in North America. When a software product is introduced in a new geographical location, the language and conventions have to be changed to ensure that the people in that area can use the software product properly. Localization testing is done to address this aspect.

When a user clicks a Submit button, the software product needs to perform the requested operation. If the response takes more time than the acceptable limit, then it is a waste of user time. Performance testing is used to ensure that the software product responds to the user interactions within the stipulated time. If this does not happen, then it is a performance defect and needs to be fixed. Software developers can then tweak their source code. Performance problems can happen because of faulty query techniques used to access the database or because of messages that go through a large number of software components.

### *9.9.3 Regression Tests*

In agile and spiral software development methodologies, many software products are being developed using incremental integration so that new features can be added to the existing software products continuously. This process poses a grave risk because the existing functionalities of the product may be broken in the new versions. This happens because when new features are added to the existing software product, many times, the existing source code will be changed to accommodate the new features. In fact, sometimes a lot of existing source code is rewritten to create a new version because the old source code was not scalable to allow the addition of new features. Similarly, any old source code can be refactored to make it scalable to the addition of new features. This results in changes to the old (existing) functionality of the product. In these scenarios, once the new features are added, many old features may not work at all or may work very differently from what they are supposed to do. It is important for the old features to continue to work in the same way for the existing customers of the software product. Because of this, during software testing, old features are also tested during the new release of the software product. This is known as regression testing.

Regression tests are a permanent feature of agile projects. In each iteration, the suite of regression tests are run to ensure that the existing software product features are still working after the addition of new features during that iteration. In traditional Waterfall projects, regression tests are done during maintenance.

In the case of legacy software products, the software needs to be maintained and supported while it is operated by the end users. During maintenance, some changes to the source code are made. Thus, regression tests may be needed to ensure that the existing features of the software product are working properly.

Regression tests are performed for unit tests, integration tests, and system tests. Regression tests are needed for all these tests because the changes to the source code happen at the class level and can affect the entire source code of the software product under test.

## 9.10 User Acceptance Testing

Software development is such a long process that by the time the software is ready to be deployed (i.e., used by the end users), it is very much possible that what was intended by the customer or the end users is totally lost and what is being given to the customer is altogether a different entity. This is because a lot of different people are involved at different stages of software development and what is stated at one stage by one set of people is translated differently by the next set of people at the next stage, and they do their part of the project as per their understanding (which, most of the time, is different from what was intended). Thus, it is possible that the end result is quite different from what the customer expected. This is why there is a need for user acceptance testing of the software product when it is ready to be handed over to the customer. The customer runs the software product to see if all the software features are as per his or her expectations.

User acceptance testing should always be based on the SRS. It is also of interest for the test team to do pre–user acceptance testing before actually handing the software product over to the customer. Pre–user acceptance testing can be carried out after all the code fixes (that occurred during the system testing stage) have been done. At the pre–user acceptance testing stage, the test team will not perform testing to search for defects that need to be fixed. Rather, they will see if all the features of the software product are as per the SRS. In the pre–user acceptance testing, the testers should observe all the known defects, create walk-arounds (a walk-around is an alternative way of performing a transaction when the primary method of performing that transaction is not possible because of some software defect; more details are available in Chapter 10) wherever possible, and document them properly. These will help end users when they actually carry out their user acceptance testing. Thus, end users not only compare the software product with the SRS but also consider the list of known defects and walk-arounds while performing the user acceptance testing.

After the user acceptance testing is found to be satisfactory, the software system can be used by the end users.

## 9.11 Other Important Tests

We have discussed different kinds of tests based on different criteria such as level, life cycle, hierarchy, and needs. These are the important tests that are performed on most of the software projects. However, there are more types of tests. Let us discuss some of them.

### *9.11.1 Sanity Tests*

From a software life cycle point of view, we can see that testing is needed during software development. However, software testing is also needed during maintenance and during the operation of the software product. We already discussed that regression testing has to be done after the maintenance of the software product. However, testing is also needed during everyday operations of the software product to ensure that the product is available to the users and that it is working as usual (so that the end users do not suffer). This is called sanity testing. In the past, all software products were inside the firewalls of the organizations and there was no outside connection (through the Internet or any other extranet). However, nowadays, a large number of applications are accessed over the Internet; thus, they are prone to security threats. In such scenarios, these applications must be continuously checked to ensure that they are safe and running smoothly and that the users are using them without any problems. This is the purpose of sanity testing.

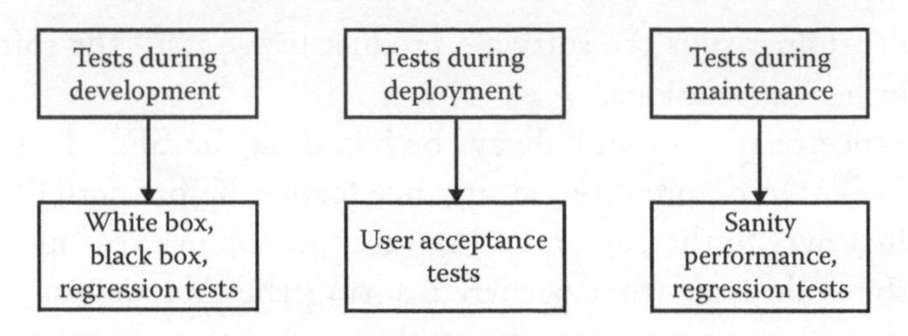


**Figure 9.9** Software testing types during software development, deployment, and maintenance.

In Figure 9.9, you can see various types of tests done when a software product is in the development phase, deployment phase, and maintenance phase. We also need to perform tests in the production phase when a software product is in actual use (by the users). Note that the current discussion is more about the tests that are conducted on a software product and not on other software artifacts that are generated during the software development process. Thus, we have not shown verification tests in Figure 9.9.

Sanity tests are even more important for Software as a Service (SaaS) vendors. SaaS vendors earn their revenue by allowing their software product to be used by their customers on a subscription service basis, and they charge on a monthly or weekly basis. If, for some reason, any functionality of the product or the entire product is not available or not working properly, then the vendors lose their revenue. Because of this, SaaS vendors continue to monitor their product for its availability and for any defects caused by the environment in which the product is working.

Sanity testing is used on a daily basis for SaaS products that are in production. SaaS products, when they are in production, are used by a large number of users. The same product instance (e.g., a single website where the SaaS product is hosted) may also be used by the end users belonging to different customers. This may sometimes lead to the website not working properly because of wrong data pushed in by the end users. Sometimes, it can also happen that because of some defect in administration privileges, some software product features are disabled by some end users. In a nutshell, a lot of things may go wrong with a SaaS product in production.

A sanity test done every day ensures that the critical software product features are working properly.

### *9.11.2 Smoke Testing*

Smoke testing is done for the purpose of testing whether a piece of source code integrates with an existing source code base (software product build). If a piece of source code has integration issues with the existing source code build, then the build will fail when that piece of source code is integrated with that existing source code build. A software build fails when it does not compile. A failed build creates problems for all the software developers who are using that build because they will not be able to integrate their newly developed source code with that failed build.

Smoke testing is done to determine whether a piece of source code, when integrated with the central build, will result in a breakage of the build.

On agile projects, continuous integration of the source code with the central build takes place. This means that the frequency of integration of the pieces of source code with the build is very high, sometimes as high as 20 times per day. Doing a manual smoke test in such a scenario is not possible. There are software products available to perform smoke testing automatically each time a piece of source code is integrated with the central build.

### *9.11.3 Cyclomatic Test*

Cyclomatic tests are done to see the complexity of the software product and its components. These tests are static. Cyclomatic tests are done to assess the size of the software product (to be created) and the effort required to create the software product. The complexity of a software product increases the required effort exponentially. The complexity of a software product or any of its components is determined by assessing the number of logic statements that contain too many decision points. If a class or a procedure contains too many if–else statements or too many loop statements, then it will be treated as a complex class or procedure.

### *9.11.4 Incremental Testing*

Incremental testing is very similar to regression testing. We have seen regression testing in an earlier section. Regression testing can be done for any software product after changes happen in the software product for reasons such as increments, enhancements, and incorporation of customer change requests. Incremental testing is meant for testing only when a software product is being built incrementally.

### *9.11.5 Exploratory Testing*

The most popular method of software testing is to follow a structured approach. This triggers the generation of a list of test cases based on the SRS. The drawback of this approach is that the software product cannot be tested thoroughly because the number of test cases will be too big if you want to cover all aspects of the software product. For example, it is customary to check the main business flow in any transaction in a typical testing effort. Nowadays, in most software products, there are many alternative ways (paths) to do a transaction. Because of time constraints, in any testing, it is not possible to check all these alternative ways of doing transactions.

To overcome this limitation, the software product should be tested without following a structured approach. This approach is known as exploratory testing. However, exploratory testing should be restricted to a time limit to make sure that the testing tasks will be completed as per the planned schedule. Exploratory testing should be done by the domain experts who have knowledge of the nature of inputs and the nature of expected outputs. The domain experts have business knowledge and know how the end users are going to use the software product. Hence, they know which alternative ways (of doing a transaction) will be used by the end users. Domain experts are also able to test a software product because they are well acquainted with the way the transactions will be performed. Thus, the domain experts may not need test cases to test the software product. In fact, writing and maintaining test cases is a laborious and time-consuming task. Doing away with the test cases will allow the domain experts to test the software product really quickly. This will allow the domain experts to test the software product comprehensively within a short span of time. The domain expert will test the software product and whenever a defect is found, only then will he or she write a test case against that defect to log that defect for fixing. The same speed of testing cannot be achieved by a structured testing approach that uses test cases.

Exploratory testing is primarily used on agile projects as it helps reduce the testing time during the product development iterations.

### *9.11.6 Coverage-Based Testing*

When a large software product needs to be tested, it is often not possible or advisable to test the entire system because of the associated cost and available time for testing. In such cases, priority test cases should be identified and executed to make sure that the critical functionality of the software product is tested within that short time allocated for testing. Priority test cases for a product are the ones that test the crucial aspects of that product. Some of the test cases (that are noncritical) may not be tested because of time constraints. In such situations, test coverage is not 100%. However, testing using priority test cases is still a good practice because critical test cases were executed; thus, testing was effective. The remaining noncritical test cases can be kept and used for testing later on by the test engineers.

*Code coverage* is another term that is often used in testing. It refers to the portion of the source code (of the product) that has been tested or not. If some part of the product has not been tested, as revealed by a code coverage analysis, then that part can be taken for testing.

### *9.11.7 Alpha Testing*

A large number of customers use a software product; therefore, if the software product has many defects, attending to all these customers for product support is almost impossible for the vendor. At the same time, because of cost and time constraints, performing a thorough test of a new release (of the software product) using a large testing team cannot be done in-house for a long period. Even if the testing team has done a good job, the software product may still have some critical defects because the product is not thoroughly tested because of cost and time constraints. What are the options the software vendor has in order to launch his or her new release into the market within a reasonable time frame?

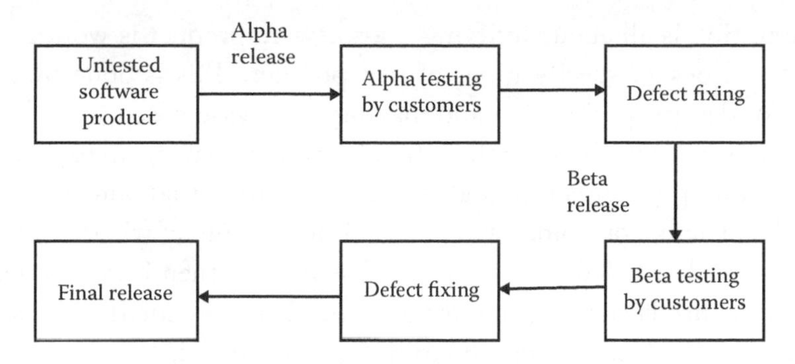


**Figure 9.10** Alpha and beta release cycle of a software product.

One option is to release an alpha version or alpha release (as depicted in Figure 9.10) of the product to a select list of customers for free for a limited time. These are enthusiastic customers who are willing to play with the new version of the product. They play around with the alpha version, and whenever they find defects, they report them to the vendor. This way, within 2 to 6 months, the product would have been tested by thousands of end users and a good number of defects would have been reported.

#### *9.11.8 Beta Testing*

After the alpha release of a product, a large number of critical defects will be identified and subsequently removed (by the vendor) from that product. Still, there may be more critical defects in that software product. If the vendor feels that the product needs more testing, then he or she may opt for a beta version or beta release of that software product instead of going for a regular version release (as depicted in Figure 9.10). In the beta release, the vendor distributes the release of the product free of charge to a select list of customers. Nowadays, both alpha and beta versions are placed on the website of the vendor where any user can download them for free. Once the beta version of the product is tested thoroughly by the end users and if the vendor feels confident of a regular release, then he or she will fix all the defects found in that beta release and then release the regular version of the product into the market.

In fact, alpha and beta testing are the most cost-effective way for any vendor to get its products tested thoroughly within a short time span by thousands of end users. The vendor saves time and testing costs by having the end users themselves test the product for free.

## 9.12 Test Case Design

A test case is a statement that defines what needs to be tested, how it will be tested, what is the precondition for that testing, and what is the expected output from that testing. Before you can write a test case, you need to think about how you should design your test cases to make your testing effective.

Software testing is all about verifying if a software product is working as per the requirement and design specifications of that product. This is done by taking each requirement or design specification and making a test case based on it. The test case consists of providing some input to the software product, having an expected outcome when the software product is run, and then finding the actual outcome. Comparing the expected and actual outcomes will provide information on whether the particular test case has passed or failed. If the test case has passed, then it means that the software product feature is running fine for what has been tested. If the test case fails, then the software product feature is not running as expected.

A test case may consist of many fields such as the serial number of the test case, test statement (i.e., test case description), entry criteria, exit criteria, expected result, actual result, tested by, test date, and supporting facts. Table 9.1 shows a test case template. This kind of template can be used to create the test cases.

A serial number for a test case can be any number that will clearly identify the test case for managing the testing activities related to that test case. The test statement defines what operations will be carried out during that test, for example, clicking on the Submit button on the main web page of the software product that is being tested. The entry criteria for a test are the requirements that need to be fulfilled before the test can run. For example, if the main web page has a textbox to fill out, then part of the entry criteria is filling out that textbox before clicking the Submit button. The exit criteria for a test case are a set of conditions based on which you can determine that the test case execution is finished. For example, after clicking the Submit button on the web page, if the web page navigates to the results page, then this is the exit criteria.

The actual result is the output or the result received from the product after supplying all the required inputs. The tester will provide the required inputs and check the output. If the expected result is being computed by the software product (i.e., if the expected result matches the actual result), then the tester will mark the test case as passed, and if the expected result is not being computed, then the tester will mark the test case as failed. The expected result for a test is derived from the requirement specification that may state what can be expected from the software product after the user performs an action on it.

After the test has been conducted, each test case should be signed by the tester who did the testing. The tester should provide the test date and any specific observation that is made when the test was conducted.

Each kind of testing needs a different kind of test case. In case of black box testing, tests are performed by executing the source code. In contrast, the static tests are performed without executing the source code.

**Table 9.1** Test Case Template

| Serial number | Test case description | Entry criteria | Exit criteria | Expected result | Actual result | Test date | Test pass/fail |
|---|---|---|---|---|---|---|---|
| | | | | | | | |

Unit and integration testing are performed with the knowledge of the source code so that any defects that are found will be fixed by making changes in the source code. These tests are commonly known as white box tests as they are done with knowledge of the internal structure of the system. In contrast, system and user acceptance testing are done without knowledge of the internal structure of the system. This is why these tests are known as black box testing. For white box testing, knowledge of the internal structure of the system is used to derive the test cases. For black box testing, software specifications (software design or SRS) are used for deriving the test cases.

Apart from white box and black box testing, experience-based testing is also used extensively and it is explained later.

At the most basic level of test case design, the test team should analyze what kinds of tests (white box or black box, functional or nonfunctional, etc.) are needed on the project. This will help in designing the appropriate test cases. Major tasks in test case design and analysis include reviewing the requirements, reviewing the software design, reviewing the interfaces, reviewing the architecture of the software product, prioritizing the test areas, creating the test data, and setting up the test environment (test bed preparation).

Writing test cases requires a lot of consideration. What type of testing needs to be done (whether functional, performance, usability, security, etc.), what kind of inputs may be required, what kind of testing is to be carried out (positive, negative, boundary value analysis, etc.), and so on are some of the concerns before the test cases are actually written.

The best way to write functional test cases is to prepare the test scenarios first. Test scenarios will capture the business logic and business flow that are required to be tested. Each test scenario will have many execution paths. These paths are to be identified. Out of these paths, the critical paths should be chosen to write the test cases.

For security testing, the parts of the application that could be prone to security threats should be identified. User registration, user authentication, roles, features related to the roles, and the accessibility of different parts of the application to different roles should be tested. Similarly, unauthorized access needs to be checked to make sure that the security mechanism works properly. The application should also be checked for data theft, impersonation, and so on.

In sanity tests, critical parts of the application are tested in the production environment to make sure that they are working fine and to ensure that the application is running smoothly and that the end users are not facing any difficulty while using the application in their daily work.

### *9.12.1 White Box (Unit and Integration) Test Case Design*

White box testing is also known as structural testing. White box testing includes unit testing and integration testing. Generally, unit testing is done to ensure that the source code statements, transaction paths, decision trees, and so on are correct as per the design specifications. The source code statements, transaction paths, and decision trees are programming constructs that implement the business logic (see Chapter 7 for these topics). During execution, depending on the user inputs or through an event, a specific code will be executed depending on the programming constructs. During unit and integration testing, all the paths of the executing code are checked. To verify if the tests have checked all of these programming constructs, coverage analysis is performed. Let us discuss some of these aspects of structural testing in detail.

*9.12.1.1 Test Coverage* The source code should be assessed to see if it covers the entire software design. There are some techniques such as equivalence partitioning and boundary value analysis that are used to assess the code coverage of a given piece of source code. Test coverage is expressed as follows:

$$\text{Test coverage} = \frac{\text{number of coverage items exercised}}{\text{total number of coverage items}} \times 100\%.$$

Here, coverage item is any item that we have been able to count.

When test-driven development is adopted, each class and its methods are tested first and only then is the source code written. In such cases, test coverage will naturally be high. All the decision paths are also likely to be covered. However, because of either time constraints or oversight, some decision paths may not be covered. At the same time, it is also possible that all negative, positive, and boundary values could not be tested because of the same reasons.

Test coverage report depends on the type of coverage to be assessed. The test coverage equation given above will thus change for each type of coverage to be done. Thus, if the type of coverage is the source code statements, then the equation will look something like below:

$$\text{Statement coverage} = \frac{\text{number of statements covered in testing}}{\text{total number of statements}} \times 100\%.$$

The types of coverage are explained next.

*9.12.1.2 Types of Coverage* Earlier, we discussed unit testing, integration testing, and acceptance testing. Test coverage assessment is done at all stages to see if all the parts of the software product have been tested sufficiently. Software product parts that are critical for business use should be tested more thoroughly to make sure that they do not contain any critical defects that may hamper the use of the software product.

At the unit level, transaction paths, statement coverage, branch coverage, and decision coverage are some of the things that are tested. At the integration level, we test for the validity of the interfaces and interactions among the components. Calls to procedures, objects, or modules should also be covered here. At the system and acceptance level, requirement specifications, user screens (user interfaces), performance, security, and so on need to be tested. Specifications that can be tested here include boundary values, equivalence partitions, decision tables, state transitions, and so on.

Therefore, we can see that the test coverage for different levels of testing is different. At the structural level (unit and integration testing), test coverage refers to the amount of source code that has been covered through some type of testing. At the system or acceptance level, test coverage is really concerned with whether the requirement specifications have been fully tested. In fact, at the system level, software design is covered. At the acceptance level, requirement specifications are covered.

Now, let us see some specific structural test case design techniques.

*9.12.1.3 Statement Tests* Statement tests are done to test a set of statements that belong to a unit of decision or logic in the source code. In a statement test, it is ensured that each statement is executed at least once. For instance, if the piece of source code being tested contains conditional statements (if–else, switch, go-to, etc.), then obviously all the statements will not be executed; only the statements that are in the path of control flow will be executed. In this case, finding the correct control flow is the objective of testing. Correct control flow can be understood by the following example:

```
Integer a, b, d, f
a = 2
b = 1
if b < a
then a = d
print ("a is equal to d")
else
a = f
print ("a is equal to f")
```

If you test the above code, then you always get "*a* is equal to *d*." This means that the else part of the statement will never be executed. Obviously, the code is not written properly. Since a part of the statement will never be executed, there is a defect.

Similarly, if a piece of source code is not referenced anywhere, then that piece will not be executed. Unused variables will also not be executed. If some pieces of a source code cannot be executed, it is a defect known as dead source code. Finding these defects is the objective of the statement tests (these defects can be corrected later).

*9.12.1.4 Decision Tree Tests* In the source code, only some specific statements of the conditional statements (such as if–else or switch) are executed when the source code is run. The flow of execution is determined by the conditions (decision points) that are placed before these conditional statements and the execution follows accordingly after meeting these conditions. Such cases can be represented in the form of a decision tree. During decision tree testing, the flow of execution is tested to see if the flow is happening as per the conditions. It is easy to understand the flow of execution by representing various conditions in the form of a decision tree. More details and an example are provided in Section 9.12.2.4.

### *9.12.2 Black Box (System and User Acceptance) Test Case Design*

As mentioned previously, white box testing cannot be used to test whether the software design specifications or requirement specifications are met by the product. Therefore, to perform these tests, the source code is executed and the tests are performed in the run mode. Test cases are designed by referring to the requirement specifications and software design specifications.

*9.12.2.1 Test Case Workflow* In case of exploratory testing (explained earlier in this chapter), there is no fixed workflow to be followed because the testers do not create a test case beforehand. Only upon finding a defect will they create a test case against that defect and log that defect for fixing. However, with test case–based testing, there is a definite workflow involved from requirement specification analysis to test case execution to defect logging to defect closing. Test case–based testing is a structured method of testing and a predefined workflow is followed. We will see this workflow in a later section.

*9.12.2.2 Equivalence Partitioning* When the test cases are designed for black box testing, the test data should be created in such a way that they belong to the correct input in some situations and to the incorrect input in other situations. This way, it will be ensured that the code is working correctly for both scenarios: in the incorrect input situation, the test case fails (if the output matches the expected wrong output), and in the correct input situation, the test case passes (if the output matches the expected right output). Sometimes, these test cases are also known as negative and positive test cases, respectively.

*9.12.2.3 Boundary Value Analysis* Many parts of a software product consist of decision trees. A decision tree is used to assign some values to a variable depending on some criteria. For example, suppose a software feature provides a discount to its customers. Depending on the purchase value, the discount rate varies. The discount rate is 5% for purchase values up to US$100, 10% for purchase values more than US$100 and up to US$200, and 15% for purchase values beyond US$200. For this scenario, there are three critical paths.

1. Purchase values: US$0 to US$100
2. Purchase values: US$101 to US$200
3. Purchase values: US$201 and beyond

How many test cases will you need to test this scenario? What is the minimum number of test cases required for this scenario? Considering positive and negative tests as well as boundary values, in the initial assessment, it seems you may need more than 10 test cases. For the boundary value, you will have to make a test case for US$0 as well as US$100 for the US$0 to US$100 range because the requirement specification says up to US$100 purchasing value. The test thus should include a test for US$0 purchase value as well as US$100. The negative test case should include a test for some value below zero purchasing value (e.g., US$-1) and one for more than US$100 (e.g., US$101). You will also need to make a positive test (e.g., US$50). Thus, five test cases will be needed for US$0 to US$100. Out of these five, one negative test case (for US$101) can be used as a positive test for the next range of purchases. Therefore, only four test cases will be needed to test the US$101 to US$200 range. You will not need to create any test case for the US$201 and beyond range because one negative test case from the US$101 to US$200 range will cover a test for a value above US$200. Hence, you will need nine test cases to test the entire functionality of this example.

*9.12.2.4 Decision Table Testing* The system design specification may contain some complex business rules that need to be tested at the system level. This testing is done by decision table testing. Many times, these rules can be hierarchical (one rule is below some other rule) so that if the conditions for the first rule are met, then the execution path proceeds to the second rule and so on. Otherwise, the second rule is never executed. When such complex sets of rules are to be tested, a good decision tree needs to be created using these rules. Once created, the test cases should cover all possible paths in that decision tree. These situations can be represented as a decision tree (which we discussed earlier). Decision table testing and decision tree testing are two alternative ways of testing the flow of execution in such situations. The only difference between these two is whether you represent the test cases in the form of a tree, that is, a flowchart (for decision tree testing), or in the form of a table (for decision table testing). Except for this difference, they will both do the same job.

Generally, decision table testing (or decision tree testing) is required for cases where a range of input values is not sufficient to test a software product. As we know, combining the boundary values and partitioning the input values for a range of values are sufficient for testing most software systems where the output follows a definite pattern, for example, when a requirement specification specifies that the accrued amount (the principal amount plus the earned interest on that principal amount) has to be paid for a fixed deposit (i.e., certificate of deposit) account after its maturity. In this case, the tester will provide the input values in a range for the interest rates and then find the accrued amount based on the principal amount and the interest rate.

However, assume a different scenario in which the rate of interest is different for different principal amounts; then the tester cannot test this software component by just using a range for interest rates. The tester must also use a table (or a tree) consisting of a range of principal amounts and their corresponding interest rates. The tester will need to compute the accrued amount and compare this amount with the amount given as output. For example, suppose the rate of interest is 4% for the principal amount ranging between US$400 and US$1000; the rate of interest is 5% for the principal amount ranging between US$1001 and US$5000. In this case, the tester will create a decision table where he or she will have to provide the principal amount ranges and the corresponding interest rates. Based on the range into which the principal amount belongs, the rate of interest will be determined and the accrued amount calculation done accordingly. This scenario can also be tested by using boundary value analysis.

This scenario can be more complex. Imagine if the requirement specification also states that if the accrued amount is more than a certain amount, then there will be income tax applied. In addition, imagine if the income tax rate is determined based on the range within which the accrued amount falls. In such a scenario, the accrued amount will be calculated by determining the interest rate first. Later, depending on the range within which the accrued amount falls, an income tax rate will need to be applied. Thus, you can see that such functionality cannot be tested by just using a simple rule such as boundary value analysis. You must create a decision table (or a tree) and then carry out your system testing accordingly.

*9.12.2.5 State Transition Testing* In some instances, the software changes its state based on a condition or a transaction. In different states, the software behavior is different. For instance, there could be a program to calculate the number of wine bottles to be filled from a full casket. Initially, when a new bottle is placed, it is empty. When the wine is poured to the bottle, then it starts getting filled. A program is needed to check if the bottle is full. Once the bottle is full, a cap is placed on the bottle. When this program runs, the following states of a bottle can be observed:

- The bottle is empty.
- The bottle is being filled.
- The bottle is full.
- A cap is placed on the bottle.

For example, the transition "wine is being poured into the bottle" performed on an empty bottle will change the state of that bottle from the "The bottle is empty" to "The bottle is being filled." When the state is changed, the behavior of the system should also change. For example, if the amount of wine in the bottle represents the behavior of the system, then, as we know, there is a difference in behavior when the bottle is empty and when the bottle is full of wine. When you test this program, you will find that any bottle will be in one of these four states. The test cases should capture each of the states. In addition, for each state, the test cases should capture all possible transitions from that state. This kind of testing is state transition testing. State transition testing is used if the behavior of the software can be described using a finite state machine. State transition is used in many embedded and automation systems.

*9.12.2.6 Use Case Testing* In many projects, use cases are made as part of software design. On those projects, use cases can be used for testing purposes. Use cases represent business and workflows well. Therefore, these flows can be tested by performing use case testing. The test cases should be designed in such a way that each actor and the activities performed by that actor are covered.

*9.12.2.7 Experience-Based Testing* Complex and large enterprise systems cannot be tested using simple testing methods. Such systems are composed of complex business rules, elaborate workflows, many features attached to different roles, and so on. The scope of many such systems is global. This means that such a system allows multi-currency transactions, country- and area-specific taxes, different languages, complex organization structures, and so on. Such a system consists of a system of systems integrated with each other. Testing such a system is a huge task. Creating and maintaining test cases to test each and every aspect of that system is simply impossible. In such cases, exploratory testing is a good solution. Business (domain)-specific testing should also be deployed. This kind of testing comes under experience-based testing. Experienced domain experts, subject matter experts, and business analysts do these tests because they have knowledge of the industry and know how the software is used in that specific industry. They also know what the end users expect from the system as well as how the tasks are performed in the field and how the system can help in accomplishing those tasks. Experience-based testing is also known as domain or scenario testing.

*9.12.2.8 Choosing the Right Testing Techniques* We can see from the ensuing discussion that there are many techniques (such as experience-based testing, use case testing, and decision table testing) available for doing system-level testing. Which of these techniques will be suitable for testing our project? It depends on the kind of project and the customer requirements. The most significant factor will be the development model adopted for the project.

At the unit or integration level, testing will be much less complex because you only need to worry about a few methods or classes. However, at the system level, things are different. The entire system is there before you to test. The output results during a test can surprise you because the output could be influenced by many components comprising the software system. Testing at the system level definitely requires good requirement specification and design documents. At the same time, a good testing technique ensures that no unnecessary testing is performed, thereby ensuring that effective testing is done and there is no time and effort wasted with that testing.

### *9.12.3 Test Case Design for New Technologies*

If you come across software features that are made using new technologies, then some of your old test design techniques may not work. For instance, if you want to test a web page with Asynchronous JavaScript (AJAX) components, then your old technique of page refresh will not work. AJAX components load asynchronously to your web page.

Suppose you have a form on a web page in which the user can enter his or her personal information. In traditional web forms, when a user completes the form and submits, then the application server receives this information and updates the database. However, if the user tries to submit the form containing some data that are identical to the data that are already in the database, then the database will not be updated but a message will be displayed to the user. Generally, usernames are such fields with a unique key defined on them. Suppose the user enters the username "Matt" and submits the form (to create a new account) and there already exists a record with the username "Matt," then that user will get a message like "Username Matt already exists. Please use a different username." The point here is that it is cumbersome for the user to keep entering a different username and try to see if the application accepts it each time. Wouldn't it be a nice feature if the username field is checked simultaneously when the user enters a username? In such a case, when the user enters a username in the username field, it is confirmed whether this username is valid and there is no identical username that already exists in the database. This validation is done even without the user clicking the Submit button. This kind of functionality can be achieved by an AJAX component. In this case, just when the focus shifts from this username component (i.e., when the user tries to move to another input field such as a zip code, after filling in the username field), the AJAX component connects to the database and checks the validity of the input in that field. This functionality takes place because the AJAX component works asynchronously from the web page. To test such a component, you need to take care of the asynchronous nature of AJAX components. Your test case should include the steps to capture the focus shift information from the username field (to another field such as zip code) and how much time it takes to validate the username information.

When you are testing new kinds of components, you will need to take care of special requirements like the one described here for an AJAX component. Most of the time, domain knowledge of the new technology is essential to conduct any meaningful testing.

### 9.13 Test Preparation

To test a software system, a testing environment must be created. If testing is performed in an unsuitable environment, then there will be issues such as difficulty replicating that defect later (for the purpose of verification of that defect). If the conditions under which a test is performed are not stable, then it will be difficult to replicate a defect. In that situation, if a software developer made a request to see if the reported defect is actually a defect, then it will be difficult to reproduce the defect. Therefore, a good testing environment is a must. The computer, on which the testing environment is created, should be separate and should not be used for any other purpose. A copy of the software product to be tested should be installed on that computer. Master data and sample transaction data (which are explained in Chapter 8) need to be created in the software product. For software defect tracking, a good tracking tool should be used.

### *9.13.1 Test Environment (Test Bed) Preparation*

A test environment is a hosting of the software product (that is under test) on a dedicated computer. The test environment should have the same hardware and software configuration as that of the server (production server) on which the software is supposed to work for the clients or users once it is ready. This is necessary to mimic the production environment as closely as possible while testing. When the application is tested on such an environment, there will be fewer chances for the application to behave differently when it is deployed on the production server. Thus, the application will have fewer chances of undetected defects when it is hosted at the customer's premises and when users start using the application.

The other reason for a dedicated test environment is that when a defect is logged, it can be reproduced when the developers want to verify the defect for fixing. If it is not a dedicated environment, then the defect may appear differently or may not be found (when the developers want to verify it) because of the changes in the environment. Many times, the defect may not be found on some other application environment if a different version of the application is used for verification.

### *9.13.2 Test Data Preparation*

Careful and planned test data preparation is an area that is often neglected by test teams. Imagine that you need to test a large software product and you have not thought about the test data beforehand. Do you think you will be able to do a good job of testing the application? Suppose you understand how boundary analysis works but you do not know what test data will be needed to do the boundary analysis for the application you are testing. You simply will not be able to test your application. You must understand various test cases. You must understand what should be the test data and how to populate it in the application. In addition to the usability, robustness, or operability tests, other tests also need data to do the testing.

Generally, large-scale systems have two types of data, master data and transaction data, as explained in Chapter 8.

It is indeed important to create meaningful and testable data for your testing.

### *9.13.3 Configuration Management Preparation*

Configuration management is very important in testing. All the artifacts (design artifacts, requirement artifacts, etc.) that are delivered to the test team should be managed well. If changes are made to the artifacts (by the development team) but not relayed to the testing team, then it will mislead the testing team. This will result in a waste of time and resources because the testing team may be working on the wrong artifact and thus lose all the effort made on that wrong artifact. Apart from the development artifacts, the test team itself has its own artifacts in the form of test plans, test cases, test case execution reports, and so on. Not only does the test team need to keep these artifacts up to date, but also these artifacts need to be in sync with the development artifacts. For this reason, the test team should always be in sync with the development team and should keep all versions of the artifacts in sync.

Unit testing and integration testing are performed by the software developers themselves (or sometimes by the software testers who are an integral part of the development team) on most software development projects. However, system tests are always performed by software testers who are not part of the development team. For this system testing team to work effectively, a system test plan is created. This test plan outlines the activities such as test data preparation, test case creation, test case execution, defect logging, and defect retesting that need to be performed. Test case execution results (i.e., outcome of the tests) in the test reports also need to be archived. A good configuration and version management strategy will enable all the documents related to the testing to be kept in a safe and accessible manner. Configuration management is explained in Chapter 12.

## 9.14 Test Life Cycle

A test life cycle (Figure 9.11) is used for system testing. For unit and integration testing, this kind of a formal test life cycle is not maintained because unit and integration testing is considered more a part of the development life cycle.

The test life cycle consists of several stages. The specific stages of the test life cycle are shown in Figure 9.11. As we have already outlined, unit and integration level testing is not considered part of the test life cycle. This means that only testing under the system-level tests and user acceptance–level tests is considered for the test life cycle.

In this section, we use Figures 9.11 and 9.12 to explain various sections. However, the entire section is not fully based on these figures. Only some of the section names are shown in them. Sections 9.14.1 through 9.14.3 are covered in Figure 9.11. Section 9.14.4 is completely covered in Figure 9.12. Sections 9.14.5 and 9.14.6 are not covered in either of these figures.

### *9.14.1 Test Planning*

The test life cycle starts when a test plan is made for a software product. In test planning, the type of testing that needs to be done is determined, that is, whether we need

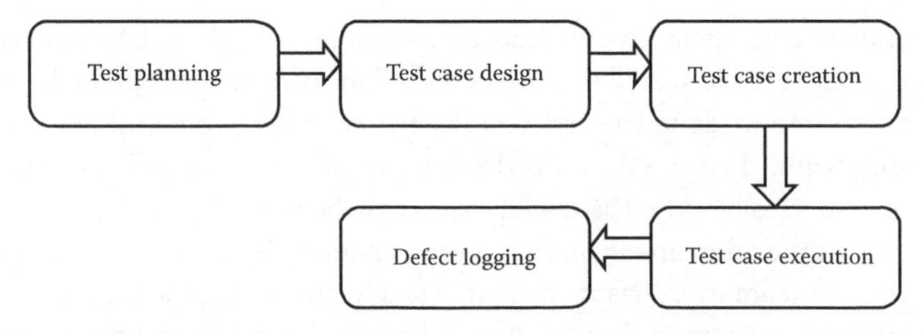


**Figure 9.11** Test life cycle.

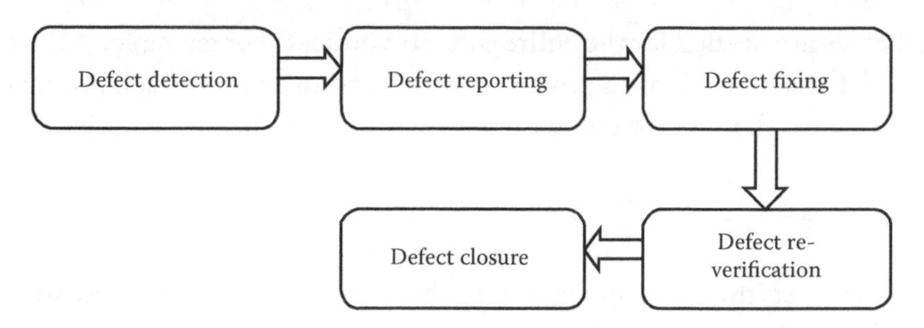


**Figure 9.12** Defect life cycle.

functional testing only or nonfunctional testing also needs to be performed. We need to find out which parts of the software product need to be tested. We also need to make an estimate of the number of test cases that need to be made.

Once this test planning is done, each tester is assigned a testing area (the test type and the specific part of the software product that the tester will test). Each tester will then design and create the test cases. The testers will then execute these test cases. During test case execution, testers may find that some test cases pass while others fail. The failed test cases will point to the software defects in that product. These software defects may need to be fixed. The entire workflow of finding and fixing the software defects is included in the defect life cycle.

The defect life cycle (Figure 9.12) is part of the "Test case execution" stage of the test life cycle. Once a test case enters the test life cycle as given in Figure 9.11 and reaches the Test case execution stage and if that test case passes that stage, then that test case is done with the test life cycle. However, if it fails at the Test case execution stage, then the defect life cycle starts. Once it starts, defect detection, defect reporting, and other stages shown in Figure 9.12 will be executed. Once all these stages shown in Figure 9.12 are complete, the "Defect logging" stage of the test life cycle (the last stage in Figure 9.11) is executed. More details on the defect life cycle are provided in Section 9.14.4.

When all the test cases are executed and all the defects are fixed, a test report that depicts the performance of the testing activity is created.

The testing process continues until the complete system reaches a satisfactory quality level.

### *9.14.2 Test Case Design and Creation*

We have discussed test case design techniques in several previous sections. These techniques include decision tree design, equivalence partitioning, and boundary value analysis. Once the test case design is done, test case creation can be started. Testers will write or create all the test cases as per the test case design chosen for various parts of the software product.

When we discussed test case design techniques, we also discussed how many test cases are needed to fully test a software feature. Thus, we can easily figure out how many test cases are needed for the entire software project. For example, if the software product under test consists of 10 product features and on average each product feature needs 8 test cases, then we have to write 80 test cases for the entire project.

#### *9.14.3 Test Case Execution*

Once we have everything set up and are ready to go with our test cases, we can start executing them. Depending on the test cycle level, different kinds of test cases will be executed. If we are at the structure level of testing, we will execute unit tests and integration tests in parallel with the development cycle (in fact, unit tests are executed even before the actual source code is written). System-level testing is performed after the development cycle is complete. If the system is ready for deployment, then user acceptance testing is carried out by the end users (sometimes, it is better to have pre–user acceptance testing to make sure that no defects pass through the deployment or production stages).

For regression tests, we check whether the existing set of regression test cases are valid. If they are not, then we make changes to them. Once all the test cases are executed and the test cycle is completed, a report is made about the number of test cases executed and the number of passed and failed test cases.

After performing a cycle of test execution, if it is found that the level of product quality (the customer decides the quality level of the software product) is below the expected level, then the test cycle can be repeated. The expected level is the ratio between the number of defects found in the test cycle and the expected number of defects. If too many defects (more than the expected number of defects) are found in the first test cycle, then this may pose problems not only for the testing team but also for the entire project team. This will indicate that the software product has not been built correctly and has serious design or construction faults. In such a case, it is better to overhaul the software design or construction and stop further testing.

#### *9.14.4 Defect Life Cycle/Tracking*

As explained earlier, the defect life cycle is part of the test life cycle and it starts during the test case execution. When the testing team starts executing the test cases, they may realize that some of the test cases do not pass; thus, there are some defects. These defects must be declared, reported to the concerned person(s), fixed, and finally closed after verification (Figure 9.12).

The defect life cycle (Figure 9.12) starts with defect detection. If a test case fails, then the tester knows that a defect has been detected. The tester will file a defect in a defect tracking system. This is a system where defect logging, defect fixing, verification reports, and finally defect closure are done. Once a defect is logged in the defect tracking system, defect fixing will be assigned to a software developer. The software developer will make changes in the source code and see to it that the defect is fixed. Once the defect is fixed, the tester will be notified. The tester will run the test case to check if the test case now passes or if it fails again (defect reverification). If the test case passes, then it means the defect has been successfully fixed. Once the defect has been fixed, the defect will be closed by the tester.

Defects are the most important part of testing. Whether the money spent on a software testing team is justified is determined by the number and quality of the defects detected by that testing team. If too many critical defects escape to the production stage, then the effort made during testing cannot be justified.

### *9.14.5 Test Reporting*

Test reporting (i.e., reporting of the test results) is an important aspect of testing because only through these reports will the management and the customer know how effectively the testing life cycle has been executed. There are many metrics that are associated with testing projects. Some of the most important metrics include requirement traceability, defect injection rate, defect removal rate, expected defect density, number of test cases, and test case execution rate (they are not explained in this book because of space limitations). The reports should address the defects, if any, associated with each of these metrics.

Several reports are generated once the test execution cycle starts, one of which is the daily reports. The number of test cases run, the number of passed test cases, the number of failed test cases, the number of reported defects, the number of fixed defects, the number of failed defect fixings, and so on are some of the reports that are generated daily. At the end of the test cycle, a consolidated report is generated on the basis of all these daily reports. Apart from these reports, there are also reports about the criticality of the defects. Therefore, we have reports about how many critical defects were reported, how many medium severity defects were reported, how many mildly severe defects were reported, and so on. On top of these reports, there are also consolidated reports for each module of the software application that is under test. Therefore, we can have module-wise defect reports as well.

### *9.14.6 Exit Criteria and Test Closure*

When should one exit from the software testing phase so that product deployment activities can start? Some of the exit criteria from testing include the successful completion of the test objective (e.g., all the test cases have passed through the testing life cycle successfully), infeasibility of carrying out further testing (immature software product under testing), and completion of testing as per the contract. If any of these milestones is achieved, software testing activities can be stopped.

Before closure of the testing phase, some closure activities also need to be performed. All the test reports, defect reports (both closed and still open defects), test bed with all the test data, and so on need to be handed over to the project manager. If any of these items are needed for the next round of testing, then they need to be procured. An important activity to be performed before test closure is the preparation of the lessons learned from that project. Every project is a challenge for the test team because many new things and issues are generally encountered in every new project. If these things are documented well, then that information can be vital for future projects.

## 9.15 Case Study

The OBAAS that is being built as part of the case study needs to be tested. What test strategy can be adopted for this system? Let us find out.

### *9.15.1 System Parts*

- User interface: Web browser.
- Middle tier: Written using object-oriented language.
- Back end: A database is attached to save the records related to customers and their bank transactions.
- Glue: Scripting language is used to tie the three tiers.
- Client-side validation is used to validate user inputs.

### *9.15.2 Test Strategy*

It was decided that unit tests, integration tests, and system tests should be used. System tests include functional and nonfunctional tests. Nonfunctional tests include performance and security tests.

### *9.15.3 User Interface Testing*

A usability test was conducted for the user interfaces to ensure that the navigation links and menus were working fine. The labels, text fields, and drop-down menus were also tested for any problems. No defects were reported because all these elements were working properly.

User interface testing was done at the system level.

### *9.15.4 Client-Side User Input Validation*

The requirement specifications stated that whenever a textbox was left empty and the user clicked the Submit button, a dialog box should inform the user to fill that empty textbox. Testing was done by leaving textboxes empty and clicking the Submit button. The system recognized this validation and promptly displayed a dialog box to fill the empty textboxes. Hence, this validation was found to be working fine.

Client-side input validation testing was done at the system level.

### *9.15.5 Unit Testing*

The architecture of OBAAS is based on Model–View–Controller (MVC). One class named "createservlet" works as a controller. However, this implementation of MVC works only for one model response at a time. This is because, in the application, no session management is involved. The user information is lost after each response from the server. This means that whatever business logic is run, it always connects to the database and the data are either queried or manipulated. These data from the database are then passed on to the view. Thus, in OBAAS, unit testing always involves a connection to the database and determining whether the business logic is working properly against that database.

Unit testing is done using a unit testing tool. All the test scripts and test details are saved in the tool. Thus, when a new release of OBAAS is planned, regression testing for all the unit tests can be performed automatically.

A sample unit testing done for account money transfer in OBAAS is included here. The sample code to be tested follows:

```
class money_transfer {
      connect with database();
      amount_transfer float (float amount_account1, float amount_
account2, float amount_trans) {
      if amount_account1 > amount_trans
      then amount_account1 = amount_account1 - amount_trans;
      amount_account2 = amount_account2 + amount_trans;
      update database with amount_account1 and amount_account2;
      disconnect from database;
      return amount_account1;
      else
      print ("the account does not have sufficient funds");
      disconnect from database;
      }
}
```

To test this piece of code, here is a sample test script:

```
class test_transfer {
test database connection();
float x;
if database connection() successful
then
x = money_transfer.amount_transfer(100, 200, 50);
print (x);
}
```

The unit test described above will tell you the balance in your bank account after the money transfer. If money transfer does not happen (because of problems in OBAAS), then you will get a message.

### 9.15.6 *Integration Testing*

A continuous integration approach was used to build OBAAS. Hence, integration testing has always involved the continuous integration aspect.

Each class was developed and unit testing was performed to check the business logic. Once the class passed all the unit tests, it was integrated with the build (main build) on a software developer's local copy that is available on his or her local computer. If any build failed, class interfaces were inspected and then integration was tried again. If the class successfully integrated, then there was no build failure. This meant that the class passed the integration tests.

### 9.15.7 *System Testing*

System testing involves functional and nonfunctional testing. Functional testing involves testing the functionality for account balance, service request, bill payment, account

**Table 9.2** Sample Test Cases for the System Testing of OBAAS

| Serial number | Test case description | Entry criteria | Exit criteria | Expected result | Actual result | Test date | Test pass/fail |
|---|---|---|---|---|---|---|---|
| 1 | Create new account in OBAAS system. The user points his/her browser to http://www.abc.com. The user clicks on new account button. The user enters username, password, account number, etc. in the textboxes provided. | The user computer is connected to the Internet. The user computer has a web browser installed. | The user clicks the submit button on the create account page and the browser navigates. | The system should display a web page stating that an account has been created. | The system displayed a web page stating that an account has been created. | | Pass |
| 2 | Performace: When user clicks the Submit button, the system responds within 5 s by navigating to the response page | The user computer is connected to the Internet. The user computer has a web browser installed. | The user clicks the Submit button on any user screen and the browser navigates. | The system should navigate to response page within 5 s | The system navigated to the response page in 3 s | | Pass |
| 3 | Usability: Text fields on a user screen are ready for input | The user computer is connected to the Internet. The user computer has a web browser installed. | The user enters text in a text field. | The user is able to input in a given text field on a user screen. | The user was able to input in a given text field on a user screen. | | Pass |

creation, money transfer (for OBAAS 1.2 only), and account closure. Nonfunctional testing includes performance, usability, and security testing. Table 9.2 describes some test results for the system testing performed on OBAAS.

### 9.16 Chapter Summary

Software testing is important because only through rigorous testing can software defects be identified and removed and software products be of adequate quality. Traditionally, for projects that follow the Waterfall model, testing used to be done only after the software product was fully constructed. This meant that the artifacts that were generated in upstream processes like requirement specifications, software architecture and design, and source code development were never tested before the completion of the product construction. This implies that the defects that were present in these artifacts remained with them for further artifact creation down the line. This practice used to lead to defective software products that required excessive effort for testing and removing all those defects. Agile models recognized this fact and introduced concepts such as test-driven development and peer programming. These efforts ensured that all the produced artifacts were checked from the very beginning when the artifacts were started for development.

The current practice is that all the artifacts including requirement specifications, software architecture and design, and software source code are checked before they are released for further activities. Tests are conducted to check the business logic even before the source code is written in the test-driven development. These initiatives have ensured that the software product is defect-free. Further testing is done at the integration and system level to ensure that defects, if any, are detected and fixed. Before a software product is released, the software product is thoroughly tested, ensuring that it does not contain defects.

Many kinds of artifacts need to be tested for various aspects of the software product. Therefore, many types of testing are conducted on these artifacts during the software development life cycle. Thus, we have static tests done on the requirement specifications and software design documents. Source code inspection (review) is also performed. All these types of testing are known altogether as software verification. Note that the source code is not executed in verification. In validation, the source code is executed. Then, we have unit testing, which is done to check the business logic for each class. After a class is developed, it is integrated with the main build of the software product. At this stage, integration testing is performed to check if the interfaces of the class perfectly match the calling methods in other classes. This way of integrating each class (one by one) is known as continuous integration.

Once all the components of a software product are integrated, the software product is ready to be tested as a whole. At this stage, system testing is performed. System testing is a black box testing technique, which means that it does not need the source code for testing (i.e., executable code is enough). In system testing, the testers take the requirement specifications and design documents to create the test cases and then perform their test to see if any software defects are present. If defects are found, then they are fixed. Once all those defects are fixed, the system is ready to be released to the customers. At this stage, the customer can perform user acceptance testing. Unit testing, integration testing, system testing, and acceptance testing are all part of validation.

System testing is actually a group of many testing types. These testing types can be broadly categorized as functional and nonfunctional testing. Functional testing tests the business logic to determine if it works properly. Nonfunctional testing includes tests on performance, security, and usability.

There are many approaches to effective testing. Some of them include boundary value analysis, equivalence partitioning, decision tree testing, and test coverage. These techniques should be used to ensure that effective testing rather than comprehensive testing is carried out. This is because if useless testing is carried out, then it will not result in finding the defects despite spending a significant amount of time. What it boils down to is that if there is a range of values involved in business logic implementation, then there will only be three scenarios that needed to be tested; thus, only three test cases need to be prepared for testing. You only need to ensure that decision tree testing is not involved when testing this product feature. These three scenarios include an input value below the boundary of that range, an input value within the boundary, and another input value above the boundary of that range. There is no need to create an unlimited number of test cases and conduct further tests as these will only result in a waste of time and money.

## QUESTIONS

1. At what levels is software testing performed?
2. What is black box testing? Why is it known as black box testing?
3. What is white box testing? Why is it known as white box testing?
4. What is a unit test? Please provide an example.
5. What is an integration test? Please provide an example.
6. Why are testing tools needed for unit and integration testing?
7. What is the verification part of verification and validation?
8. What is the validation part of verification and validation?
9. What is functional testing? Please provide an example.
10. What is nonfunctional testing? Why is it needed?
11. What are the steps involved in testing?
12. What is a test case?
13. What is continuous integration?
14. How do you perform integration testing in a continuous integration environment?
15. What is alpha testing?
16. What is beta testing?

## Recommended Reading


Paul Amman, Jeff Offutt (2008), *Introduction to Software Testing*, Cambridge University Press, USA.

Boris Beizer (2009), *Software Testing Techniques*, Coriolis Group, USA.

Lee Copeland (2004), *A Practitioner's Guide to Software Test Design*, Artech House, USA.

Cem Kaner, James Bach, Bret Pettichord (2002), *Lessons Learned in Software Testing: A Context Driven Approach*, John Wiley & Sons Limited, USA.

Anne Mette, Jonassen Haas (2008), *Guide to Advanced Software Testing*, Artech House, USA.

Thomas A. Sudkamp (2005), *Languages and Machines: An Introduction to the Theory of Computer Science*, 3rd Edition, Pearson, USA.

# 10
# SOFTWARE RELEASE

**In Chapter 9, we learned**

- **What software testing is**
- **What software verification and validation is**
- **What the levels of testing are**
- **What unit testing and integration testing are**
- **What system testing and user acceptance testing are**
- **What functional testing and nonfunctional testing are**
- **What alpha testing and beta testing are**
- **How testing can be done in maintenance and production**
- **What the steps involved in testing are**

**In Chapter 10, we will learn**

- **What a software release is**
- **What a release cycle is for Waterfall model–based projects**
- **What a release cycle is for agile model–based projects**
- **What activities are performed in a software release**

## 10.1 Introduction

Software release is the stage when the software product has been fully developed and tested and is ready to be deployed at the customer's site. However, before that happens, you need to ensure that everything has been checked and there are no loose ends that need to be tied.

During product development, many versions of the software product could have been developed. Many versions of the software product are developed for reasons such as fixing a defect or simply because the same product has been developed to cater to different customers with some customer-specific features added. Then, there could also be different versions of the software product for different operating systems. The project team will need to decide which version(s) will be deployed at the customer's site.

Some of the activities performed during software release include the following: creating end user and technical manuals, providing end user training, and deploying the software product at the customer's site. Deployment at the customer's site may include creating the master data and some transaction data as well. Sanity testing also needs to be performed after the deployment to ensure that the software product is working well in the production environment.

### 10.2 Software Release and Software Engineering Methodology

In traditional Waterfall models, software release is the last activity in a project. The software under construction is fully built and is now ready to be deployed at the customer's site. The project team develops the user and technical manuals for the software product. If necessary, the project team also provides training to the end users. If there is any integration of the developed product with other external systems required, then that integration is also carried out. When everything is finished, the deployment of the complete software product at the customer's site is completed. Master and transaction data (refer to Chapter 8 for details) are also created in the installed system. Once all these software release activities are completed, the users can use the software product.

Software products developed using the Waterfall model may have alpha, beta, and other releases. This kind of release strategy is used to test the product in the market for customer response as well as to find any defects so that such defects can be fixed in the final release.

In agile and incremental methodology, release management can be an ongoing activity. On Scrum or XP projects, an iteration cycle lasts for 1 to 2 weeks. After each iteration (sprint), there could be a minor release of the software product. If the software product is installed as a web-based product and is used by the users, then these minor releases can be implemented in the software product seamlessly. The users may not experience much change in their usage of the software product. When major releases are planned, the changes in the software product can be announced to the users well in advance so that they are ready for the changes. It is also a good idea for the company to keep the old functionality (in the older release) intact and ask users if they want to use the new functionality or stick to the existing (i.e., old) functionality. On the basis of the individual user preference, the users should be able to use the versions they like.

### 10.3 Integration

The newly developed software product may need to be integrated with some existing software products. Integration matters need to be addressed before releasing the product. If the integration is elaborate and needs to be performed at many integration points, then the integration efforts may be large. However, if the software product needs to be integrated with only a limited number of integration points, then the integration efforts will be much less. For example, if your software product needs to get share prices in real time from another system, then it will be just one-point integration. This kind of integration will not involve much effort.

## 10.4 Documentation

Documentation is very important for any software product. Any improper usage of the software product may cause damage to the product, data, and other systems connected to the product. In most cases, improper usage of a software product may stem from the lack of good user manuals. User manuals provide a complete reference on how to use the software product. Users generally spend time reading the user manuals to learn all the functionality provided by the software product. If good documentation is not available, then the users will not be able to use the software product properly and completely.

Technical manuals for the software product are equally important. If good technical manuals are not available, then, during maintenance, the software maintenance project team may find it difficult to do their work. A technical manual provides details about the software architecture and how the software design is implemented. A good technical manual is definitely a valuable resource for the software maintenance project teams.

### *10.4.1 User Manual*

A good user manual should contain complete step-by-step instructions on how to use each and every feature of the software product. For example, if a sales transaction involves five steps, then all these steps and available options should be described in the user manual. There could be walk-arounds (explained below) for some transactions because of some unfixed software defects. These walk-arounds should be described in detail in the user manual. Screenshots should also be provided so that the users do not have any doubts about using any feature of the software product.

There are some automation tools available from companies such as Activedocs and Report Lab to develop the user manuals. These tools help increase the productivity of the tasks related to user manual preparation. When such tools are used, care should be taken to ensure that the generated manual is user-friendly. This can be achieved by first creating the user manual using the automation tools and then making the changes manually wherever they are needed.

**Walk-arounds:** A software defect may prevent a user from using some of the features of that software product. However, it is quite possible that such a software defect may not be fixed immediately. In such cases, the project team tries to find out if it is possible that the user can perform all his or her activities by using some alternative way. For example, the user has to calculate income taxes using a software product, and for doing it, the user needs to get the customer names in the customer name boxes of the user interface. Because of some defect, the customer names do not appear in the customer name boxes of the user interface. The project team was informed and they found that the user can find the names of the customers from a search function provided in some other user interface and then copy and paste them in the customer name boxes of the initial user interface. This alternative method works. This alternative technique is explained in the user manual and thus the user was able to calculate his or her income taxes despite the fact that there was a defect in the software product. This alternative way of performing a task with a software product by bypassing an existing defect in that software product is known as a walk-around.

#### *10.4.2 Technical Manual*

Technical manuals are important for the maintenance of any software product. Even the comments that are made inside the source code of the product are very important. In the source code of the product, calls are made to classes, methods, and variables that are defined somewhere else in that source code. A comment at the point where a call is made can provide an idea of why that call was made. Similarly, if there is a large chunk of source code, then it may be difficult to understand its logic. If some complex logic is used somewhere in the source code, then, again, comments will immensely help the maintenance team understand that complex logic.

In general, it is difficult for a person to understand the logic of the source code written by another person. Sometimes, it is difficult for a person to recall the logic that he or she has written long ago. Thus, documenting the source code is very important for the maintenance of a product.

Technical documents are a must if your software product is supposed to be available for integration with other software products. In that case, you need to provide complete Application Programming Interface (API) documentation of your product. A software product can be integrated with any other software product through the exposed interfaces. An exposed interface of a software product is an open interface through which another software product can easily integrate with that product. API documentation describes all the details of the exposed interfaces of a software product so that if a software vendor wants to integrate its software products with the exposed interface of that software product, then the vendor will have all the information on how the integration can be done.

### 10.5 Release Cycles

A software product can undergo many releases. It can have alpha, beta, and other types of releases apart from the standard release, which is also known as the final release. In the case of incremental product development, there are many minor and major releases of the software product. An incrementally developed software product can also have a final release if that product will not be developed further.

#### *10.5.1 Release Cycle for Waterfall Projects*

For Waterfall projects, software is released after the software product is fully developed. Therefore, in most cases, there is no release cycle involved. In a few cases, a release cycle may be involved. For example, we can have an alpha, beta, and final release.

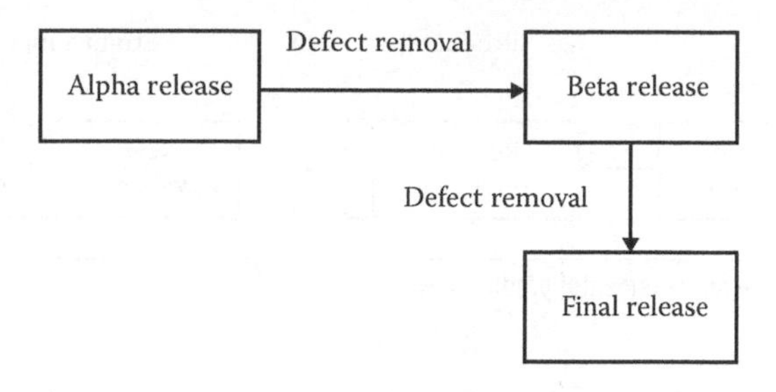


**Figure 10.1** Alpha–beta–final release cycle.

In the alpha release, a software product is released generally free of cost to the customers. This version of the software product is not thoroughly tested; thus, it may have some software defects. Since it is free, a large number of customers install and use the software product. During usage, customers may encounter some software defects in the alpha release of the product. Customers report the defects to the software vendor. The software vendor promptly repairs all such reported defects.

This way, the software product is tested free of cost for the software vendor. Imagine the cost the testing task may have incurred had the software vendor hired a large testing team to do the task. However, by giving to its customers for free, the software vendor had its product tested free of cost in exchange.

The alpha release of a software product is generally an untested product and is tested by the customers. If the software vendor finds that there are a large number of defects found in the alpha release, then the vendor may release a beta version of the software product after fixing all the defects that were found in the alpha release. Again, the beta version of the product will be offered for free to the customers.

After the beta release, all the reported defects will be rectified by the software vendor. Finally, the vendor may release the final version, which may be a paid version and will have almost no software defects (Figure 10.1).

### *10.5.2 Release Cycles for Incremental Building of Products*

When software products are developed incrementally, there will be a large number of releases. These releases can be both minor and major. All these minor and major releases are dictated by the market/customer demands. The marketing team or the customer gets market feedback and provides their input to the vendor about the required features of the product. The project team then develops these new features in a timeboxed (iteration) plan. The release date is fixed by the marketing team. Thus, the project team is left with a fixed time to develop the required product features. The release cycle and different versions that are built in an incremental development model are depicted in Figure 10.2.

Most web-based software products are developed using this timeboxed concept. Project teams that develop the software products incrementally are always under pressure because of this timeboxed product development concept. These teams work in a

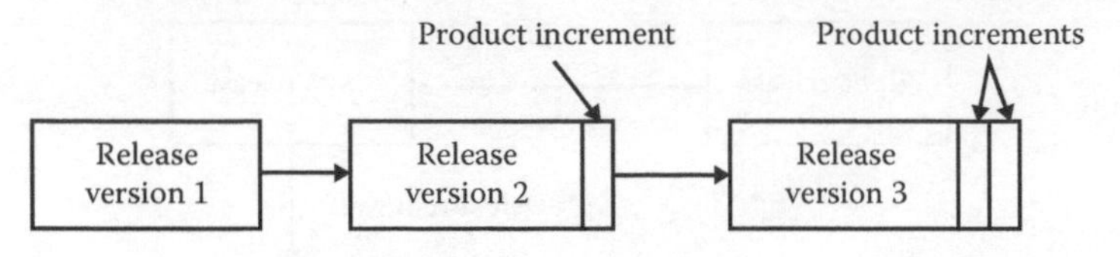


**Figure 10.2** Release cycle for incremental product development.

very different environment compared to their counterparts who work on Waterfall projects. The project team working on building a software product incrementally will keep on working on the same project, thus developing a single product with its new release each time. In contrast, in Waterfall projects, the focus is on projects. In a Waterfall project, once a product is built, the project team will have to move to some other project where they will work on building another software product.

Like in the Waterfall model, alpha and beta releases of a software product can be done during the incremental building of products as well if the software vendor wants to sell its products almost defect-free and wants to save time and money on testing the software products.

#### *10.5.3 Difference between Alpha–Beta–Final Release and Incremental Release*

In traditional Waterfall projects, there is just one release of a software product after the software product is completely built. If any maintenance (software defect removal or enhancements) is required to the software product after its release, it can be done in a maintenance project. Maintenance projects are good when the vendors have just a few customers. Also, maintenance projects are suitable only for large software products. It is not possible to maintain small software products for individual customers for financial reasons. The best business model for small software products is to maintain just one or only a few versions of the software product rather than keeping several customer-specific versions. Customer-specific versions are explained in Chapter 11.

In traditional Waterfall projects, software vendors want to bring their software products to the market as quickly as possible. One of the best options for them is to use the alpha–beta–final release route. The alpha or beta release allows the software vendor to test its software product faster than any other way such as testing it in-house.

The very idea of incremental product development is that the customer will have a software product (even if it has only a few of the features of the total features ultimately required for the product) as early as possible. Because of this, an initial release of the software product will become available to the customers with some product features that are useful to the customers. In Figure 10.2, each successive release of the software product has more features than the previous release. The product size increases as the increments are added to the product in each release.

From the above discussion, it became clear that while the alpha–beta–final release of a software product is all about discovering and removing the software defects (thus

**Table 10.1** Some Comparisons between Alpha–Beta–Final Release and Incremental Release

| Comparison parameter | Alpha – beta – final release | Incremental release |
|---|---|---|
| Purpose | To improve quality | To build product incrementally |
| Product size change | Remains constant | Increases in each release |
| Product quality | Improves | Remains same |
| Development model | Waterfall | Agile with incremental product building |
| Time to market | Fast | Fast |
| Product size | Small to even large | Small to even large |
| Product versions | Few | Few to many |

improving the product quality) as quickly as possible, the incremental release of software product is all about increasing the product size and adding more product features in each successive release. Some other factors such as product size and purpose of release, and so on are shown in Table 10.1.

## 10.6 User Training

User training is a core activity during software product release. Users need to be trained so that they can use the software product effectively and efficiently. If the organization size is small and is located in a single city, then the training can be provided to all the end users of that organization. If the organization size is large and scattered geographically, then providing training to all the end users will be time-consuming and expensive. To tackle this challenge, some key users from that organization are identified and are provided the required training. Later, those key users provide training to the other end users of that organization.

For user training, video presentations can also be helpful, specifically if a software product is being distributed or deployed over the web. Video and image creating tools that can capture user screenshots and create a video depicting the user screen are a good way to create and provide product documentation. These videos show how to use the product effectively.

## 10.7 Deployment

The new software product needs to be deployed at the client site once it is ready. Deployment activities include software installation, master and transaction data creation, and software configuration requirements, if any. Software configuration during software release is all about making sure that the right version of the software product is used at the customer's site along with all the material that is needed to run this version of the software product at the customer's site. The material may include files (e.g., a template to be used by the users) and other software or hardware components.

When the new software product needs to be distributed over the web, deployment involves placing a copy of the software product on a web location and providing a link to it so that the users can download a copy of it and install it themselves. Digital copyrights may be enforced by licensing agreements. The fee, if the software product is not free, can be charged through credit card. User manuals also need to be provided at the site so that users can download them as well.

If the new software product itself will be deployed on the web and users will access it through a URL, then the software product needs to be installed on a web server. User manuals for that software product can be provided on the same web server (where that product is installed) so that users can learn how to use that software product.

### 10.8 Software Migration Strategies

Software migration is the process of switching from a software product that has become outdated or is no longer needed to a new software product. Software migration is generally used for software products such as Enterprise Resource Planning (ERP) or any enterprise-level software product. For example, a customer may have been using a legacy custom-made ERP system but has now decided to use SAP ERP in place of the existing system.

Software migration can be a challenging job. This is because there could be some critical data that need to be migrated to the new system. These data, which reside in the old system and need to be ported to the new system, pose some challenges. The challenges include the difference in data format between the old system and the new system (and the additional data that are required in the new system).

To address these challenges, some migration strategies are used. Here are some tasks that are performed in a typical migration strategy:

- Data from the legacy system need to be ported to the new system. Data migration is a complex task requiring specialized skills. Some of the tasks performed during data migration include data field mapping and master and transaction data creation in the new system. Data field mapping is done because the format in the old system is different from that of the new system. For example, the "date" fields in the old system are in MM/DD/YYYY format, but in the new system, the "date" fields are stored in the DD/MM/YYYY format. Data field mapping will help convert the existing data to the new format; thus, the data can be correctly ported to the new system.
- If there are any open transactions in the old system, either they need to be closed or a special arrangement needs to be made to port the open transaction data into the new system. Open transaction data are a set of data belonging to a transaction that is still not completed. For example, suppose a purchase order was raised in the system but the vendor has yet to deliver the goods, then this purchase order is still open. Typically, a purchase order is closed after the goods are delivered by the vendor and the receipts are issued for this purchase order. If an open transaction is attempted to be ported from an old system to the new system, then some mismatches can happen in the data (this topic is

very specialized and cannot be discussed here further; for further information, you can read a book on ERP systems). If such open purchase orders do exist in the old system, then they need to be ported after they are completed or an arrangement needs to be made in the new system for porting such open transactions.

- Data migration is associated with cut-over activities. During data migration, users will be using a mix of legacy and new systems, and the legacy system is connected with the new system. Although users might be using the user interface of the new system, some data may be saved in the legacy system because not all data have been completely migrated to the new system at that stage. Once data migration to the new system is complete, the legacy system should be cut out from the new system. Afterward, all the transactions will be carried out on the new system only. The data will also be saved in the new database created for the new system.

Migration strategies are important for software products that are developed for business use. Business users should not stop doing their everyday business transactions just because an old (existing) system will be replaced by a new system. Business users continue their work during software migration while the project team slowly transfers data from the old system to the new system.

### 10.9 Software Product Release Checklist

A complete software product consists of many artifacts. Before its release, the project team has to verify the deliverable artifacts. All the deliverable artifacts need to be handed over to the customer at the time of software release. Here is a list of artifacts:

- The complete source code of the product.
- The binary executable code of the product.
- Hardware or software parts that are not part of the software development project but still may be part of a larger contract with the customer.
- The user manuals of the product.
- The technical manuals of the product.
- Downloadable files if they exist. A downloadable file could be a file template that is needed by the users to create some documents. These prepared documents may need to be uploaded into the system (to complete a transaction or prepare a report).

Some or all of the artifacts from the above list could be included in the deliverables. The exact list of deliverable artifacts depends on the contract that was signed by the customer and the developer before the start of the project.

To make sure that all the deliverables are handed over to the customer, you may create a checklist. You also need to ensure the version of the artifacts that need to be delivered to the customer. Thus, the version of each artifact is also important. If you deliver a wrong version to the customer, then it may cause problems to the customer or to the end users.

### 10.10 Chapter Summary

After a software product has been developed in a software project, it needs to be released so that the customer or the users can use it. Software release is a phase in any software development life cycle where the software product is already complete and ready to be deployed. A software release involves handing over the following artifacts to the customer: software product, user and technical manuals, and any other supporting material. End user training is also part of software release.

Apart from the regular release of a software product, other kinds of software release are also possible, namely, alpha and beta releases. The primary goal of alpha and beta releases is to get the software product tested for free by the customers.

There are many incremental releases for projects that are done using agile methodologies. These releases could be minor or major. Each new release of the software product will have some new product features added.

In cases where a Commercial Off-the-Shelf (COTS) or Software as a Service (SaaS) product needs to be implemented, data migration from an existing legacy system needs to be performed as well (if such a legacy system exists). Data migration involves the migration of the existing customer data from the legacy system to the newly implemented COTS or SaaS product.

## QUESTIONS

1. What is software release?
2. How is the release cycle carried out for agile projects?
3. What is an alpha release?
4. What is an incremental release?
5. What items are there in a software release checklist?
6. What is software migration?

## Recommended Reading

Michael E. Bays (1999), *Software Release Management*, 1st Edition, Prentice Hall, Upper Saddle River, NJ.

Louis J. Taborda (2011), *Enterprise Release Management: Agile Delivery of a Strategic Change Portfolio*, Artech House, Norwood, MA.

# 11
# SOFTWARE MAINTENANCE

**In Chapter 10, we learned**

- **What a software release is**
- **What a release cycle is for Waterfall model–based projects**
- **What a release cycle is for agile model–based projects**
- **What activities are performed in a software release**

**In Chapter 11, we will learn**

- **What software maintenance is**
- **What the steps in a maintenance process are**
- **What the strategies of software maintenance are**
- **What software maintenance types are**
- **What reverse engineering is**

## 11.1 Introduction

Software maintenance is performed on a software product after the end users start using it. Software maintenance is performed to rectify any software defects or to enhance the software product. Techniques such as reverse engineering are used to perform the maintenance of a software product when its source code is not available.

Monitoring and maintaining a software product after it is taken into a production (i.e., working) environment is also part of software maintenance. A software product is said to be in the production environment if it is deployed at the customer site and is being used by the end users to perform their regular work.

Software maintenance is costly. It is estimated that the maintenance cost is more than 80% of the total cost incurred in the development and maintenance of a software product. This fact highlights the importance of software maintenance.

To maintain a software product, a software maintenance project can be planned and then executed much like a software development project. During its life, a software product might have undergone many such maintenance projects performed either for defect fixing or for enhancement purposes.

### 11.2 Software Maintenance and Software Engineering Methodology

A software product developed using the Waterfall model will be in maintenance mode after it has been released and used by the end users. Any changes to the software product have to be done by taking a maintenance project. For Waterfall model–based software products, a maintenance project is always a separate project from the software development project.

When a software product is built using incremental development models, a maintenance project and the software development project can occur at the same time. A development team may be working on building the next version of the software product while, at the same time, a maintenance team is working on fixing some defects for an older version of the software product. This is in contrast to the situation where a development project and a maintenance project for the same software product are always taken up separately for software products in the Waterfall model.

Generally, software vendors who build a software product incrementally provide support services for the older versions of that software product. These services may include providing production environment monitoring support and maintenance of the software product. Previously, these jobs were handled either by the internal IT teams of the customer or by some third-party software service providers.

### 11.3 Production Environment Maintenance

What is a production system? A production system is a software product that is being used by the end users (do not get confused with the definition of production system in artificial intelligence, which has a different meaning). The end users will be performing some transactions using the software product. A lot of transaction data will be generated and then saved in the database. Such products cannot be changed without making a plan beforehand and getting approval from the customer (who owns the product) because the end users are using that product round the clock. The goal of the maintenance project team is to make the required changes to the production environment without affecting the end users. The production system should not be down during maintenance. The product as well as the master and transaction data should not be changed. Taking on a maintenance project in a production environment is a real challenge.

To do a maintenance project on a product that is in a production environment, a parallel (i.e., new) system should be set up and all the work related to the maintenance project should be performed on that parallel system. After making all the changes, the parallel system should be tested to verify if it is working properly. Once things look fine with this parallel system, the changes performed on it can be replicated in the production system.

While performing production environment maintenance, the version of the software product is also an important consideration. Suppose Version 7 of the software product is under development and both Versions 5 and 6 are being used in the production system. If maintenance needs to be performed on Version 6 of the software product, then care should be taken that other versions (e.g., Version 5) of the software product do not get affected by the changes made to Version 6. It is quite possible that both Versions 5 and 6 are installed on the same computer; thus, there is a possibility that changes will be made to the wrong version. Keeping the versions separate from each other requires careful handling. If the company is also maintaining some customer-specific versions in the production environment, then, again, version control is a must. A customer-specific version of a software product contains software features that are specifically built for a particular customer. Those features are not available in the standard version of the software product. If a software vendor has many customers, then it is possible that the software vendor has built many customer-specific versions of the software product. In such a scenario, the software vendor must take care of maintaining these different versions. A version and configuration control software may be required for this. We will learn about configuration and version control in Chapter 12.

### 11.4 Production Environment Monitoring

Companies that run their business using mission-critical software products want to make sure that the uptime of their software production environment is very high. For a company, a mission-critical software product is the one that is used by the users to perform tasks that are considered critical by that company. For example, if a software product is used to take care of purchase orders and purchase-related transactions, then that can be considered mission critical. If, for any reason, the production environment is down, then those companies may incur heavy monetary losses. The downtime of a production environment can be attributed to any reason such as network failure, hacking, hardware failure, data loss, or software defect. To minimize such downtime, these companies invest heavily in technologies that help minimize such occurrences. To prevent downtime caused by software defects, these companies may have a contract with the software vendors or any other service providers to monitor the production environment for finding and fixing the defects.

The production monitoring team does sanity tests on a daily basis. The sanity test includes running the test scripts that check if all the critical functionality of the software is fine. If any errors are detected, then they are reported immediately. A software developer or a project manager from the vendor or the service provider may be available to fix or to contact the right person to fix the problem immediately. This kind of monitoring of the production environment is done round the clock every day.

### 11.5 Maintenance Process

Software maintenance is a process involving many tasks. The maintenance process can be depicted by a continuous series of subprocesses where these subprocesses can be planned and then executed. As shown in Figure 11.1, the process starts when a request is made for the maintenance of a product. The tasks that are included in the maintenance are specified in a list. Each task in that list is also assigned a priority. The requirements are classified and identified in that list. Classification puts all the related requirements in one place so that all of them can be taken together. Otherwise, some of them could be missed. Identification of maintenance requirements deals with finding out the pieces of source codes (in the existing product) that need to be changed.

The next step in the maintenance process is to do an analysis on how the changes to the existing source code or the addition of a new source code will achieve the required functionality specified in the maintenance request. Then, a design of the software components that require changes as per the identification of maintenance requirements will be made. This software design may be a new design for the addition of new source code or may be a redesign for changing a piece of existing source code. Once the design is complete, then the implementation work begins. The project team will write the new source code or change the existing source code as per the design.

Once all the source code writing is completed, the system will be tested for validation (software testing). The tests will determine whether the changes made to the system are fine and the system is usable. If any defects are found, then they will be fixed. Once the system is certified to be reasonably good, then the end users will be asked to perform acceptance testing. If acceptance testing is successful, then the system can be taken for deployment.

This maintenance process can be used whenever there is a need for software maintenance. Thus, we can see that there may be many maintenance cycles occurring on a software product during its life. In each maintenance cycle, care should be taken to ensure that the documentation is completed properly as per the maintenance. Without proper documentation, later, it will be extremely difficult to understand how, why, and where the changes were made to the system.

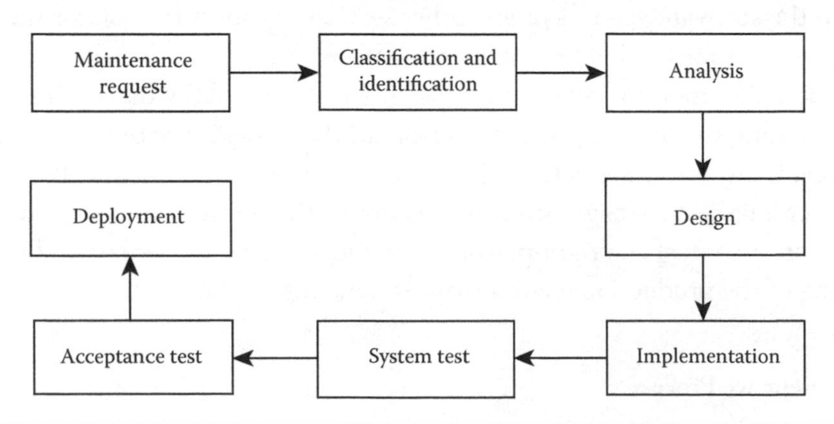


**Figure 11.1** Maintenance project life cycle.

### 11.6 Types of Software Maintenance

There are different types of software maintenance and any of them may be performed on a software product. Some of them include the following:

- Adaptive maintenance: Adaptive maintenance is done when the environment in which you use your software product is changed and you want your software product to work with this changed environment. For example, your software product is built to run on a specific operating system. This operating system is no longer supported and you are forced to use your software product on some other operating system. In this case, adaptive maintenance needs to be performed to ensure that the software product is successfully ported to the new operating system.
- Perfective maintenance: Perfective maintenance is done to incorporate changed user requirements. For example, an order management system is currently able to take customer orders for automobile service parts. Now, the customer has added a new business line that deals with selling computer accessories. The current system is not designed or configured for it to be used to take orders for computer accessories. Hence, a new set of user requirements has arisen and the software product needs to be able to handle these changed requirements. In this case, perfective maintenance is needed.
- Corrective maintenance: If the end users found some defects in the software product while using it, then the maintenance that is conducted on the product is corrective maintenance. In this maintenance, all the reported defects are fixed.
- Preventive maintenance: Preventive maintenance is done in advance even before any defects are reported. In most cases, it is done during software development itself. Many types of errors that may occur when you run the product can be caught through error handling routines in the source code. For example, in your code, you want to connect to a database and you have provided the database connection information such as database name and database driver to be used. If, for some reason, the database is not available, then if your source code is run, it will raise an error and the software product will stop responding to any user requests. If you have provided an exception handling source code to such a situation, then when the database connection fails during the operation of the software product, the user will be provided with a friendly message regarding the problem and the user can work on some other task where the connection with the database is not needed. Exception handling is a very good way of allowing users to continue their work with the software product even after some errors occur while running the software product. This kind of maintenance is known as preventive maintenance.

### 11.7 Software Maintenance Strategies

When end users use a newly released software product, they may encounter some software defects. The question is, when will these software defects be fixed? Taking a software maintenance project involves effort similar to that for a software development project. You need to have software engineers and a project team who can take the maintenance assignments.

You need to devise strategies on how you will do maintenance for a software product. Sometimes, you may need to fix the software defects urgently, while at other times, a software defect can be fixed at a later time. Similarly, a defect can be fixed individually or at times it can be fixed together with other defects.

- Fix immediately: In some environments, fixing the individual software defects immediately is possible. If a software product is being developed incrementally and, at the same time, a previous version of that software product is already in production, then fixing the software defects (in the production version) immediately is possible. If the software development project team has access to the production environment and is also responsible for maintaining the production environment, then individual software defects can be fixed immediately. This is especially true if the software product is web based. Fixing a defect in a web-based software product is easier and faster because it allows the project team to work remotely on the software product and fix the problem. In contrast, if a software product is hosted in-house, then the project team has to physically arrive at the customer site and then fix the defect. It is not possible for a project team to regularly visit a customer site physically and fix a defect each time.
- Fix periodically: If the software development project team is responsible for the production environment but has no access to the production environment on a regular basis, then fixing the individual software defects is not feasible. This will be the case when the production environment is behind a firewall (hosted in-house). In these cases, the project team can record all the reported defects and then do the maintenance work periodically.
- Fix as a separate maintenance project: If the software development project team is not responsible for the production environment, then a separate maintenance project team is required to fix the defects. In this case, the customer can invite the developer or any other service provider for the product maintenance. All the reported defects will be recorded and presented to this project team. This team will study the defects and make a maintenance project plan. The team will then make a software design to fix the problems. On the basis of the design, additional source code may be needed or the existing source code may need to be modified. The system will be tested after making changes to the source code. Once the testing results are fine, then the modified product will be deployed in the production environment. In this kind of project,

it is better to do the maintenance work (i.e., development work related to the maintenance project) in a separate environment. For example, the project team will design and build the software product on a separate computer. After testing, if the modified software product is found to be working fine, then this build of the software product can be replicated on the production environment.

- Using a patch: Many software vendors create a patch to fix the software defects and then make this patch available to their customers. Customers or users can themselves apply the patch to the product. Patches are developed for standard software products. For example, software vendors like Microsoft, Oracle, and SAP AG provide such patches for their standard software products. A patch to fix a software defect may not work for a specific customer if the original source code of the software product is modified to suit the needs of that specific customer. For example, SAP ERP is often implemented at the customer sites with some changes in the original source code of the software product to suit the specific needs of a customer. In such cases, when SAP releases new patches to fix the known defects in their product, these patches may not work with the installation of SAP ERP at some customer sites because the source code is changed. In such cases, SAP AG (the company that builds the SAP ERP) provides instructions, along with the release of each such patch, on how to apply that patch when changes have been made in the source code.

### 11.8 Reverse Engineering

A software product is built by writing the source code. Once the software product is compiled, the source code is converted into binary code. Actually, this is the binary code that runs on the users' computers and with which the users interact. All software vendors who sell proprietary software products sell only the binary code. They keep the source code of their products to themselves as a trade secret so that nobody can use that source code and build a competing or better product.

Suppose the binary code of a software product exists but there is no source code available. If someone wants to build the same software product or wants to enhance that software product, how will they be able to do it? What if the software vendor who built that software product is no longer in business or does not support that product? In such cases, reverse engineering can be used. A project team can be set up to study the features of that existing software product and this team may make a design to recreate that software product. On the basis of that design, the project team can write the source code and build the software product (Figure 11.2).

An important aspect of reverse engineering for software products is licensing and copyright issues. Generally, if you want to build a copyrighted software product, then it will be an infringement of the copyright laws. However, if you build a software product using reverse engineering and if the final product is significantly different

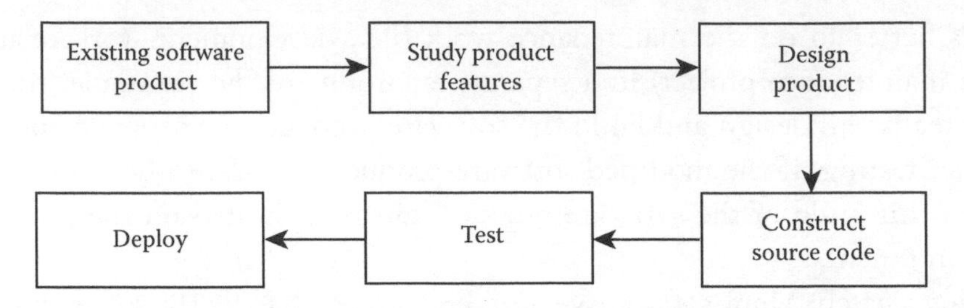


**Figure 11.2** Reverse engineering process.

from the original product, then copyright law may not be applicable. You should make sure that you are not violating any copyright laws before taking any reverse engineering initiative.

### 11.9 Chapter Summary

Software maintenance is a process that is used to either fix a software defect or enhance an existing software product, after the software product has been developed and has been used by the users. When a software product is deployed at a customer site, it is known as a production instance of the software product. When a software product is modified by changing its source code to suit the needs of a specific customer, that version of the software product is known as a customer-specific version. When this is not the case and the software product is installed without changing any source code, that version of the software product is known as the standard version.

Maintenance of a software product can be carried out by creating a project that is known as a software maintenance project. The project team will study all the known defects and determine the problem areas in the source code from where a software defect is originating. Once this analysis is done, then a design for the software product can be made. This design incorporates the changes that are needed to fix the defects. Based on this design, the existing source code may be changed or a fresh source code will be written. Once all the changes (that are supposed to fix all the defects) are made to the source code, that version of the software product will be tested. If it works fine, then it will be replicated in the production environment.

Fixing individual defects one at time is not always possible. In such cases, a maintenance project can be taken up to fix all the known defects by changing the source code. If a project team is responsible for fixing the defects in a production environment on a continuous basis, then fixing the individual defects is possible. In such a scenario, the project team has access to the production environment and the team can modify any source code to fix a problem at any time. Depending on the options available, a maintenance strategy can be made to fix the software defects.

Sometimes, for some reason, the source code of a software product is not available. In such cases, if a software defect has to be removed or an enhancement needs to be made on that software product, then reverse engineering is used. In reverse engineering, the software product is studied for its behavior during user interaction or for its response to some events. On the basis of this study, a project can be instituted to construct a software product that looks and behaves exactly like the original software product.

## QUESTIONS

1. What is software maintenance?
2. What strategies can be used to maintain a software product?
3. What is reverse engineering? How is it used for software products?
4. What is a production instance of a software product?
5. What is a customer-specific version of a software product?
6. What are the phases in a software maintenance project? How are they different from a software development project?

# Recommended Reading

Macario Polo, Mario Piattini, Francisco Ruiz (2003), *Advances in Software Maintenance Management: Technologies and Solutions*, Idea Group Publishing, Hershey, PA.

Donald J. Reifer (2012), *Software Maintenance Success Recipes*, Auerbach Publications, Boca Raton, FL.

# 12 Configuration and Version Management

**In Chapter 11, we learned**

- **What software maintenance is**
- **What the steps in a maintenance process are**
- **What the strategies of software maintenance are**
- **What software maintenance types are**
- **What reverse engineering is**

**In Chapter 12, we will learn**

- **What software configuration management is**
- **What software version control is**
- **What a software build is**
- **How configuration and version control systems work**
- **What continuous integration of source code is**

## 12.1 Introduction

In this book, so far, we have learned various activities that take place on a software development project. We have also learned all the activities that take place on a maintenance project. When a software development or maintenance project is in progress, many artifacts are generated. These include project plans, software requirement specifications, software design documents, software source code, test scripts, and test results. We have learned what all these artifacts are about. We also know that many versions of the same artifact are generated for many reasons. For example, when a requirement is changed based on the end user request, the changed requirement, along with the corresponding changed design and source code, will have a new version. Generally, a software development team keeps the old version as well as the new version because, in the future, any changes in the new version may have to be reverted. Keeping the old version is important for reference purposes as well. Keeping these different versions results in the generation of several versions of various artifacts.

Managing all these artifacts and their versions is a difficult task because there are continuous changes to these artifacts as the project teams continue to update them (when the project work progresses).

The artifacts on a project can be in two states: work in progress and finished. Work-in-progress artifacts are the ones that are not complete at the time of their assessment. For example, if the project team is currently involved in software design, then the design documents are in work-in-progress status. When the design work is completed, these design documents will be in finished status.

Special skills and tools are required to manage different versions of the artifacts. These tools are known as configuration management and version control systems (CVS systems). There are two parts in managing artifacts: configuration management and version control. Let us look at some details of each of these parts.

Version control is all about managing different versions of the same artifact. For example, the source code of a software component can have more than one version. How are you going to manage these versions? It is possible that, by mistake, a wrong version could be used for further development. To minimize such errors, a version control system is used. A version control system will use branching of different versions and manage all the artifacts of the same version on the same branch. For example, all the artifacts with Version 5 are on the same branch. Branching in version control is similar to a folder in the file system in any operating system. A file system has folders. In each folder, there can be subfolders and files. When you need to refer to a file or a folder, you need to refer to the parent folder and travel inside that tree structure until you reach the file or folder that you need.

Configuration management, on the other hand, keeps all the required files or artifacts in one place so that a particular version of a software product can run. For example, suppose a software product needs a database connection for manipulating the data in a database. If the software product is not connected to the right database, then this product will not work (connection to a database may be based on a connection string that can be specified in a file and this file needs to be located in the filing system from where it can be read correctly). Similarly, if the software product needs some Excel files to be located at a particular location in the file system and if these Excel files are missing, then the software product will not run. Therefore, for example, if Version 3 of a software product needs Version 4 of the Excel file to run, then Version 3 of that software product and Version 4 of that Excel file must be present at their respective places.

Suppose we have a software application inside the file abc.exe (i.e., the name of the software application is abc.exe) located at c:\my-folder and this application needs a file, calculation.xls (i.e., an Excel file), to run. In that application, it is defined that this Excel file should be available at the location \compute relative to the location of that application file itself. You then need to ensure that this calculation.xls file is available at c:\my-folder\compute. The other aspect of configuration management deals with the right version of the files in conjunction with version control. Thus, if abc.exe needed Version 4 of the calculation.xls file, then the location c:\my-folder\compute must contain Version 4 of the calculation.xls file. If any other version of the calculation.xls file is located at c:\my-folder\compute, then the software application will not run.

Configuration management is more related to a software build. A software build is made up of compiled source code files along with other required files (as seen in the previous paragraph). A software build is thus the repository of all the abovementioned files. The files in a build must be at the right place and the right versions of the files should be used in a particular build. Only then can the build run. If a file is not at the right place or a wrong version of a file is used in making that build, then that build will not run.

A software build is the most important artifact in any software development project. The process of making a software build starts when a software developer creates a new file. This file contains some programming source code. Now, this file should be integrated with the software build. To do this, the software developer checks his or her file in the software build. If more than one version of that software product is maintained during the project, then there could be many versions of such software builds. When there are many software builds available, the project manager decides which build to use for further development of the software product. Generally, when a new version of a software product starts to be used for the project, a new branch in the configuration and version control system is created. All the new versions of the project artifacts are then developed and saved in this new branch. The new software build is also saved in this new branch.

We will learn about branching and configuration and version control concepts in the following sections of this chapter.

**Usage notes:** Currently, CVS systems are categorized as centralized CVS systems and distributed CVS systems. The naming conventions used for CVS systems are somewhat confusing. In this book, we will use the term *centralized CVS system* for both centralized and distributed CVS systems. This is because, in software development projects, truly distributed CVS systems are not used. In essence, distributed CVS systems are an improvement over the old-type centralized CVS systems.

## 12.2 Configuration Management and Version Control Concepts

Configuration management and version control are carried out using a CVS system. A CVS system can be automated or manual. Nowadays, automated CVS systems are used on most software projects because several CVS tools are available in the market. Doing configuration and version control manually is very difficult and prone to mistakes, whereas CVS tools provide a foolproof and fast mechanism for all tasks involved in configuration and version control on any software project. Let us discuss CVS systems and related concepts.

When a software product is built, many versions of various artifacts are generated. Hence, we end up having many versions of the artifacts including requirement specifications, design specifications, source code, and test documents. To incorporate any change request for an artifact, a copy of the work-in-progress artifact (if that artifact is not yet finished) or a copy of the finished artifact is first saved (with a different name) to create another copy of that artifact. Now, the requested change is incorporated into the newly created copy of that artifact. Thus, now we have two copies (versions) of the same artifact: one is without that change request and another is with that change incorporated. In the same way, as time passes by, too many versions of most of the artifacts are thus generated for various reasons. By the end of the project, a large number of versions for each of the artifacts and possibly several builds will be created.

All the artifacts that are generated during the software development project need to be organized in such a way that they are easily accessible whenever they are needed by a software project team member. Generally, artifacts on a software project are saved as an electronic file. For example, if a software designer has created a software design and now wants to keep this design in a place where it can be accessed by any of the software team members, then what options does he or she have? He or she can make a copy of the file containing the design on a server that is accessible to everyone. On a server, there are many places where this file can be placed. One option is that a file system (or a directory) is created on the server specifically for the design document and other artifacts at specific places defined inside the file system of the server. However, there could still be a problem. The problem is related to different versions of the same artifact as well as many copies of the same version of the artifact with different team members. It is important to differentiate between a copy and a version of a file. When a file is updated and then saved as a new version, we end up having two different versions of that file and the contents inside the new version of that file will be different from the contents inside the earlier version of that file. On the other hand, if we copy a file and save it in a file system (without making any changes to that file), then we will end up having two different copies of that file and the contents inside these two files are exactly the same. We will learn more about this problem in Section 12.2.1.

### *12.2.1 Artifact Versions and Many Copies*

If you look at Figure 12.1, you will realize that there could be many versions of the same artifact located both on the local computer of the developer and on the server. Many versions of the same artifact are generated on almost all projects because of the incorporation of change requests into that artifact. Generally, the original artifact (e.g., Version 1) is saved on the local computer (when that artifact was created and then saved for the first time on the local computer of a software developer) as well as on the server (when that software developer saved that artifact on the server as well). After some time, when the change requirement is incorporated into that original artifact, again, a copy (e.g., Version 2) is saved on the local computer of that software developer and then on the server. After some time, another change request is received by that software developer and a third version (Version 3) is also created and saved

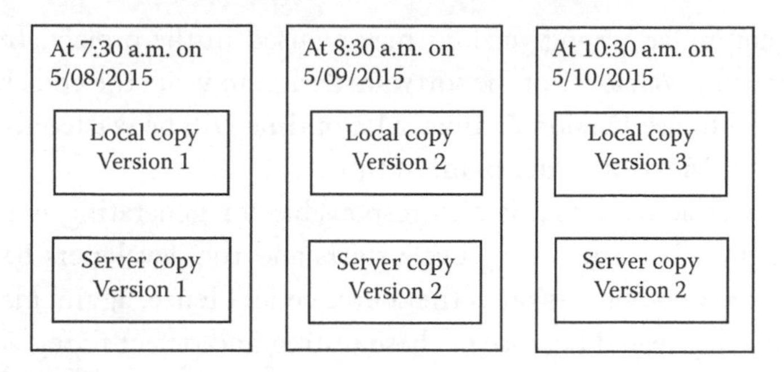


**Figure 12.1** Many copies of an artifact on a local computer and on the server.

on the local computer. However, this time, by mistake, that software developer saved Version 2 on the server. In Figure 12.1, you can see that Version 2 of that artifact (i.e., the same copy) is saved on two different dates (and at two different places) on the server. This kind of mistake is possible because the files containing various versions of the artifacts are being saved in different folders. Hence, although the file names are the same, they have been saved in different folders. All these versions of the same artifact are confusing; therefore, a mistake can be made in choosing the right version of that artifact for further development.

This problem of having many versions and copies of the artifact can happen for the following reasons:

- A wrong file name was used at the time of saving the file: Suppose a designer was working on Version 2 of the design and by mistake he or she saved the file as Version 3.
- Wrong folder: By mistake, a right version of an artifact is saved in a wrong folder.
- Many copies of an artifact version: There is one copy of an artifact at the software developer's local computer. When the developer copies this file on the server, we have two copies of the same file: one on the local computer and another on the server.
- Version confusion: Suppose the software developer has made some changes in the file (but did not create a new version of that file) and kept it in his or her local computer. However, the software developer forgot to update this file on the server. Now, although both files have the same name and version number, the content inside those files is not the same.
- Same file with different names: Suppose there is a file on the server and, by some mistake, its name is changed. Now, it will be difficult to find out the version of that file.

There is another problem regarding the files and their versions. Some people do not remember where they have previously saved a copy of their artifact in the file system; thus, they end up making more than one copy of the same version of the artifact on their local machines as well as on the server. Once there is a mix of versions and their copies on any computer, wrong versions may be used in the project. Imagine a software developer using Version 1 of the software design to write the code but actually he or she is supposed to use Version 2. Thus, a lot of time will be wasted for rework upon discovery of this mistake at a later point in time.

There is another activity that is also responsible for generating many versions of the same artifact. When the testing cycle starts and the developers fix the software defects, they create another version of the source code. Hence, again, many versions of that source code are created because of these testing and defect fixing activities.

### *12.2.2 Limitations of the File System of an Operating System*

Although the file system of any operating system provides many good features for the configuration and version control of the files in a project, this kind of system has some limitations:

- It is difficult to provide synchronization between a file on a local computer and a file on the server. A note on synchronization: Two different files, if synchronized, will have exactly the same contents after synchronization. Synchronization of files between the local computer of a software engineer and the server is required because the software engineer keeps working on the same file at his local computer to make necessary changes when required. First, he or she makes changes to the file on his or her local computer and then synchronizes this file with the existing file on the server so that the copy of the file on the server is also updated with the changes he or she has done to the local file. If the software engineer has created a new file (an entirely new file or a new version of an existing file) on his local computer, then, first, he or she saves it on his or her local computer and then uploads this file on the server so that the configuration of the file system on the local computer becomes the same as that on the server. A file management system of an operating system cannot provide foolproof management of synchronization of files.
- If the files need to be managed across many computers, then a filing system that can effectively manage all these files across the network of computers is needed. However, a filing system of an operating system is meant to work only on the computer on which the operating system is installed.
- The file system of an operating system offers only version control facilities. It does not offer configuration management facilities.

A file management system of an operating system of a computer is not capable of looking inside the files that may reside on other computers, to find out if those files are the same as the files located in its own file management system. This weakness results in the problems we discussed in Section 12.2.1. A good system should be able to find the differences between the files. This functionality is achieved through CVS systems.

*12.2.3 Software Build*

When a new file (source code file) is saved on a software build, this new file will be integrated with the existing files residing on that software build. When a software developer checks in (submits) his or her source code (in the form of a file) on a software build, the code is first compiled; only then it is integrated with the software build. The software build should always be in a compiled state. If, for some reason, a new file integration (with the software build) results in compiling problems, then the software build status changes to a failed (or broken) status. This creates problems because other developers will not be able to integrate their files (containing the source code) to that failed software build. This is because once a software build is in a failed state, even if those files to be integrated have no compilation problems, integrating them with the failed software build will not change the software build's (failed) status. Thus, the developers submitting their files on that failed software build do not know whether their files are good enough (i.e., error-free compilation).

Maintaining a software build is done through a specialized arrangement. A software build system is used to manage the software builds. However, these software build systems do not integrate with the file system of any operating system. When a software developer wants to check a file into the software build maintained at the server, some checks are necessary before doing this operation. We will learn about these checks in Section 12.2.4.

*12.2.4 File Locking*

A CVS system is used by many users. A file on a CVS server can be accessed by more than one user. In such a situation, changes to the file can be made by more than one person at the same time. This can lead to unpredictable file editing. To control this situation, file locking is used by CVS systems. When a user checks out a file from the server, this file will be locked by the CVS system. Now, if another user tries to check out this file from the server, then the system will display a message that this file is locked and cannot be edited. The other user can only view that file because if he or she makes any changes to that file, the system will not allow him or her to save those changes.

When the first user makes changes in the file and now wants to check in that modified file on the server, the system will allow him or her to check in that modified file. Once the first user has checked in that (modified) file, it can be checked out by another user to make some changes. This is because, after check-in by the first user, there is no lock on the file; thus, it can be checked out for making any changes.

**12.3 CVS Systems**

CVS systems are used on software projects to manage all the artifacts that are generated on the project. We have learned about the limitations of a file system of the operating systems. To overcome these limitations, CVS systems are used for these projects. Configuration and version control for a project can be done centrally or can be distributed. If the team is using a distributed CVS system, then care should be taken to synchronize all these distributed systems.

### *12.3.1 Software Build on CVS Systems*

The older centralized CVS systems used to have a central repository for keeping all the files and folders (i.e., all the different artifacts with different versions) at a central server. The software build (with which the currently developed source code should be integrated) would also be located on this server.

In this older scheme of things, any team member can create an artifact on his or her local computer and upload it on the CVS server. If other team members want to modify this file, then they can download a copy of that artifact and use it. This arrangement is good for all the artifacts except for the files containing the source code. If a file (containing the source code) is checked in a software build and this file is not integrated with the software build, then the compiler will throw compilation errors. When this happens, the software build is said to be in a failed state. This software build, which is in the failed state, will prevent the rest of the team members from integrating their files (containing the source code) with that software build, as explained earlier. This is why a software build should always be in a passed state so that the source code files can be integrated with it.

The problem with having just a central software build (i.e., just the software build on the server) is that there is no way for a software developer to check if the source code file that he or she has written will be compiled neatly and there is no chance of the software build getting broken. In general, there are several software developers in a team; therefore, the central software build is in a broken state very often. If the software build is broken, then it will be difficult for any software developer to check in his or her source code files to that software build. Since most software development projects use continuous integration of source code (we learned about continuous integration in Chapter 9; more details on continuous integration are provided in a later section of this chapter), a better way of managing the software build is needed.

To overcome this problem, a better CVS system has evolved. In this better CVS system, each software developer has a local software build on his or her computer. This local software build is synchronized with the central software build. Although this arrangement is popularly known as a distributed CVS system in the software engineering community, it is almost similar to a client–server architecture and it still needs a central server; all the local software builds need to be synchronized with the central software build. In effect, this type of distributed CVS system works more like a central CVS system than a peer-to-peer CVS system.

In a peer-to-peer arrangement, any file can be kept at any one of the peers and all other members of that peer-to-peer network can have access to this file. Any peer in this arrangement can also check in and check out any file located on any of the peers.

We can clearly see that the distributed CVS systems used on software projects do not behave like a peer-to-peer system. They behave more like a central CVS system.

### *12.3.2 Branching in CVS Systems*

Creating a new branch in a CVS system is very similar to creating a new directory in the file system of an operating system. For example, in a Windows operating system, you can create a new directory to house all the files you want to keep in one place. Inside that directory, you can create subdirectories to house related files. The directory and subdirectories form a tree structure. The files (inside this directory and subdirectories) are like leaves. A CVS system with all its branches, subbranches, and files forms a similar tree structure.

In Figure 12.2 you can see this tree structure. There is the main trunk on which the branches (Branches 1–3) are located (the trunk is the vertical straight line, shown at the left-hand side of this figure, to which these branches are attached). On each of these branches, there are files or documents (Design documents version 1, version 2, etc.). These documents are like the leaves (i.e., leaf nodes) of that tree.

When you want to build a new version of a software product, you create new versions of the artifacts for the software product. In CVS, a new branch will be created to house all the new versions of the artifacts. When a new branch (in the tree structure) is created in a CVS, you have two choices: create it without copying anything from the previous version or create it by copying everything from the previous version. Which choice is better? Creating by copying all the artifacts from the previous version is practical. This is because there can be thousands of existing artifacts on an

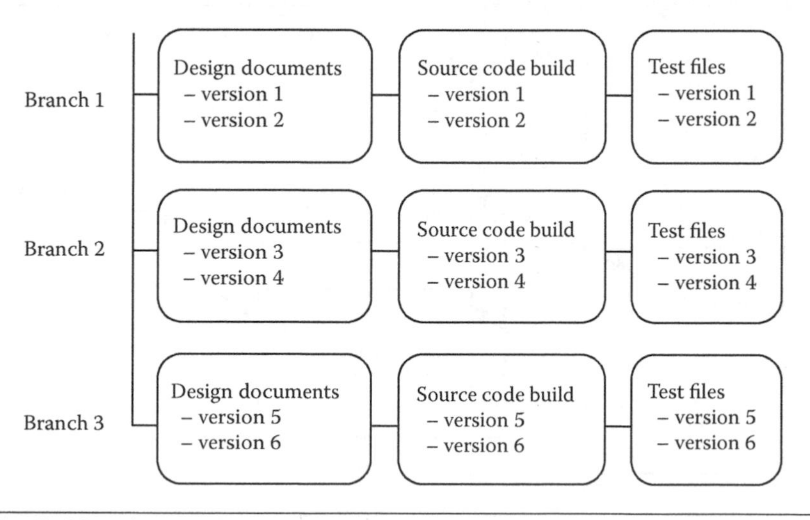


**Figure 12.2** Each branch represents one or more versions of all the artifacts in CVS.

ongoing project. Existing artifacts can be used as the baseline artifacts for creating the new version of the product. If required, you can modify these artifacts or insert new information into these artifacts, or you can remove some existing artifacts or add some new artifacts. The biggest advantage of having copies of the previous version comes in the area of software source code. The entire source code can be copied into the new branch intact and be modified based on the new version of the design artifact. This approach is specifically useful when the source code is added incrementally to the existing source code.

Figure 12.3 depicts a centralized CVS system. This figure also shows the server part of the CVS system and its branches as well as the client CVS systems connected to the server CVS system. How the file synchronization happens between a branch of a server CVS system and a client CVS system will be explained later. A branch can contain many versions of the same file or a single version of each file. It depends entirely on the need of the project.

In the centralized CVS systems, file synchronization happens between a client CVS system and the server of the centralized CVS system (or with a branch of the server CVS with which that client CVS system is working as a client).

Branches are created in a centralized CVS system to keep different versions of the same software product. If a person is working on some file from a specific branch (say Branch 1) of the server CVS system, then his or her client CVS system should have the same files (with the latest updated copies) as those on that specific branch (Branch 1) of the server CVS system if that person wants to check in (submit) the source code of a file (that he is currently working) on the software build on Branch 1. This means that the files in Branch 1 and the client CVS system of that person should be synchronized. The files in that person's client CVS system do not need to be synchronized with the files in another branch (say Branch 2) of the server CVS system.

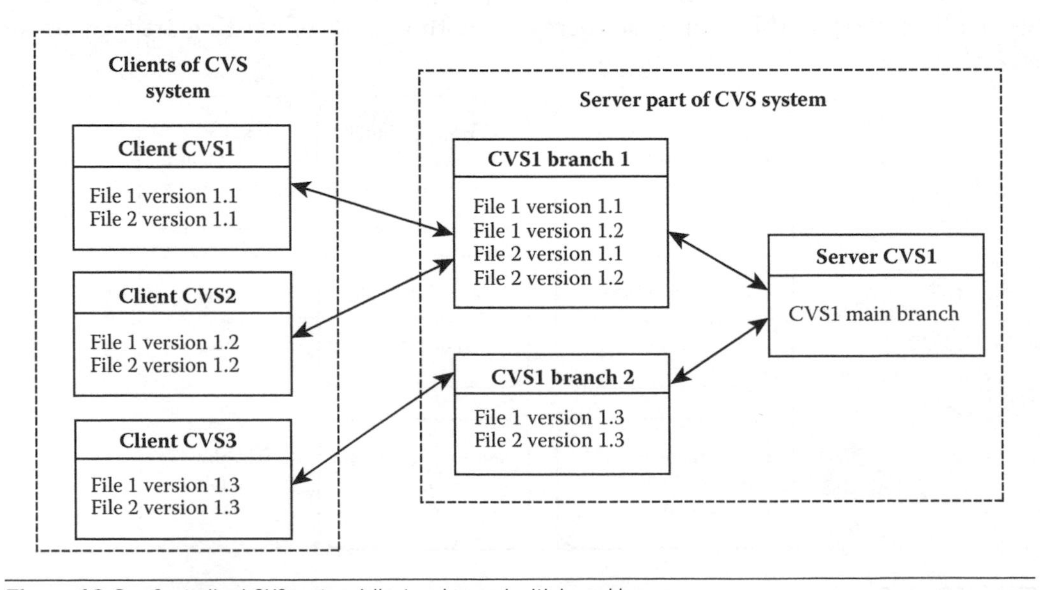


**Figure 12.3** Centralized CVS system (client and server) with branching.

How a centralized CVS system works can be understood by referring to Figure 12.3. Figure 12.3 depicts a centralized CVS system. It shows a CVS system: CVS1. CVS1 has three branches: the CVS1 main branch, CVS1 branch 1, and CVS1 branch 2. Two clients, Client CVS1 and Client CVS2, are currently synchronized with CVS1 branch 1 of Server CVS1. Client CVS3 is synchronized with CVS1 branch 2 of Server CVS1.

From the current configuration of the CVS system, you can see that Client CVS1 can synchronize files on CVS1 branch 1. Client CVS2 can also do the same with CVS1 branch 1. However, Client CVS3 is configured to synchronize only with CVS1 branch 2 of Server CVS1. Currently, the files File 1 version 1.1 and File 2 version 1.1 reside on Client CVS1. Similarly, File 1 version 1.2 and File 2 version 1.2 reside on Client CVS2.

Suppose all the files on CVS1 branch 1 belong to a software build. That is, assume that all the files residing on this branch (File 1 version 1.1, File 2 version 1.1, File 1 version 1.2, and File 2 version 1.2) are part of this software build. Assume that currently the software build on this CVS1 branch 1 is in the passed state. In addition, assume that currently the software build on Client CVS1 is in the passed state and it comprises files File 1 version 1.1 and File 2 version 1.1. Suppose a software developer who is using Client CVS1 has created a new file, "File version 3.1." The software developer has checked in this file with the local software build on his or her local computer (Client CVS1). The software developer has found that the new file has integrated with that local software build without any compilation problems. What could happen when the software developer tries to integrate (i.e., check in) the newly created file (File version 3.1) with the central build on CVS1 branch 1? If the software build (i.e., central build on CVS1 branch 1) is configured for continuous integration, then the software build will be broken after the newly created file (File version 3.1) is integrated with this software build. Why? This is because the local build on Client CVS1 and the central build on CVS1 branch 1 are not synchronized. If they are synchronized, then both of them would have contained the same number of files with exactly the same version of files at each place. However, you can see that the central build CVS1 branch 1 contains four files (File 1 version 1.1, File 2 version 1.1, File 1 version 1.2, and File 2 version 1.2) but the local build on Client CVS1 only contains two files (File 1 version 1.1 and File 2 version 1.1). An important requirement for a software build in a distributed (client–server) environment for continuous integration is that the local (client) build and the central (server) build must have exactly the same configuration (i.e., the same number of files with exactly the same version of files). If there is any difference between these two builds, then the build will fail when a new file is integrated.

To overcome this problem of failing build, the software developer must synchronize his or her local build with the central build before integrating any of his newly created file with the central build.

### *12.3.3 Branches and Different Product Releases*

As mentioned earlier, to take care of the problem of having more than one copy of the same version of an artifact, there is a branch concept in CVS. A branch in a CVS system contains all the artifacts related to the same version of the software product that is being developed (note: this is somewhat different for agile projects, as explained in the next paragraph). For example, in Branch 1 of the CVS system, all the artifacts belonging to Version 1 of the software product are maintained. Similarly, in Branch 2, all the artifacts of Version 2 of the software product are maintained, and so on.

If you look at Figure 12.2, you will notice that there is more than one version of the same artifact in the same branch. This is in contrast to the fact mentioned in the previous paragraph that states that only one version of any artifact should be present in one branch. Actually, both statements are correct. In agile environments, software product release is done frequently because the new features are added after each iteration. Hence, at any given point in time, there are many releases of the same software product. The source code, along with all the project artifacts, is maintained on one (i.e., on the same) branch of the CVS system for one release of the software product. However, there could be minor releases of the software product as well. This means that there could be many versions of the source code and project artifacts in any major software product release. Each branch in the CVS system can be mapped to a major release of the software product. This scenario is depicted in Figure 12.2. In this figure, you can see that there is more than one version of the same artifact inside one branch in the CVS system. For example, Branch 1 contains two different versions (Versions 1 and 2) of the source code in the "Source code build." This means that this branch of the CVS system is holding one major version (e.g., Version 1) and one minor version (e.g., Version 2) of the software product.

### *12.3.4 File Synchronization in a CVS System*

CVS systems have been built to take care of all the requirements for managing the files and software builds on a software project. Although the software build systems are separate from a CVS system, most CVS systems integrate well with these software build systems and provide a seamless operation. File synchronization and version management of files are the essential parts of any CVS system.

Let us understand how the file synchronization and version management of files are handled in a CVS system.

A CVS system consists of two parts: a client part and a server part. The client part is installed on each of the local computers of the software developers (and on the local computers of all people working on the software project). The server part is installed on a server that is centrally located and all the project team members have easy access to that server.

The technique to keep all the files on the server part of a CVS system up to date with the files located on a client part of that CVS of a local computer is achieved by keeping all the files on that local computer always synchronized with the files on the server. A check-in and check-out mechanism is used as explained in the previous section to ensure that only one person can work at a time on a file. This ensures integrity of all files on both the server and local clients of the CVS system.

To ensure that the synchronization of files between the server and local computers is maintained, the complete filing structure and the files should be up to date on each client. If a person wants to check in a file on the server and if his or her local CVS system is not synchronized with the server CVS system, then the server may not accept his or her request. This is because some files may have changed on the server CVS system; thus, the client CVS system (i.e., the files in that client CVS system) on that person's local computer may be outdated. Hence, before a person checks in a file on the server CVS system, he or she needs to synchronize the files in his client CVS system with the files on the server CVS system. Synchronization of all the files in the client CVS system (i.e., on the local computer) happens when the user chooses to update all the files (in the client CVS system) that have been changed on the server CVS system since that user checked out a file from the server or checked in a file to the server.

All the CVS systems allow automatic update of changed files in the client CVS system. Synchronization of files at the server and client CVS systems ensures that the version of each file in the CVS system is always maintained.

In a CVS system, if synchronization of files is required, then doing this synchronization manually will be laborious and error-prone. It is difficult to manually compare two or more documents (or files) to verify if they are the same. Automated tools are available with any CVS system for comparing the documents. Once the files are compared, synchronization can be done. Apart from comparison of files, some tools can automatically synchronize the CVS systems. When synchronization of a complete CVS system is done, all the files residing on the client CVS system will be synchronized with those on the server CVS system (here, server CVS system means the branch of the server CVS system with which that client CVS system is connected). For example, in Figure 12.3, if synchronization is done, then Client CVS1 and CVS1 branch 1 will have the same number of files and the same version of files. However, Client CVS1 is not configured to be synchronized with CVS1 branch 2; therefore, the software developer who is using Client CVS1 will not be able to synchronize his or her Client CVS1 (i.e., the files and other artifacts in his or her Client CVS1) with CVS1 branch 2 (i.e., the files and other artifacts in CVS1 branch 2).

Figure 12.4 depicts the document synchronization in a CVS system.

Figure 12.4 depicts the file synchronization process. If a file needs to be submitted at a location in the CVS system where there is already a file of the same name, then a file comparison will take place. If both files have the same content, then it can be said that both files are synchronized. On the other hand, if contents inside these files are different, then the contents will be merged.

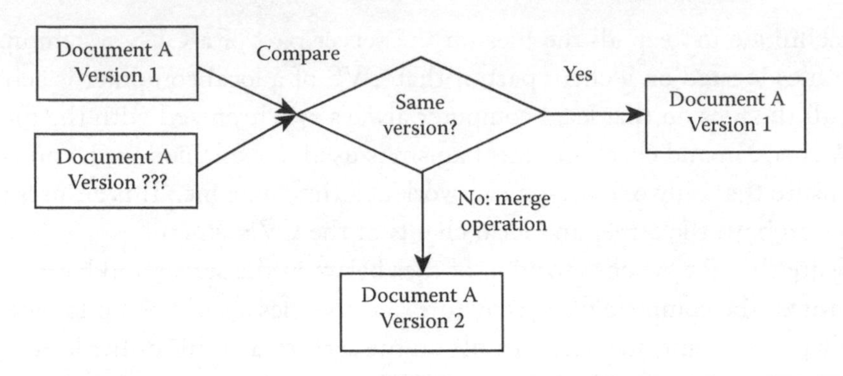


**Figure 12.4** Document comparison and synchronization in a decentralized CVS system.

### *12.3.5 Roles in CVS Systems*

CVS systems are specialized software products. The users use a CVS system for synchronizing the client CVS system files with the server CVS system (or with a branch of the server CVS), checking out a file from the server CVS system (or from a branch of the server CVS) for update, checking in a file in the server CVS system (or in a branch of the server CVS) after file update, checking in a newly created file in the server CVS system (or in a branch of the server CVS), checking out a file from the server CVS system (or from a branch of the server CVS) for viewing it, and so on.

Most CVS systems have at least three types of roles for the people having access to those systems. An account created in a CVS system (for a person) is assigned one of these roles. There is a role for the users of the system. The users can create, update, delete, or view their own files but cannot update or delete the files of the other users. Then, there is a role for the managers of the system. The managers can create, update, view, or delete any files. The system administrator plays the topmost role. The system administrator has the ability to do anything with the system. The administrator can create new users, modify existing users, delete existing users, or modify or update or delete the files. The administrator can also create new branches in the system.

### *12.3.6 Continuous Integration of Source Code*

Most of the software development methods these days are of the agile type. The software product is built incrementally using agile methods. Therefore, continuous build of the software source code is required in agile methods. A software build system should provide this functionality so that the software developers can check in their source code whenever they finish their piece of source code.

Generally, a software build system is divided into two parts: client software build and server software build. First, a developer checks in his or her source code on the client software build that is installed on his or her local computer. If the client software build compiles and integrates the piece of the newly developed source code without any error, then the developer will check in his or her source code on the server software build.

It is important that the client software build is up to date (i.e., in sync with its server software build) to ensure that the piece of source code will compile successfully on both client and server software build. Thus, developers synchronize their client software build with the server software build regularly.

A good software build system will never allow the addition of source code to a failed build. If a build fails, then the configuration management system should allow the build to revert to the last successful version of the build. The developer should modify and check the piece of that source code (because of which the build has failed earlier) and then add it to the build. If the source code is good, then the new build will not throw any error messages.

In Figure 12.5, we can see the smoke test process for software build. The smoke test process starts when a developer checks in his or her source code on the server software build. The server software build will compile after integrating this piece of source code. If the compilation is successful, then the software build sends a message with a passed build status. If the build has passed, then the software build will again be ready for submission of new pieces of source code. On the other hand, if the build fails, then the software build sends a failed status message. A good software build system will generate a detailed message showing the cause of failure. Using this message, the source code can be fixed by fixing the root cause of the failure. The developer will once again check in the fixed source code. This process can be continued if the software build fails again.

When the build is in a failed state, no other developers can check in their pieces of source codes for integration. In this case, no new versions will be created. If the build is in a failed state, then immediately a message will be generated by the smoke testing software.

Developers keep on adding new source code files to a software build until the build fulfills the specified requirements. Once all the source code files have been successfully added and the software build is in a passed state, this software build will be

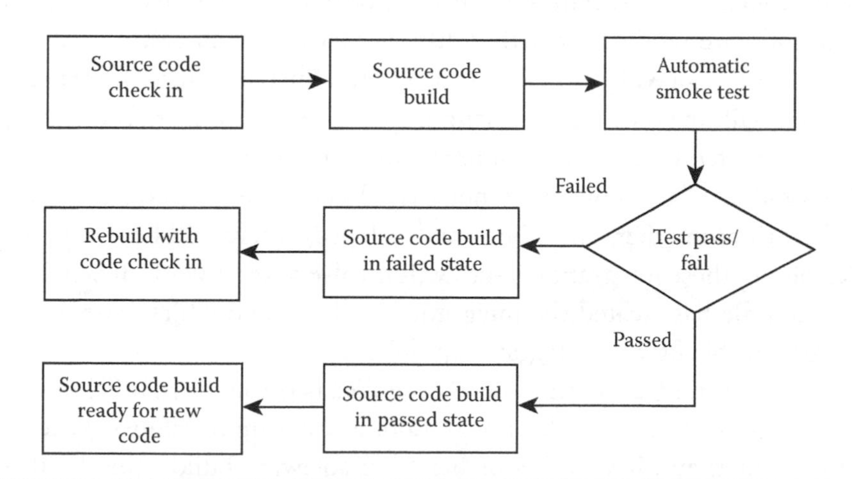


**Figure 12.5** Continuous integration with automatic smoke test.

known as a release of the software product. If this is a minor release of the software product, then this software build can be copied in the same branch of the CVS system with a new name (so that another version of the software product, which will correspond to another minor release of the software product, can be started). The developers will then start adding or changing the source code files to this new software build.

Nowadays, distributed configuration and version control systems (distributed CVS systems) are used on software development projects. In a distributed software system (or a distributed system), all the files and directories that need to be managed are scattered over the computers that are part of that distributed system. Does this same definition hold true for distributed CVS systems as well?

Although a CVS system is classified as a distributed system, its nature is different from that of a true distributed software system. These CVS systems have a central server and local computers and they work like a client–server system. However, the main distinguishing factor for these CVS systems is that all the files residing on a local (client) system must be synchronized with the files residing on the central server.

### *12.3.7 Why Continuous Integration of Source Code?*

Why do files residing on a local system need to be synchronized with files on the central server system of any CVS system in a continuous integration environment? Let us find out.

Suppose a software developer has created a new source code file and tested it on his or her local CVS system. If this file integrates neatly, then it means that the software build on this local CVS system is in the passed state and the build is healthy. This healthy software build is now ready to take/accept other source code files to be integrated.

Each of the software developers can create their own source code files and keep integrating them with their local software build. However, at some point in time, these separate local software builds will need to be integrated with each other to create a complete software product (note that different parts of a software product may be simultaneously developed by different people). This means that each software developer has been building only a component of the software product. These components need to be integrated to create the complete software product.

The integration of components is not easy. When the software components are integrated (with each other, once they are developed), there will be many integration issues. Resolving these integration issues will involve a lot of work. It will be difficult to trace which file has created the integration problem and which part of the source code was responsible for that integration problem.

A better way of integrating the source code files is through what is known as continuous integration. In continuous integration, a software developer writes a source code file and integrates it with his or her local software build. This local software build is installed on the local part of the CVS system on the developer's computer. If the source code file integrates well and there are no compilation errors encountered, then the developer will integrate this file with the software build on the central server (or a branch of the server). In the same way, all other software developers continue writing the source code files and integrating them with the central build. If any of the source code files creates compilation errors, then these errors are fixed immediately. Since the compilation errors during the integration are fixed for each file, fixing these errors is not a big problem. The complete software product is built on the central server by accepting one file at a time. At the same time, there is no need for the integration of the software components (developed by different people to create a product) later because all the source code files belonging to all the components of that product are integrated continuously.

If the configurations on the local software build and the central software build are different, then compilation errors can occur when a developer integrates a source code file on the central software build even if this same file did not create any compilation errors on the local software build. Configuration of the local build (representing a software developer) and the central build can be different if, for instance, some other developer has checked in a file in the central build after the first software developer synchronized his local CVS system with the central sever and then started adding more source code to a file. Because the current configuration of the central build and local build of the first software developer is different, the file being checked in with the central build by the first developer will not integrate and the central build will fail. To make sure that this phenomenon does not happen, the local build (of the first software developer) must be synchronized with the central build before a file (from that first software developer) can be integrated with the central build. If such synchronization is done, then the newly added file by the second developer on the central build will also be copied in the local build of the first developer.

Now, you can realize the point we made early in this section about the importance of continuous integration of software build in a CVS system.

## 12.4 Chapter Summary

Configuration management for a software product is the process of managing the files, artifacts, and source codes so that the software product can run. Maintaining the required files (e.g., supporting files such as an Excel file) and source code for each version (of the software product) separately is difficult because, generally, more than one version of the software product is maintained during development. A configuration management tool helps manage each version of a software product in isolation from the other versions of that product.

Version control is required on software products because a large number of versions for each artifact are generated during the software development project. A good tool is needed to manage different versions of various artifacts.

Software projects use some tools for configuration management and version control purposes. These tools also integrate with the software build management software.

For continuous integration, a central software build is required. Developers can write the source code and then integrate it with the central software build.

## QUESTIONS

1. What is configuration management for software projects?
2. What is version control for software projects?
3. What is a software build?
4. What is a centralized CVS system?
5. What is synchronization of a file?
6. What are the weaknesses of a file system for version control?

## Recommended Reading

Jessica Keyes (2004), *Software Configuration Management*, 1st Edition, Auerbach Publications, Boca Raton, FL.

Frank B. Watts (2009), *Configuration Management Metrics*, 1st Edition, Elsevier Inc., Norwich, NY.

# 13

# Software Project Management

**In Chapter 12, we learned**

- **What software configuration management is**
- **What software version control is**
- **What a software build is**
- **How configuration and version control systems work**
- **What continuous integration of source code is**

**In Chapter 13, we will learn**

- **What software project management is**
- **Project management in Waterfall model–based projects**
- **What a project schedule is**
- **What are the techniques to plan, monitor, and control software projects**
- **Project management in agile model–based projects**
- **What team management is on software projects**
- **CPM/PERT, Gantt chart, and earned value management in project management**
- **What customer management is for software projects**

## 13.1 Introduction

A project is a series of related activities with the purpose of providing a service or building a product. A project is characterized by a definite start date and an end date. A project consumes resources such as materials, time, and manpower. For example, a civil engineering project may require machines to construct a building. A civil engineering project will also need skilled people to plan and execute different tasks. A civil engineering project manager is deployed to plan and manage the project. The same is true with software projects. A software project needs skilled people to design and construct the software product. To manage a software project, a project manager may be deployed. The project manager will be responsible for managing all the activities related to the project.

To build a software product, you need skilled people who can gather and prepare the end user or customer requirements, create software design, write the source code, and test the software product that is being built. You may also need infrastructure such as personal computers, servers, networking equipment, Internet connections, and office equipment.

A software project manager creates a software project plan and monitors and controls the entire project. A large software project may need hundreds of skilled people to perform various project activities. On the other hand, a small project may have just a bunch of people.

## 13.2 Project Management and Software Engineering Methodologies

The software engineering methodology that is used to develop a software project plays a vital role in the management of that software project. All aspects of project planning, project monitoring, and project measurements (these terms are explained in the latter sections of this chapter) are changed based on the software engineering methodology used.

### *13.2.1 Project Management for Waterfall Projects*

Project planning is done meticulously for software projects that follow the Waterfall methodology. Project planning is used to achieve a complete overview of the budget and schedule aspects of the project. The customer and the project stakeholders can measure project progress at any time to determine if the project is progressing as per the plan or is deviating from the baseline (i.e., planned) schedule (i.e., time) or budget.

Project monitoring is done using techniques such as Gantt charts, earned value management (EVM), or critical chain method. We will learn about these in this chapter.

### *13.2.2 Project Management for Agile Projects*

Ideal project planning is a difficult task for the software projects that are based on the Waterfall methodology. The fundamental risk with such software projects is that when the software product is completed, after spending money and allotting time, customers complain that it is not the product they were looking for. This happens because throughout the project execution, the customers are never on board and have no idea what kind of software product is being developed. Thus, after completion of the project and when the software product is delivered to the customers, the customers find the product unsuitable for their needs.

The agile methodology provides a way to develop a software product incrementally. The risks of developing the wrong product are reduced in agile projects because only a small increment of the software product is developed at a time and the customer's feedback is addressed after each incremental development. This means that elaborate project planning is not needed but full attention is provided in developing and shipping the small increments of the product. Project planning and monitoring efforts on such projects become small. We will learn about software project management in agile environments in the latter sections of this chapter.

## 13.3 Project Planning

Projects are invariably large in nature; thus, they require careful planning before they are executed. If project planning is not carefully done, resources and time will be wasted, in which case the quality of the project will be affected because of troubleshooting and ad hoc quality management arrangements.

### *13.3.1 Project Planning for Waterfall Projects*

The traditional project planning process comprises the following steps or tasks:

- Define the scope of the project.
- Create a work breakdown structure (WBS) of the tasks.
- Estimate the effort for the tasks.
- Calculate the needed resources.
- Map the logical dependencies between the tasks and identify the critical path.
- Plot the tasks on a project schedule.
- Calculate the total cost.
- Find out ways to trade off between resource usage and cost.
- Baseline the project plan.

Let us study each of these project planning tasks.

*13.3.1.1 Define the Scope of the Project* Scope is the total amount of work to be completed during the project. For software projects, scope is the number of software product features to be developed. For example, an inventory management system may include software product features such as inventory count, inventory movement, inventory staging for receiving, and inventory staging for shipment. Now, if the customer also wants to include a packaging management feature in the system, then the project scope will increase. Increased project scope means increased project work. Hence, it is very important for the project manager to define the project scope at a minute level to make an exact scope of the project.

Scope also includes the quality level to be achieved. Generally, quality level is measured in terms of the defect density per testing cycle. For example, if 90 defects were found in the first testing cycle of a software product of size 100 lines of source code, then the defect density at this stage is 90 defects. In the second round of testing, if 40 defects were found (after fixing the first 90 defects), then the defect density at this stage is 40. If the required quality level to be achieved is five defects per 100 lines of source code, then further testing will be needed after the second round of testing. This means that extra effort is needed in testing the software product. This extra effort will

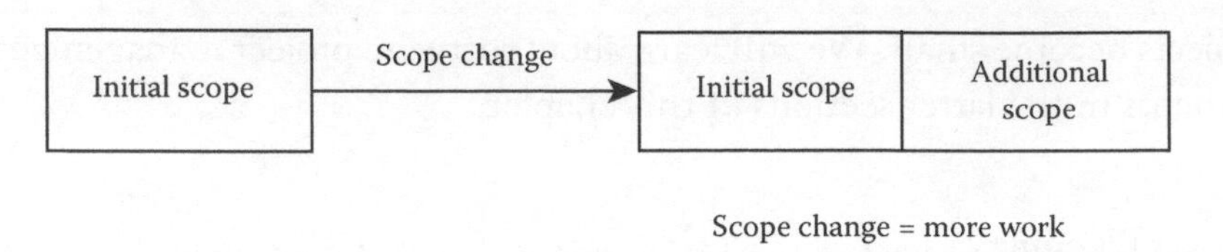


**Figure 13.1** Scope change results in more work.

result in an increased project budget and schedule. As shown in Figure 13.1, if the initial scope of the project changes, then more work is needed to complete the project.

*13.3.1.2 Create a WBS of the Tasks* We can divide the entire project into many tasks. For example, the project can be divided into project initiation, project planning, project monitoring, and project closure tasks, or it can be divided into requirement gathering, software design, software construction, and software testing tasks. Each of these tasks can be further divided into many subtasks. For example, the design task can be divided into two subtasks, namely, Module 1 design and Module 2 design.

We need to divide a project into many manageable tasks. If we do not do this, then it will be extremely difficult to manage the project. You can imagine something similar when you plan a hiking trip to a faraway place. If you do not make a good plan for the trip and have not broken down all the tasks well, then your hiking trip will be difficult.

The best way to create manageable tasks for a project is by using WBS. The WBS is a complete listing of all the project tasks. The WBS does not provide scheduling information but provides a complete list of tasks that needs to be worked on during the project. In the WBS, you should try to break down each task into subtasks as much as possible. This will help in looking for opportunities to do many tasks in parallel, thereby reducing the duration of the project.

WBS should be created by considering the schedule and resource conflicts. You should not assign the same resource to two or more different tasks at the same time unless the resource is used by these tasks part-time, which is rare anyway.

Table 13.1 shows a project schedule. The columns WBS no., Summary task, and Subtask are part of the WBS. All the dates are in month/day/year format. The

**Table 13.1** Project Schedule Table

| WBS no. | Summary task | Subtask | Start date | End date | Resource | Dependency |
|---|---|---|---|---|---|---|
| 1 | Design | | 10/10/2015 | 12/12/2015 | | |
| 2 | | Module 1 design | 10/10/2015 | 11/11/2015 | Andy | |
| 3 | | Module 2 design | 10/12/2015 | 12/12/2015 | Rachel | |
| 4 | Source code | | 11/12/2015 | 02/05/2016 | | |
| 5 | | Module 1 build | 11/12/2015 | 01/05/2016 | Larry | 2 |
| 6 | | Module 2 build | 12/13/2015 | 02/05/2016 | Ashley | 3 |

remaining columns in this table are added to create a project schedule. Note that these columns are not part of the WBS. They are in fact part of the project schedule. We will learn about project schedules later.

*13.3.1.3 Estimate the Effort for the Tasks* Once you divide the project into different tasks, effort is required to complete each of them. Before a person starts working on a task, the project manager needs to find out how much time is needed to complete it. The project manager may be able to create a schedule for this task by knowing this estimated time in advance. The required effort will be influenced by the size of the task and its productivity level. For example, suppose it is estimated that a software product contains 1000 lines of source code. It also needs to be tested. If the productivity level of the software developer who is assigned this task is estimated to be 10 lines of source code per hour, then this task will take 100 (1000 divided by 10) h of effort. If this product is allocated to a person who works 8 h per day, then it will take 12.5 (100 divided by 8) days to complete. Please note that if this task is assigned to some other software developer, the required time needed to complete this task will be different since the productivity level of this developer may be different. The project manager may create the schedule for this task accordingly.

*13.3.1.4 Calculate the Needed Resources* How many resources will be required to complete a task is determined by the total effort required for that task as well as if the task can be divided among many people. Some tasks cannot be divided; hence, they need to be completed by one person. If a task can be divided among many people, then that task can be completed in a shorter amount of time because the people involved will work on the subtasks that are independent of each other and can do the work in parallel. If the work can be done in parallel, then additional resources (such as additional computers and workspace) may be needed. A goal of any project is to complete it as early as possible, that is, divide and assign the tasks if possible for faster completion.

Figure 13.2 shows the duration of a complete task and the reduction in the duration of that task after dividing it into two subtasks. In the first case, when the task was not divided, only one person was working on it. In the second case, after dividing that task into two subtasks, one person worked on each of these subtasks.

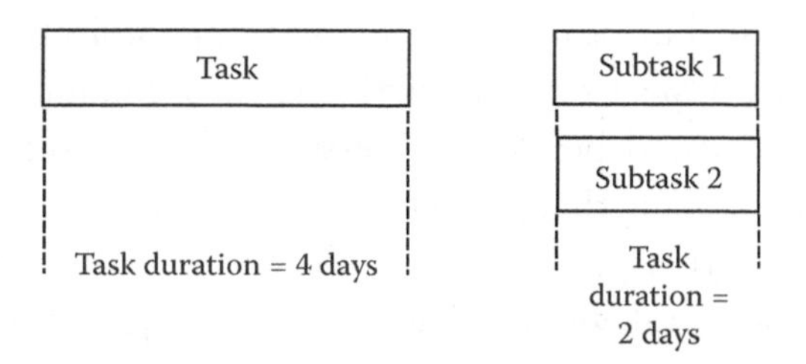


**Figure 13.2** Task duration for a complete task and the task duration after division of that task.

*13.3.1.5 Map the Logical Dependencies between the Tasks and Identify the Critical Path* You need to consider the dependencies between the tasks and subtasks to make the schedule reasonable. For example, in Table 13.1, Module 1 build is dependent on the completion of Module 1 design. The critical path is the shortest path that connects the task having the earliest start date with the task having the latest end date in the project schedule, by using zero or more tasks in between. For the project schedule given in Table 13.1, the critical path starts from Module 1 design (start date, October 10, 2015) to Module 2 build (end date, February 5, 2016). Finding the critical path is important because it will give you the total duration of the project. In Table 13.1, the total project duration is around 3 months and 25 days (the time gap between October 10, 2015, and February 5, 2016). More details on critical path are provided in Section 13.4.1.3.

*13.3.1.6 Plot the Tasks on a Project Schedule* A project schedule can be plotted in the form of a schedule table (Table 13.1) or a Gantt chart (Figure 13.3). There are many automation tools available, such as Microsoft Project and Primavera, to create and maintain the project schedule.

In Table 13.1, all the tasks have a WBS number. A summary task can be defined at the aggregate level of two or more subtasks. For example, design is a summary task because it has two subtasks: Module 1 design and Module 2 design. Each subtask has a start date and an end date. Thus, the summary task has a start date equal to the earliest date among the start dates of all its subtasks and an end date equal to the latest date among the end dates of all its subtasks. You can assign resources to each task. Finally, you can define the dependencies among the tasks. We have defined that the task with WBS no. 5 (Module 1 build) is dependent on the completion of the task with WBS no. 2 (Module 1 design). Similarly, the task with WBS no. 6 is dependent on the completion of the task with WBS no. 3. Unless the task with WBS no. 2 is completed, you cannot start the task with WBS no. 5. Similarly, you cannot start the task with WBS no. 6 until the task with WBS no. 3 is completed.

The project schedule can also be plotted in the form of a Gantt chart. Figure 13.3 depicts a Gantt chart.

Figure 13.3 contains a Gantt chart in which the tasks are plotted against time. This Gantt chart is an alternative way of showing a project schedule. The benefit of a Gantt chart is that it is graphical; thus, understanding a project schedule is much easier with it. Figure 13.3 shows the four project tasks for a sample project.

If you have created a project schedule with the start and end dates for each of the tasks (in a project) and the dependencies between those tasks, as well as all other information necessary to make that project schedule complete, then it is called a baseline project schedule for that project. Generally, the baseline project schedule needs the approval of the project stakeholders (i.e., customer or project owner). The baselines are created during the planning stage of the project (i.e., before the project starts).

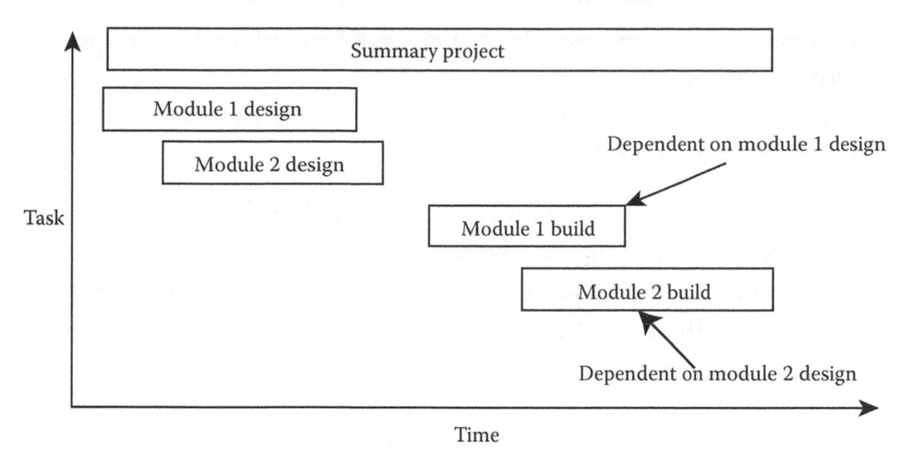


**Figure 13.3** Gantt chart for project scheduling.

Once the project work is started, the project manager uses these baselines to measure the progress of the project work. Therefore, the baselines are used as the reference points to measure the progress of different tasks in the project.

A project schedule provides complete information as to when each project task starts and ends. It also provides links between various tasks in the project. For example, if there is a requirement (or a dependency) that a project task (say Task 1) must start after the completion of another task (say Task 2), then Task 1 cannot be started as long as Task 2 is not complete. This means that if Task 2 is delayed, then Task 1 will also be delayed. In Waterfall projects, all the software development tasks are dependent on each other. Writing the source code is not possible until the software design is complete. Similarly, testing the software product is not possible until the source code is fully written and compiled. For example, in Table 13.1, Module 1 build is dependent on Module 1 design and Module 2 build is dependent on Module 2 design. Module 1 build cannot start before Module 1 design is completed. The same is true with Module 2 design and Module 2 build.

*13.3.1.7 Calculate the Total Cost* Once you have determined the effort required for each of the tasks in a project, you can calculate the cost for each task. Adding the costs for all the tasks will give you the total cost for the project. Suppose you have the following scenario:

- Two business analysts are scheduled to work for 2 months each. Their average salary is US$5000 per month.
- One software designer is scheduled to work for 1 month. Salary = US$6000 per month.
- Three software developers are scheduled to work for 3 months each. Average salary = US$5000 per month.

- Two software testers are scheduled to work for 1 month each. Average salary = US$4000 per month.
- A project manager is scheduled to work for the entire project duration of 7 months. Salary = US$7000 per month.

You can now easily calculate the project cost by adding the costs for each task and the number of resources assigned to those tasks. This cost is for development alone. There could also be costs for hardware purchases, office rent, management expenses, and so on. Generally, these costs put together are below 20% of the development cost.

*13.3.1.8 Find Out Ways to Trade Off between Resource Usage and Cost* Resource usage costs money. If you keep a person in a project for 1 month, then his or her salary for that month will be counted as part of that project cost. Cost is a factor that needs to be considered while assigning the resources to a project. To minimize the project costs, never keep a resource on the project more than it is required. For example, a software developer is needed on a project for 2 months; once his or her assignment is complete, he or she should be released immediately so that he or she can be placed on some other ongoing project.

People hired for a project have different skill levels and thus salaries. It is also possible that a person with the same skill set and experience as that of another person has a lower salary. In such a scenario, work can be assigned to the person with the lower salary and thus the company will save money. The person having a higher salary may have some other skills that can be utilized for some other suitable project in that company. This kind of scenario is prevalent in most large software services companies having a large number of ongoing projects at any point in time.

There is another way to reduce project costs. Sometimes, a task has enough time to complete. Therefore, that task is less important in terms of schedule requirements, in which case you can assign that task to a junior member of the team. Suppose, for software testing, you have 100 test cases and you estimate that 50 of these test cases can be taken care of by a junior tester. Assigning them to the junior tester will save you a lot of money because the junior tester's salary is much less than that of any experienced tester. Although the speed at which the junior tester will be working on those test cases will be slower, you can still fit the schedule of testing by the junior tester in such a way that the overall schedule for testing is not hampered. Junior members of a project team gain valuable experience when they work alongside experienced team members. Assigning them this way is a great way to enhance your team's capability.

*13.3.1.9 Baseline the Project Plan* When you need to track the progress of a project, you must have some baseline dates that you can use as a reference. The baseline dates compared to the actual completion dates are thus maintained for all project tasks. If you do not maintain the baseline dates, you will not be able to track the progress of the project.

You will assign the start date for the first task in the project. From there, you can assign subsequent dates to other tasks in the project schedule plan. These dates are your planned or baseline dates for various tasks. You freeze these dates once they are set up and agreed upon by different parties involved in the project.

Providing good baseline dates is extremely important. For example, Task 1 is dependent on the completion of Task 2 and you forgot about this dependency when you assigned the baseline dates for these two tasks; therefore, you assigned the same start date for these two tasks. On project execution (i.e., once the project is started), it was found that Task 1 will not meet the desired schedule and only then did you realize that it cannot start before the completion of Task 2. Correcting this error in the project schedule is too late now because once the project execution starts, you cannot change the baseline dates. Baseline dates have to be properly set up when the schedule is made and before the project execution starts.

### *13.3.2 Project Planning for Agile Projects*

All projects have to deal with time, cost, scope, and quality constraints. You need to fix the scope on traditional projects because the project size is directly dependent on the size of the source code of the software product. The larger the software product size is, the larger is the software project. The project stakeholders also need to consider the time limit to develop the project, as well as cap the project cost. Finally, the software product's quality also needs to be considered.

In traditional projects, all the terms related to time, cost, scope, and quality are considered and agreed upon before the project starts. However, in agile projects, things are a bit different. The cost, quality, and time are fixed. However, the scope is kept open. This is because the agile product development is supposed to be responsive to the customer needs.

Since agile projects rely heavily on incrementally developing the software products and, at the same time, the project scope is kept open, most traditional project planning techniques do not work on those projects. For example, you cannot create a WBS for the project tasks because the tasks on the agile project are not fixed (scope can change). Similarly, a project schedule cannot be drawn using techniques such as the Gantt chart because of the varying scope.

Instead of using traditional tools for creating a project plan, a new set of tools have evolved for agile projects. These include timeboxing, a many-layered project plan, user stories, and scenarios.

In Figure 13.4, we have depicted many layers of project planning on agile projects. The topmost-level planning, "Project plan," is the planning for the entire project. In reality, we never do project planning in agile projects because we always carry out planning at the lower levels.

The next level of planning is known as major release planning. Major release planning is done to ensure that the product reaches the customers as per the agreed upon time frame. Major release planning is always planned on the basis of the feedback from the marketing team (or from the customer, if the product is being developed for

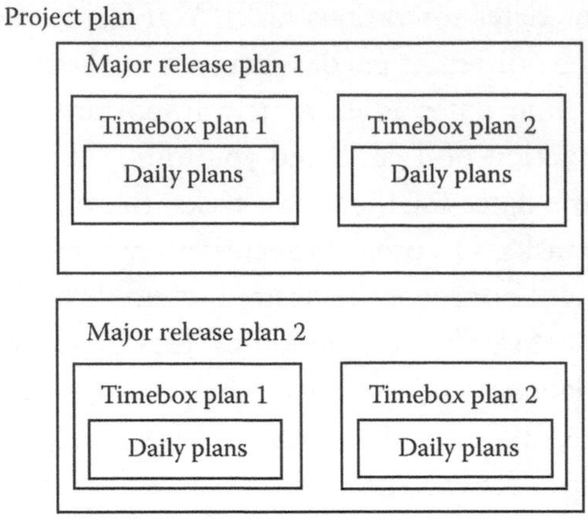

Project plan

Major release plan 1

Timebox plan 1
Daily plans

Timebox plan 2
Daily plans

Major release plan 2

Timebox plan 1
Daily plans

Timebox plan 2
Daily plans

**Figure 13.4** Many layers in project planning on agile projects.

that customer). Marketing teams analyze the market situation and advise the project team as to which product features are expected by the potential customers and at what time it will be suitable to create a product release to the market. Based on a longer-term major release plan, a series of timebox plans (i.e., iteration plans) can be chalked out. For example, a major release plan is planned ahead to be slated within the next 6 months. Product features to be included in this major release are also tentatively planned. On the basis of this major plan, some 10 timebox releases are planned every 2 to 3 weeks. All these timebox plans at this time are tentative in nature and can be changed. Firmed timebox plans can then be elaborated for the first 2 to 4 timebox plans (out of the entire 10 timebox releases that are planned). This process can be repeated until all the remaining timebox plans for that major release are accomplished.

A set of user stories form timebox planning that is done at the iteration level. Timebox planning level is also known as minor release. We learned about user stories in Chapter 2. A user story is equivalent to a use case or a requirement specification. We learned about use cases and requirement specifications in Chapter 4. Timeboxing is a concept of doing everything inside a fixed period. Time is fixed and you have a number of tasks to do. How will you do the tasks? The other limitation is that resources are also fixed. Since the scope is not fixed, it means that effort estimation is not possible (you do not know how many tasks need to be done). The best approach to be adopted for timeboxing is to assign urgency ratings to the tasks that need to be completed. For example, if a particular product feature is important and must be developed, then it should be assigned a higher urgency rating. If another product feature is not so important, then it should be assigned a low urgency rating. This way, you can prioritize all the product features. Inside the timebox, you should take the most urgent features first. Later, you can take the product features that are not so urgent. This way, during timeboxing, all important tasks will be completed. In a timebox, if time has run out and some tasks (that are planned in that timebox) could not be completed, the impact on the overall project will not be much. This is because the tasks that could not be completed in that timebox were least important.

Every day, plans are made for a project. There will be a meeting, lasting for 30 min in the morning, to plan the activities for that day. This meeting is attended by all the project team members (developers, testers, designers, project manager, etc.). This meeting is a good time to discuss the issues that affect the work of the other team members. For example, if a developer is waiting for a defect fix to be closed (after its verification by a tester), then the developer can discuss that issue during that meeting.

## 13.4 Project Monitoring and Controlling

Project monitoring involves tracking the project progress (in terms of its budget and task progress) against time as well as taking the corrective steps wherever necessary to keep the project always on track. Project monitoring and control on traditional Waterfall projects is more difficult than on agile projects for the following reasons: the duration of the development process is longer, the deliverables are available after a long time, and the risks may happen from time to time. On agile projects, project managers have to deal with low-level risks because deliverables will arrive at short durations at the end of each iteration.

Project monitoring and control consists of project schedule management, team management, supplier management, and customer management. We will learn about team management, supplier management, and customer management in the latter sections of this chapter. The remaining part of this section is all about project schedule management.

### *13.4.1 Project Monitoring and Control for Waterfall Projects*

The project schedule for a Waterfall project can be monitored and controlled using techniques such as Critical Path Method (CPM) or PERT (a French acronym for Program Evaluation and Review Technique), Goldratt's critical chain method, Gantt charts, or EVM.

*13.4.1.1 Gantt Chart* The Gantt chart can provide information for task progress in terms of its schedule. Using a Gantt chart, you can maintain the baseline dates and actual dates for any task. In Figure 13.5, a Gantt chart is shown. The tasks with solid lines are the baselined tasks. They have baseline start dates and end dates. The tasks with actual start dates and end dates are shown with dotted lines. Gantt charts also show the percentage progress of any task. The percentage progress is derived by calculating the task progress for the given date compared to the total work planned for that task. The task progress is measured as the difference between the actual progress and the planned progress. For example, suppose a task is planned for 10 days. After measurement on day 5, it was observed that only 4 days' worth of work was completed. This means that the task has slipped by 20%.

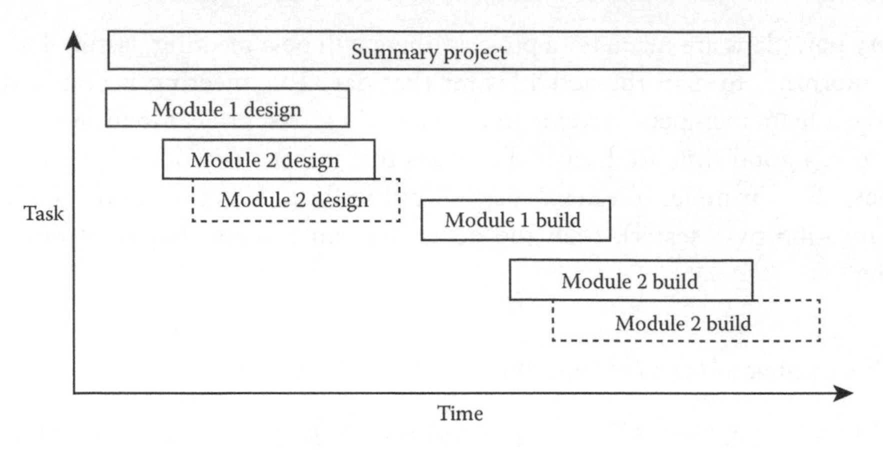


**Figure 13.5** Gantt chart with project progress measurement.

The project schedule is part of the project plan. Once the project execution commences, you start measuring the project progress. You find that a task in that project, which was supposed to start on September 15, 2015, was actually started on September 20, 2015. Here, September 15, 2015, is your baseline start date and September 20, 2015 is your actual start date. You will keep measuring the project progress in a similar fashion and record all the baseline and actual dates. A baseline date is always fixed. You cannot change a baseline date because only with baseline dates will you be able to measure the project progress and project performance. In Figure 13.5, you will see that two tasks actually strayed from their planned schedules. While Module 1 design and Module 1 build started and ended on the planned dates (i.e., baseline dates), Module 2 design and Module 2 build started and ended on dates later than originally planned. This kind of information is vital for measuring the project progress so that the project can be controlled. Controlling a project involves taking some measures so that the project schedule does not suffer further.

Milestones are the tasks with zero duration and represent some important achievements in the project and are inserted into a project schedule during the project planning stage by estimating their completion dates. For example, in civil engineering, if you are constructing a house, then "finishing the roof by the end of July 17, 2015" is a milestone, "completing the painting by August 24, 2015" is another milestone, and so on. In other words, a milestone is a point in the time horizon at which a task is supposed to complete. To make sure that the completion of each task is recorded and reported, a milestone is placed at the end point of each task in the Gantt chart. When the task is completed, it is stated that that particular milestone has been achieved. As a comparison, if you are traveling long distance by road, then the milestones could be the names of some important cities on the way along with the estimated times to reach each of those cities.

*13.4.1.2 CPM/PERT* Assume that the project plan has to be drawn by considering the project tasks. In this case, the alignment of the start and end dates of the tasks and the dependency between the tasks have to be considered. Alignment of tasks

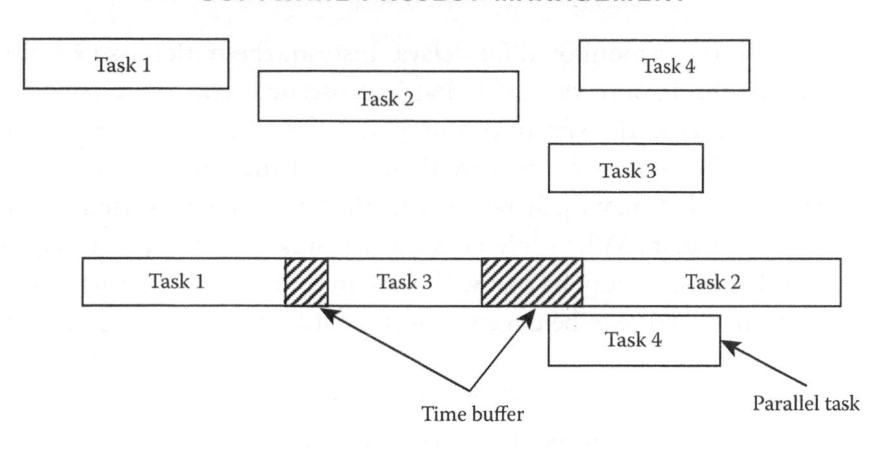


**Figure 13.6** Project tasks aligned for creating a critical path.

will result in a path composed of all the tasks. On this path, the tasks will be aligned on the basis of their start and end dates or on the dependencies on each other. For example, if Task 2 is dependent on Task 1 and the end date of Task 1 is October 10, 2015, then the start date of Task 2 can only be after October 10, 2015 but not before.

Once the tasks are aligned, you need to provide some time buffers (i.e., extra time) for risks. As a comparison, if you estimate that it takes 30 min to reach a destination from your home but you started 45 min early, then that extra 15 min is the buffer time. In Figure 13.6, you can see the time buffers provided between Tasks 1 and 3 as well as between Tasks 3 and 2. Time buffers for critical tasks allow the project team to finish the project within the schedule even when some critical tasks get delayed to some extent. Thus, the time buffers provide a mechanism in controlling a project.

*13.4.1.3 Critical Chain Method* The critical chain method uses the theory of constraints to find the bottlenecks in a project plan and provides adequate measures to tackle the constraints.

During project planning, buffers are created for the critical tasks (i.e., the tasks that are on the critical path). In Figure 13.6, Tasks 1, 3, and 2 are critical. A buffer is not created for Task 2 because it is the last task in the project and there are no other tasks that depend on it. Noncritical tasks are not given any buffers because even if they are delayed, it will not affect the project duration. For example, if the design of a component is separate from the main software product, then that design is not a critical task in the main software product. For example, Task 4 in Figure 13.6 is a noncritical task. As a comparison, in civil engineering, if the project is the construction of a house, then putting up some fence in the backyard, which is also a part of the project, is generally a noncritical task.

On the other hand, if a critical task is delayed, then it will delay the overall project. Project planning is just an estimate and cannot be accurate. The buffers are provided for the critical tasks to take care of any mistakes in those estimates. When a critical task is executed, it is not monitored for delays. Instead, the buffer associated with it is monitored and the amount of buffer that is consumed when that critical task is completed is determined. The size of consumption of the buffer will provide a guide on how the project is progressing. If more than 50% of the buffer is consumed, then it means that the project may likely be delayed. This 50% figure is used to show the criticality of the situation. When this buffer is consumed by 50%, you may need to think ahead and take some actions that will result in controlling the situation. At that juncture, a critical analysis can be done to ensure that further delays do not occur.

*13.4.1.4 Earned Value Management* EVM takes into account both the budget and the time while tracking the progress of a project. EVM is the most complete project progress tracking technique. EVM is mainly used for government contracts and large projects. EVM needs minute and accurate data to work. Therefore, if not much project data are being generated on a project, then EVM cannot be used. On small projects, the amount of project data will be low. The quality of data on such projects will also be low because the project progress measurement on such projects is not required frequently. Most project managers on small projects know most of the project progress even without keeping formal project progress data. However, on large projects, it is difficult to assess project progress without having good project progress measurement techniques. In fact, on such projects, there could be a dedicated person appointed to take the project progress measurements. Hence, on large projects, collecting data on the quality and quantity of the project completed is always good for measuring the progress of the project. Most government contracts fall into the category where project progress reports are a must even if they are small. This is because government employees need to provide justification for all the project expenses in terms of project progress reports.

To understand how EVM works, you need to understand the concept of percentage amount of work performed on a task in terms of both time elapsed and budget spent. We discuss the concept of EVM in Figures 13.7 through 13.9.

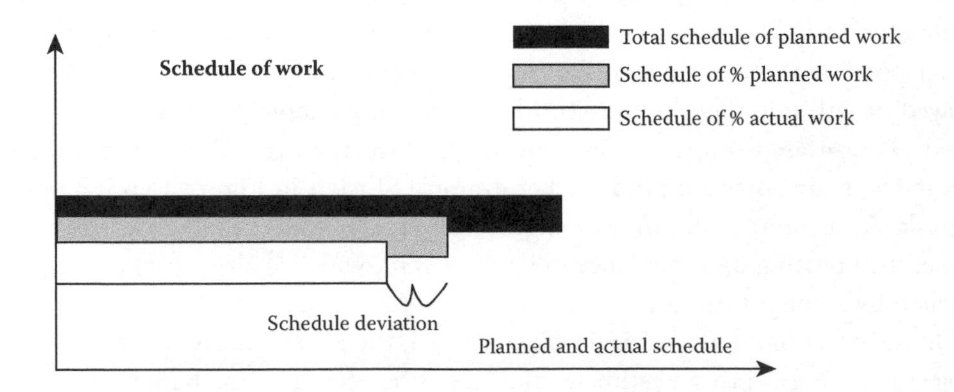


**Figure 13.7** Amount of work completed in terms of time (schedule) consumed.

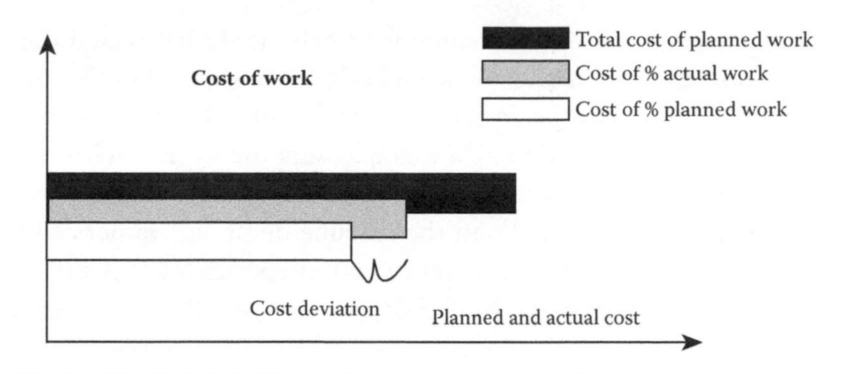


**Figure 13.8** Amount of work completed in terms of budget consumed.

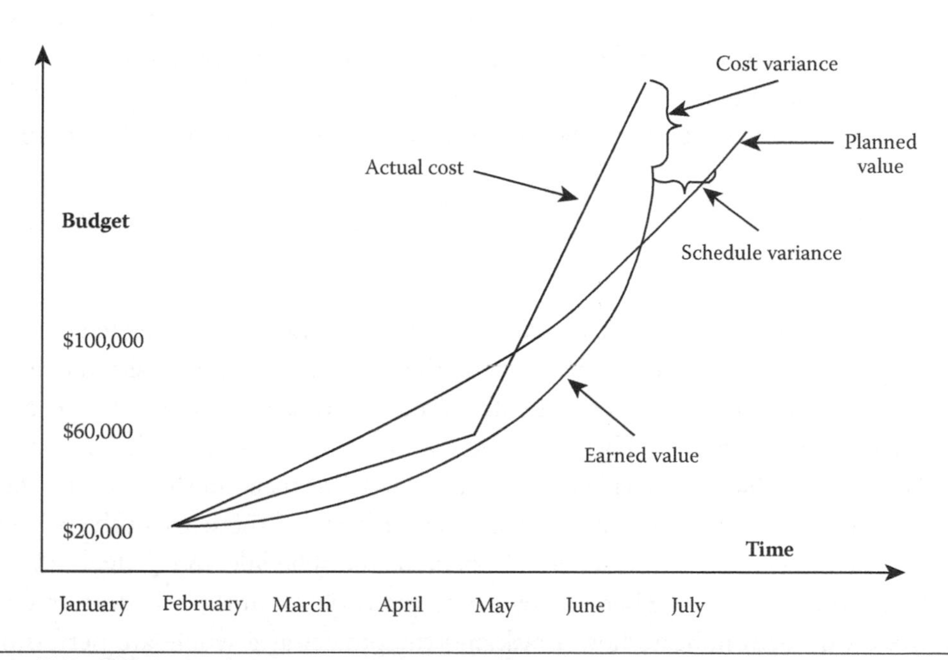


**Figure 13.9** EVM showing the budget and schedule variance.

In Figure 13.7, a schedule deviation is depicted. A schedule deviation can happen when a task slips from its baseline start or end dates. A project status report can determine whether a task has started on its scheduled (i.e., baseline) start date. If a task has started on its baseline start date but its progress lags behind its scheduled end date, then the task will be delayed despite starting on the baseline start date. If a task is measured when it is still under progress, then we can find out how much of the planned work would have been completed and how much of the actual work has actually been completed so far. Figure 13.7 presents the planned and actual completion of a task. Suppose the planned completion of a work on a certain date is 60%, but after completion status is determined, it is found that only 50% of the work has been completed, then this task has a schedule deviation of ((50 – 60) × 100/60)% or –16.66%.

Figure 13.8 depicts the budget deviation for a task. If we have fixed the baseline budget for this task and measured how much budget is consumed until a given point of time, then we can find out the deviation in the budget from the expected budget consumption until that point of time. For example, suppose we have a baseline budget of US$100 for a task. Measurement taken on a date indicates that US$50 was actually spent on this task until that date. From the baseline duration, we noticed that 40% of the duration has elapsed. Thus, the baseline budget consumption should have been US$40 and not US$50. This means that US$10 more is spent than expected until that point. Hence, the budget deviation for this task at this point is ((50 – 40) × 100/40)% or 25%.

Note that we have used the terms *cost deviation* and *schedule deviation* here. In EVM, there are terms such as *cost variance* and *schedule variance*. The formulas for calculating these terms are different from those we have used for cost deviation and schedule deviation here.

Now, we can proceed to understanding EVM. Any project has three constraints:

- Scope
- Budget
- Time

In EVM, we fix the scope of the project by setting a baseline. A baseline is the start and end points of a project. You can also set up the baselines for all the tasks in that project. For example, we can have a project with start and end dates of March 12, 2015, and June 14, 2015. These dates will be fixed. The start and end dates for all the project tasks will also be fixed. These dates will be treated as the baseline dates for the respective tasks. Once we fix the project baseline dates, it means we have fixed the scope of the project.

Once the project starts, the actual numbers for the schedule and budget are compared with the baseline numbers to determine the variance in the schedule and budget. These variances indicate how a task and the project as a whole are performing. From the cost and schedule variance figures, you can also find the projected (revised) cost and schedule variance in the future.

There is another matter related to cost variance. In isolation, cost variance can give a misleading picture about the project status. Let us consider a scenario. Project X has been approved for a duration of 1 year with budget Y. In addition, it was planned that 50% of the approved budget of the project will be spent during the first 6 months. Six months after the project started, the project manager reported that he or she had spent 50% of the budget. This may wrongly indicate that the project is as per the plan. However, this information is not sufficient to arrive at such a positive conclusion. Although 50% of the budget is already spent, what if only 30% of the work is completed in that 6 months' time? This means that the project is not doing well. EVM is used to handle these kinds of issues.

In Figure 13.9, you will see that a lot of terms used are related to EVM: earned value, planned value, actual cost, and so on. Actual cost is the money spent on a project task for the time duration for which the measurement is being taken. Planned value and earned value are calculated using the following formulas:

Planned value = (% of work planned) × (project cost).

Earned value = (% of work complete) × (project cost).

Read a book on EVM to learn more about it.

*13.4.1.5 Resource Control for Project Schedule Management* In Waterfall models, there could be a mismatch between some planned resource requirement and the actual availability of those resources. For example, the design of two modules was initially allocated to two persons. However, due to resource crunch (note that each person is a resource), the design of these modules was assigned to only one person. In this case, the project manager will not be able to stick to the original project plan. How can you manage this scenario? A solution is to find a new person who can work part-time on one of those projects instead of overloading the available person with the task of working on the design of both modules.

Care should be taken while adding a new person to a delayed project in an attempt to complete that project on time. Unless that new person is well versed with the technology and the kind of project, he or she may in fact delay the project further. This is because he or she may need the help of the project team to understand the project and the technology. This will result in time mismanagement; that is, instead of project team members working on project tasks, they will have to spend some of their time helping the new member understand the project. It is best if someone who previously worked on that project is available to join the project because that person already knows that project and can also easily mingle with the rest of the team.

Sometimes the required people are not adequately available for the project. In such cases, there is no option for a project manager but to ask some people to work overtime and take additional assignments. Overtime is also used when the project is lagging behind the schedule.

### *13.4.2 Project Monitoring and Control for Agile Projects*

In agile projects, a project is divided into several iterations. Each iteration is a mini-project in itself. You need to track the progress of each iteration in isolation from the progress of the previous iterations because, in each iteration, a fresh set of new features will be added to the existing build of the software product. However, the source code developed during each iteration needs to be integrated with the existing source code (i.e., existing build). Refactoring may be needed if the integration of new source code results in some performance or incompatibility issues.

We have learned about project planning for agile environments in Section 13.3.2. We saw that planning is done at many levels: project, major releases, timeboxes (iterations), and daily plans. We also saw that project-level planning is not done for agile projects. Project planning is derived from the aggregation of planning that is done at the lower levels. Daily plans are almost informal. They are not recorded formally. Planning for major releases is mostly a marketing plan. Project planning for major releases is also not done formally. Hence, we are left with timeboxes. Planning for timeboxes is done in agile projects to some extent, although this planning is also not very formal.

There are two reasons why project planning on agile projects is not done formally. First, agile projects are supposed to overcome the problems associated with plan-driven Waterfall models. Thus, agile projects lack any formal planning. The second reason is that these types of projects are a recent phenomenon. The project planning methods associated with agile projects are still evolving. In the future, when these methods become mature, some form of project planning methods may come out for agile projects.

Currently, some empirical methods are being used for effort estimation and project planning on agile projects. They are based on setting story points for each product feature to perform effort estimation. Once the effort estimation for a project team is established, the project schedule can be made based on the project size (number of story points to be achieved) and the effort estimate. On agile projects, effort estimation is measured in terms of the speed at which a project team can complete a story point. This is known as velocity. Once you know the capability of a project team in terms of the speed at which it can develop a software product, you can derive the project costs by finding out the total salary of the project team.

We have been discussing project planning so far in this section on project controlling and monitoring. This is because, on agile projects, project planning and project execution are not done separately. Most of the time, you make some plans for the next 10 to 15 days for each iteration (timebox) and then you do daily planning to execute (implement) the plans daily. Thus, project planning and project execution go hand in hand.

**Timeboxes:** Most agile methodologies work in timeboxing concepts. There is a start date and an end date fixed in advance and the project team is given a list of features to develop in that time frame. The priority among these features is also fixed. The features with high priority are taken up first. If time does not allow developing the low-priority features, then they are dropped from that iteration and are taken up in a later iteration.

**Story points:** Story points are a measure to determine the effort required from a project team to complete a product feature. A product feature can be complex or simple. Depending on the complexity of the feature, the required effort will vary. For example, designing, implementing, and testing a search feature can be more complex than designing, implementing, and testing a catalog page. Suppose we assigned six story points to the product search feature. Then a catalog page, which is very simple and smaller than the search feature, can get two story points. This also means that the effort required for designing, implementing, and testing the product search feature will be three times that for the catalog page.

How the story points for a product feature are derived is totally based on the previous experience of the project team. Thus, story points for a product feature will vary from one project team to another. In the future, this project planning measure can be standardized so that the story points will be the same regardless of a project team's capability.

**Velocity:** Velocity is the project planning measure that indicates how many story points a project team can design, implement, and test in a fixed time span. A time span here is a sprint if the project team is following Scrum or an iteration if the project team is following eXtreme Programming. For example, suppose a project team completes 10 story points in a sprint and then the velocity of the project team is set at 10 story points per sprint. If, on another sprint, this project team is able to do only eight story points, then this deterioration in the performance of the project team can be analyzed to find root causes.

As is the case with story points, velocity is also an empirical method. If a project team is working on a project for the first time, its velocity is unknown. Only after the project team completes two or three sprints will it be possible to determine its velocity. Presently, each project team derives its own velocity. In the future, a standard method to derive the velocity can be developed so that the velocity can be uniformly applied to all the teams.

*Example:* A problem and its solution are provided based on a hypothetical and oversimplified scenario to demonstrate how the project schedule and cost can be estimated for an agile project.

*Problem:* It is decided that Project A will be developed using Scrum methodology. There are 10 people on the project team and the average salary of a team member is US$5000 per month. It is estimated as follows: the project has 100 story points, the velocity of the project team is 4 story points per sprint, and the time duration of each sprint is 5 days. Estimate the project cost and schedule.

*Solution:* Each sprint will be completed in 5 days.

Hence, 4 story points will be completed in 5 days.

Therefore, it takes ((100 × 5)/4) = 125 days to complete 100 story points.

It takes 125 days to complete the whole project because the project has 100 story points.

Therefore, the estimated schedule for this project is 125 days.

The cost of team members' salaries for 125 days is

$$\text{US\$ } ((5000 \times 10) \times 125/30) = \text{US\$}208{,}333.33.$$

If other costs are around 15% of the team members' salaries, then the total cost of the project is

$$\text{US\$ } (208{,}333.33 + ((208{,}333.33 \times 15)/100)) = \text{US\$ } (208{,}333.33 + 31{,}249.99) = \text{US\$}239{,}583.32.$$

## 13.5 Project Team Management

A responsibility that does not often show up in the project reports of any project manager is team management. It is one of the most crucial aspects of any project. Team performance may not be up to the mark even if you have highly skilled people on your team. Managing the team members and getting the desired output from them will determine the success or failure of the project. Even a project team consisting of highly rated team members cannot perform well if the project manager fails to apply good team management techniques to the project.

A project manager needs to evaluate and match the skills of the people with the project tasks, assign suitable tasks to the people, arrange required training for the people, manage the expectations of the people, get the desired output from the people, and get the results.

Project managers are needed to manage the teams especially when the team size is large. On large teams, it is difficult to find out the performance of individual team members. On large projects, the performance of individual team members is difficult to visualize. Only the project manager can visualize this aspect; thus, his or her role as a team manager becomes indispensable. A project manager assigns tasks to the team members and ensures that the tasks are completed. Thus, a project manager knows the productivity of each team member.

On agile projects, team management is not important because team sizes are small. These small teams are mostly self-managed. Because of this, a project manager in these projects may not have a team management role.

In the Waterfall model, team management is one of the key responsibilities of any project manager. The project manager has to deal invariably with a large team on these projects; thus, he or she keeps track of each of the team members. The project manager is directly responsible for managing the team in these projects.

On some projects, a project team is managed by a technical project manager. A technical project manager is proficient in any area of software engineering. For example, a technical project manager may be proficient in coding. Generally, a technical project manager is responsible for smaller project teams. On smaller project teams, the technical project manager can manage the project as well as perform one of the software engineering tasks. For example, a technical project manager who is responsible for a project team of size 10 also performs coding. For such a small team, the technical project manager has enough time to perform both the management and development (i.e., technical) tasks simultaneously.

For large projects, it is difficult for a project manager to find time to perform any other task apart from managing the team.

## 13.6 Project Customer Management

Managing the expectations of the customers is a difficult task. Failure in properly understanding the requirements of the customers is a major reason for project failures.

At the same time, ensuring that the customer's confidence remains high during product development is also something that the project managers need to address.

If you are working on an agile project, then managing the customer's expectations is easier. The customer's representative is available in the form of a product manager or product owner. The product owner is always available with the project team to resolve any issues. Moreover, at the end of each iteration, the product owner provides immediate feedback. Thus, managing the customer's expectations on agile projects is not an issue.

However, things are very different in Waterfall projects. The customer is not available on-site with the project team. The customer is available only during the requirement gathering phase when the project team meets the customer. Once the project starts, the project manager sends project status reports to the customer. During this period, the customer never knows whether the software product is being built as per the requirements. The customer sees the product only after it is built. If the project duration is 6 months, then the testing of the product by the customer can be done only after these 6 months. This is a long time. Because of this, the customer becomes restless. Even if the project manager sends the project status reports to the customer on time, indicating that the project is going smoothly, the customer remains dissatisfied. This is because the customer may know from these reports that a software product is being developed but he or she does not know if it will match his or her expectations.

The best way to please a customer is to prepare a prototype of the product and show it to the customer at the very beginning of the project. The customer will then provide feedback and the prototype will be modified on the basis of the feedback and again be shown to the customer; this process is repeated until the customer is fully satisfied, as explained in Chapter 3. Therefore, the customer will have an idea of what to expect in the final product if the prototype approach is followed.

### 13.7 Supplier Management

For a software project, the suppliers are the companies (or individuals) to whom some or all of that software development work is outsourced. If a project or a part of it is outsourced, then managing the suppliers is a crucial task for the project manager. Aligning the deliverables from the supplier with the deliverables from his/her own team is a major challenge for the project manager. The project manager is always anxious about whether the delivery will happen at the right time from the supplier.

The best way to manage suppliers is to take great care in selecting them. If you select a supplier just because it quoted the lowest price for the work, then it may create problems later on. One should choose a supplier based on its track record, the competency level of its team, and the price that the supplier has quoted.

Suppliers can be managed best when you have service agreements defined well in your contract with the supplier. At the same time, keeping good communication with the supplier is the most important aspect of managing the supplier. The suppliers must have a clear idea of what and when things need to be delivered.

Engaging the suppliers in Waterfall projects is a common practice. Now, even some parts of agile projects are being carried out by the suppliers. Thus, the task of managing the suppliers becomes a key aspect of many software projects whether they are Waterfall or agile in nature.

### 13.8 Chapter Summary

Software project management is all about managing the resources, schedules, budgets, customers, and suppliers on a software project. A software project is managed by a software project manager. The project manager may create project plans including schedules, resource requirements, and budget requirements. Once the execution of a project starts, the project manager will monitor and control the project. Whenever problems arise during the execution of a project, the project manager will try to fix them using project management techniques.

Note that a project may need to be broken down into project tasks. This is done using WBS. A project schedule consists of project tasks. Each project task may have its dependencies on other tasks, required resources, and budget.

During the project planning stage, the project schedule is drawn using scheduling techniques such as PERT/CPM, Goldratt's critical chain method, EVM, and Gantt charts. Once the project starts executing, these same methods are used to track that project schedule. It is important to track the project schedule because if a project task is delayed, then the dependent tasks will also be delayed; thus, the entire project may be delayed. Scheduling techniques are used to track project schedules for software projects that follow the Waterfall model.

Projects those follow agile methodologies cannot be tracked using the abovementioned scheduling techniques. Agile projects do not follow firm project planning. Instead, they rely heavily on building a software product in small increments. Here, project planning in advance is not really needed or even required. Software features that need to be developed are determined 1 week in advance at the most. Only three or four software features will be designed, developed, and tested at a time. Elaborate project planning is not needed to do this. To track iterations or sprints on such projects, some concepts like story points and velocity are used.

## QUESTIONS

1. What is software project management?
2. How is CPM/PERT used in software project management?
3. How is EVM used in software project management?
4. How is a Gantt chart used in software project management?
5. What is a WBS?
6. What is a project schedule?
7. What is project planning?
8. What is project monitoring and controlling?

9. What is supplier management?
10. What is customer management?
11. What is velocity?
12. What are story points?
13. What is a timebox?
14. Why do traditional project scheduling methods not work on agile projects?

## Recommended Reading

Murali Chemutury, Thomas M. Cagley, Jr. (2010), *Mastering Software Project Management Best Practices*, J. Ross Publishing, Fort Lauderdale, FL.

Bob Hughes, Mike Cottrell (2010), *Software Project Management*, 5th Edition, Tata McGraw Hill, Delhi, India.

Andrew Stellman, Jennifer Greene (2005), *Applied Software Project Management*, 1st Edition, O'Reilly, Sebastopol, CA.

# Appendix: Answers to Questions

## Chapter 1

*Answer 1*

Software engineering is a branch of computer science that uses engineering processes to design and develop the software products. It uses the core processes of requirement management, software architecture, software design and construction, software testing, and software release management. The peripheral processes in software engineering include software maintenance, project management, configuration management, and feasibility studies.

*Answer 2*

A computer is an electronic device that uses input devices such as a mouse and a keyboard to get user input. It then processes this input and computes the required output and then displays it on a monitor. It can also send the output on a printer or any other device. The hardware of a computer includes the central processing unit and memory.

A modern computer can do a lot of highly sophisticated computing tasks.

*Answer 3*

Computers are used in almost all fields. Industries such as transportation, retail, manufacturing, entertainment, science, healthcare, and gaming use computers in one way or another. Computers are now also being used for remotely controlling smart devices such as refrigerators, televisions, and air conditioners.

Today, no large business can run without computers. Governments are using computers to better serve their citizens. There is no field where computers are not being used.

*Answer 4*

A methodology is a standard way of performing a series of activities to achieve a target. A software engineering methodology is used to develop a software product using some standard activities in the form of processes. For example, a Waterfall model is a software engineering methodology that is used to design and develop a software product. This methodology uses the processes of requirement management, software design, software implementation, software testing, software release, and so on to develop a software product.

*Answer 5*

Software development involves performing a series of processes. First, the users' requirements need to be converted into the requirement specifications. These specifications are then converted into a software design. A software design is used to write the source code. This source code needs to be tested to remove the software defects, if any. Only then is a software product produced. All these processes are labor-intensive. Software professionals who work on developing the software products are highly skilled and demand good salaries. Because of this, developing software products is costly.

*Answer 6*

Developing a software product is a challenging job for many reasons. First, the software engineering field requires a lot of creativity on the part of the software engineers to design and develop a software product. Not many people are capable of thinking at a level where they can come up with a software design or write the software programs. Finding good software engineers to develop a software product is relatively difficult.

Software engineering is a relatively young field and is still not mature. The processes used to develop the software products are still open for improvement.

*Answer 7*

Developing a software product is a challenging job. To overcome these challenges, some processes and techniques have evolved. One technique is to use established software engineering methodologies to develop the software products. Another technique is to use the existing software libraries or reuse the software components that are already built. One more technique is to use software design patterns, which are like templates. For example, we use a website template to build a website. Software design patterns work in a similar way.

*Answer 8*

Software development involves creating the requirement specifications, software designs, user interfaces, and databases, as well as performing software testing, project management, and so on. Software specifications are created by business analysts. Software design work is done by software designers. Software source code is written by software developers. Software testing is done by software testers. Software projects are managed by software project managers. Thus, several different career options are available for software engineers. A software engineer can master any of these fields and obtain employment in it.

*Answer 9*

A software project is used to develop a software product. A software project can be planned and then monitored and controlled by a software project manager. A large number of software engineering processes are involved in any software project. These software engineering processes include creating and managing the requirement specifications, creating the software design, writing the source code, and testing the software product.

**Chapter 2**

*Answer 1*

Plan-driven software engineering methodologies provide a good overview of the software development life cycle. This factor makes it the best suitable methodology for outsourcing a software development project. When a contract is to be made with an outsourcing supplier, the customer and the supplier must agree on the terms such as project duration, project cost, and quality level of the software product. Since it is possible to create a concrete plan for the entire software development project before a contract is signed for outsourcing, the customer will know when the software product will be delivered and what will be the cost.

Even for in-house software development projects, a project manager needs to provide all the information to the project stakeholders about the project to be undertaken. Using the information provided in the project plan, the stakeholders can decide whether to pursue the project or not. Once the project commences, the stakeholders can easily get all the information about project progress and can find out if the project is doing well. The project manager can monitor and control the project by comparing the baseline project plan with the actual project progress data that he or she receives.

*Answer 2*

Plan-driven software engineering methodologies are good for building the software products where the customer is very clear about the requirements. However, in most cases, the customers do not have a clear idea about what they are looking for in the proposed software product. Making a concrete plan for taking up a software project is prone to many change requests (i.e., request for changes from the customers). Adjusting the project plan to include the change requests is the most difficult task on any software project.

The other important risk in a plan-driven software project is that when the software product is built and shown to the customer, he or she may find that the software product is not as per his or her expectations. This happens because the customer is never on the project site during the entire software development process; therefore, the customer does not know if the software product being built is indeed as per his or her expectations.

*Answer 3*

In plan-driven software development, all the customer requirements are taken at the beginning of the project. The software design is based on all these requirements. This is a huge risk because, in many cases, customers are not very clear about their requirements. Even if the requirements are clear, because of the market demand, the requirements can change later on. In some extreme cases, by the time the software is built, the entire requirements may completely change and thus the developed software product will not be of much use.

Risk-driven software engineering methodologies try to avoid the risks just explained by building a software product incrementally. This will allow the customer to see the product features that are developed in almost real time (in a time gap of 10 to 15 days: this is the typical length of an iteration). Hence, the software product being developed is always in sync with the changing requirements of the customers.

*Answer 4*

Agile means the ability to cope with changed circumstances. On any software project, agility may allow the software product being built to develop product features that were not even imagined when the project started. These software product features were developed because the market demanded them. This would not have been possible if all the plans about the project were frozen at the beginning of it.

*Answer 5*

Each software engineering methodology has its own merits and demerits. A risk-driven methodology may allow a software product to be built completely in sync with the market demand, but this is not always the case. For example, a business house may need to use a software product to take care of its business transactions. The nature of the business will remain the same and will not change overnight. In such a scenario, developing a software product for this business house may not need a risk-driven methodology.

When the stakeholders want to get a clear picture about the entire project (including the cost and time involved to develop the software product), even before that project is sanctioned, a plan-driven software engineering methodology is the best fit.

*Answer 6*

Incremental building of a software product allows it to be built in sync with the changing market demands. If there is a demand for a software product feature, then it will be incorporated in the software product quickly. The software product features are ready by the completion of an iteration (the duration of an iteration is around 10 to 15 days). The software product can be released to the customers after the completion of each such iteration. Thus, the customers do not need to wait a long time to use the software product features that they were looking for.

*Answer 7*

Agile projects typically have small teams consisting of 5 to 10 people. If a proposed software product is very large in size, then a small team may not be sufficient. You will need a large team to do it. In such a scenario, agile software engineering methodologies may not be suitable.

When the project stakeholders need to know the details about the cost and time to develop a software product up front, again, agile methodology may not be the best fit.

When the proposed software product has clear requirements and these are not expected to change, then the plan-driven software engineering methodologies will be a better choice.

*Answer 8*

Software product development starts with gathering the user requirements (i.e., requirements of the users). These requirements need to be converted into requirement specifications. These specifications need to be converted into a software design. The source code needs to be written to build the software product as per the software design. The developed software needs to be tested to make sure it has adequate quality. To perform all these activities, you need to have the following processes: requirement management, software design, implementation, and testing.

*Answer 9*

A software development life cycle is the same as a software engineering methodology. The software development life cycle changes when the software engineering methodology is changed. A software engineering methodology describes how a software product is developed using the processes that are part of a software development life cycle. For example, a software development life cycle may be linear in the case of Waterfall methodology but will be cyclical in the case of Spiral methodology. In the case of agile methodology, a software development life cycle is in the form of short iterations. In the case of Rational Unified Process, the software development life cycle may involve iterations within each process of the software development life cycle.

*Answer 10*

Project management is all about project planning, project initiation, project execution, project monitoring and controlling, and project closure. In each of these project phases, there are project activities that include project tasks. For a software project, these project tasks include the gathering of software requirements, creating the requirement specifications, creating the software design, writing the source code, and testing the software product. These project tasks are essential processes of any software engineering methodology.

**Chapter 3**

*Answer 1*

A prototype is a miniature product that resembles the actual (i.e., proposed) product. A prototype is used to test or demonstrate the features of the proposed product. In the software industry, a prototype is built to show to the customers so that they will evaluate it and provide their feedback. On the basis of the feedback, the actual software product will be built.

A prototype in the software industry is generally built over iterations. The prototype is shown to the customers and their feedback is taken at the end of each iteration. On the basis of the feedback, the prototype is refined further and shown to the customers again. This process continues until the customers are fully satisfied.

*Answer 2*

The prototype can be either a throwaway prototype or an evolutionary prototype. If the project team needs to reuse the design and the source code written in building the prototype, then it will build the evolutionary prototype. If the project team does not want to reuse the design and the code (if any) written to build the prototype, then it will build a throwaway prototype.

*Answer 3*

A feasibility study is conducted to assess whether a project is technically or economically viable. If it is found that it is either technically or economically unviable, then the project is shelved. A feasibility study can be conducted by someone who has good knowledge and experience in the business or technology related to the proposed project.

*Answer 4*

A feasibility study can be either technical or economic. A technical feasibility study is done to evaluate whether the proposed project is technically viable. If the technical feasibility study is not done, then the project can get stuck during project execution because of some technical issues.

An economic feasibility study is done to assess whether the cost to be incurred on the proposed project is justified on the basis of getting returns from the developed product.

*Answer 5*

Suppose a company wants to build a new software product to replace its existing legacy system. A feasibility study is conducted to determine economic feasibility. After conducting the feasibility study, it was found that the cost of developing the new software product would be US$400,000. The feasibility study also found out that there is no real benefit of the new software product. Generally, when a software product is implemented, cost reduction is expected, but since the company was already using a legacy system, the cost reduction could not be achieved by implementing the new system. The only benefit the new system could provide was in terms of ease of use for the users.

*Answer 6*

Suppose a new software product needs to be developed for a government to provide better service to its citizens. A feasibility study was conducted for this project. During the study, it was found that the citizens are computer illiterate. It was also found that providing training to the citizens would take at least 10 years because the number of citizens was huge and the geographical area vast. This finding clearly established that the idea of introducing computer-based government services was infeasible. This finding resulted in shelving the project.

*Answer 7*

The development cost and time for a prototype should not be much. If it takes an inordinate amount of time to develop a prototype, then the project will be affected. If the prototype development requires a large amount of money and the project is finally found to be infeasible, then all the money spent on that prototype will be a waste. In these circumstances, a prototype that takes less time and money is needed. Throwaway prototypes provide these benefits.

Although the design and source code of the throwaway prototype cannot be reused in building the actual product, the benefits mentioned above make the throwaway prototype a good choice.

*Answer 8*

On some software projects, building a prototype is costly and the built prototype must be reused to justify the cost of building it. On these projects, the likelihood of the project being shelved because of any technical or economic reasons is also low. In such cases, the evolutionary prototype is used. The software design and the source code that are used in building the evolutionary prototype are reused in the actual software product.

*Answer 9*

Feasibility studies can be conducted using several methods. One method uses market surveys. Here, it is assessed whether the cost incurred on a project will provide fairly good returns. Another method is to build a prototype first and then get feedback from the customer. On the basis of the feedback, the project viability is decided.

Another method of determining the feasibility of a project is to build a pilot project. For example, if a large organization wants to implement a software product, then the organization will first test it by using it at only one of its sites. This is known as a pilot project. Once the results are good, the organization can implement (i.e., use) the software product at all of its sites.

**Chapter 4**

*Answer 1*

A software requirement is a statement made by a user who will be using the software product as to what the proposed software product is supposed to do. A software requirement (or a set of software requirements together) should include all the details that will be used to create a suitable software design. For example, a software requirement may state that the user should be able to view a graph depicting the sales from a region for each month.

This statement can be further refined to include the details about the region and the products (for which the sales graph is required). Additional refinement in the requirement may include information about what kind of graph the user should have (e.g., a pie chart or a columnar chart). Once all the details about a requirement are complete, then the requirement will help create a suitable software design.

*Answer 2*

Software requirements can be functional or nonfunctional. Functional requirements relate directly to the software product features that will help users perform their tasks. Nonfunctional requirements are related to performance, security, usability, and so on. A performance requirement may relate to how much time a software product takes to respond to a user request. A security requirement may relate to how the software product is secured against data theft, unauthorized access, and so on. A usability requirement may relate to how user-friendly the interfaces are and thus help the users perform their tasks.

*Answer 3*

Software requirement management deals with change requests (i.e., requests from the customer to change a requirement) and the modification of documents as per the changed requirement. When a change request arrives, the requirement specification related to that changed requirement needs to be updated. The dependent requirements, if any, will also need to be changed. For example, suppose a change request comes to change the monthly sales report for a product based on the category of the product instead of the sales region. Now, the sales report should not be based on a region but on a product category.

In this scenario, the related requirements such as the sales graph should also be changed accordingly.

*Answer 4*

In Waterfall methodology, all the requirements are gathered and all the requirement specifications created before the software design activities start. If a change request comes after the software design is already created, then, along with the requirement specifications, the software design also needs to be changed. Thus, a change request can have repercussions on other phases of the software development life cycle.

In agile methodology, only two or three requirements are taken for developing an increment in the software product. Because the development of these increments takes 10 to 15 days from the date on which the requirements were received from the customer, there is no possibility of requirement change requests. If the project team is keeping a product backlog, then some future requirements may need to be changed after a change request arrives, but it will not affect the development life cycle. Only the product backlog will be changed.

*Answer 5*

A use case describes the interaction between an actor and a software product. The actor could be a user, another software product, or time. Events are dependent on time. Events trigger when a certain time is reached. For example, you can run some batch input commands at certain times of a day automatically. At those certain times, the event will trigger and the software product will perform the required computations. An example is running the virus detection software at a particular time of the day. The virus detection software will run automatically at the set time daily and will detect any viruses in the operating system.

When the actor is a user, then, upon some action (from the user) on the user screen, the software product will perform some computation.

All these interactions between an actor and a software product are depicted in use cases. Use cases help in creating a software design.

*Answer 6*

Requirement specifications can be made using use cases or just plain text. Generally, a document containing all the requirement specifications is known as a software requirement specification document. On agile projects, a requirement specification can be in the form of user stories.

A requirement specification is derived from a user requirement, after refining it to make sure that it contains complete information to create a software design.

*Answer 7*

A use case can be set at aggregate, actor, or subactor levels. When a use case is set at the aggregate level, it corresponds to either a part or a complete software design. For example, a use case describing an order management system, when implemented into a software design, will show the design of the complete order management system.

If a use case is at the actor level, then it will correspond to a software component. For example, a use case describing a user login (into the system) will correspond to a software component design for the login function.

If a use case is at the subactor level, then it will correspond to a part of a software component. For example, a use case describing the user input for username will correspond to a part of a software component design for the user login function.

**Chapter 5**

*Answer 1*

The scripts that run on the application server are known as server-side scripts. A server-side script is a glue program that combines the content coming from an application server with a software product if that application server is already connected with that software product. For larger websites, keeping the business logic in compiled software components on the application server is better because it runs faster and also helps in preventing hacking.

*Answer 2*

Web architecture can be two-tier, three-tier, or *n*-tier. In a two-tier architecture, the web server is combined with the application server to form one tier. The other tier is the database. The content from the web server is displayed on a web browser. In a three-tier architecture, the web server is separated from the application server and the third tier is a database. The content from the web server is displayed on a web browser. If the website (consisting of the three tiers) is integrated with another website or a software product, then this integration layer will form another layer in the web architecture.

*Answer 3*

A high-level design depicts the software architecture. It is important to depict a high-level design for a software product because it is here that the complexity of the design of a software product needs to be reduced. Later, when lower-level software designs are made, complexity in software design is further reduced.

Designing a software product is a long process. It cannot be taken (i.e., done) at one level. It needs to be taken at many levels. A high-level software design is the first step in that direction.

*Answer 4*

Authorized access to a software product ensures that only authorized people can access that product. Unauthorized access will lead to data theft or even destruction of the software system. To ensure that only authorized people can access a software product, a software security mechanism should be built inside the software product.

*Answer 5*

Web-based architecture overcomes the limitations faced by the client–server architecture. Software development effort is reduced in web-based architecture because the client part of the software product does not need to be developed. This is because the client part is a web browser and is readily available.

In the client–server architecture, whenever there is a change in the server part of the software product, the client parts also need to be changed. This used to result in the updating of each client computer. This is a labor-intensive effort. With web-based architecture, there is no need to update client computers.

*Answer 6*

If a software product that is dependent on a particular operating system is built, then that software product can run only on the computers that have that operating system. It is desirable to build software products that can run on all operating systems so that all users can use them regardless of the operating system of their computers. To overcome this limitation, software product vendors create different versions of their software product for all these different types of operating systems. This is costly and time-consuming.

A platform-independent software product can run on any computer regardless of the operating system it operates on. This solution overcomes the problems cited earlier.

*Answer 7*

A client-side script is used for web-based software products to validate user input, disable web page source code views, disable downloading of images, and so on. Client-side scripts are also used for automatic filling of user information on a web page.

User input validation can be done either from server-side scripts or from client-side scripts. When server-side scripts are used, the script takes the user input data to the server and validates it against a database. This process takes time and may affect the performance of the software product. Client-side scripts validate the user input by running these scripts inside the browser and the validation takes place quickly.

*Answer 8*

A three-tier architecture is a type of software architecture pattern. For a client–server software product, there is a user interface layer, a business logic layer, and a database layer. For a web-based software product, there is a web server, an application server, and a database layer. The web server provides the content that is displayed on the user screen (web browser). The application server hosts the business logic layer.

*Answer 9*

Multitier architecture for a software product consists of the user interface, business logic, database, and integration with some other software products.

Multitier software architecture is used when you need to integrate your software product with some other software products. For example, if you want to allow advertisements on your website and there is a service provider that provides feeds of advertisements, then you will need to integrate your website with the website of this service provider. In this case, you will need to use multitier architecture for your software product (website).

*Answer 10*

The performance of a software product can be affected by large image sizes, frequent trips to the database, and so on. Use images with small file sizes to load them quickly on any web browser. Use database manipulation scripts with data caching so that frequent trips to the databases can be avoided.

*Answer 11*

A software architecture pattern provides templates to create software architecture for your software product. Many software architecture patterns are defined and used by the software developers. For example, we have a two-tier architecture pattern where the software product is developed using a client part and a server part. We have a three-tier architecture pattern where the software product being developed uses a user interface layer, a business logic layer, and a database layer.

*Answer 12*

Software architecture depicts a high-level design of a software product. Software architecture can be drawn for a software product using a software architecture pattern. Software architecture should show the logical design at the highest level. For example, the software architecture of a software product can show all the tiers in the software architecture.

*Answer 13*

Software components in the database layer include database tables, schemas, indexes, and so on. Components in the user interface layer include user control components, windows, and so on. Components in the business logic layer include the set of classes and their defining interfaces.

*Answer 14*

A component diagram depicts the software components and their relationships. These software components are part of the business logic layer (the middle layer) design of a software product. Although the databases and user interface layers are also composed of components, the component diagrams are not made for those layers. Database components are drawn using the entity-relationship diagrams, and the user interface components are drawn using the mockup screens.

A component diagram not only shows the components and their relationships with each other but also shows the integration points of the components with other subsystems and systems.

*Answer 15*

A data flow diagram depicts how the data pass from one component to another when an external or internal event triggers. An external event can be invoked by a user or by any other system. An internal event can be invoked by a component of the same system.

When an event triggers, the data pass from one component to another and data processing also happens along the way. Finally, the data can be saved in a database.

In data flow diagrams, an external entity that triggers the data flow is depicted. The business logic is depicted as a process and the database is depicted as a data store.

*Answer 16*

A software design pattern is a kind of template to create software component designs. Many types of design patterns are created to cater to the needs of designing various types of components. For example, you can use a software design pattern to build a large number of similar classes from a single parent class. This kind of functionality is useful because it helps in building a software product using reusable software components.

*Answer 17*

Software design patterns are broadly categorized as creational, behavioral, and structural. Some examples of creational design patterns include the abstract factory method, factory method, and builder method. Some examples of structural design patterns include adapter, bridge, decorator, and façade. Some examples of behavioral design patterns include chain of responsibility, iterator, and command.

*Answer 18*

A software component is an abstract representation of the middle-level part of a software product. A software component can be the interface, on the basis of which the classes can be implemented in object-oriented design. A web page can also be thought of as a component. Each web page contains a piece of business logic that is complete at the web page level.

**Chapter 6**

*Answer 1*

A user interface is the visible part of a software product. Using the interface, a user can provide inputs and receive outputs from the software product. A user interface consists of user control elements such as textboxes, labels, and drop-down lists. A user interface is always contained inside a container such as a window or a browser.

*Answer 2*

A graphical user interface (GUI) is a type of user interface that uses windows or similar graphical elements to be displayed to a user. A GUI is better than a character-based user interface. A character-based user interface can display information only in the form of characters, whereas a GUI can display information in the form of images, buttons, videos, sound, and so on. Thus, a GUI is very user-friendly and can help a user to increase his or her productivity.

*Answer 3*

A GUI can have many kinds of user control elements. The container elements help organize other user control elements and keep them inside their boundaries. A container can be a window, a frame, a dialog box, and so on.

A selection and display element allows a user to select or display a value on the user interface. Some examples of selection and display elements include buttons, radio buttons, and drop-down lists.

A navigation element helps users navigate between web pages or even inside a web page (among various user elements) on a user interface. Examples of navigation elements include scroll bars, links, and tabs.

A value input element helps users provide inputs to the software product so that the software product can do some computation based on these inputs. Examples of value input elements include textboxes and combo boxes.

An output element is used on user interfaces to show the result of computations done by the software product. Some examples include labels, tooltips, status bars, and progress bars.

*Answer 4*

Mockup screens are used for designing user interfaces. This way, mockup screens can be considered component diagrams for user screens. Mockup screens are drawn to show what user elements will be there on a user screen and how they will be organized. Mockup screens also depict navigation from one web page to another.

Mockup screens are great not only for designing user interfaces but also for testing the software product.

*Answer 5*

AJAX is an acronym for Asynchronous JavaScript and XML. As the name implies, AJAX components work asynchronously. An AJAX component can have events that will trigger, even if a web page is not reloaded.

AJAX components can be used for user input validation. For example, an AJAX component can be used to check whether a username is available when a user registers on a website. If no AJAX components are used on a registration web page and a username is already taken, then all the information that the user input during page reloading is lost and the user needs to fill in the registration form all over again. However, we can use an AJAX component and attach it to a move over event. Once the user fills in the username input box and proceeds to fill other information on the web page, the move over event will trigger and the username validation will be done. The web page is not reloaded while this is happening. Thus, the user does not need to keep filling in the registration form again and again.

The early implementation of AJAX components used to consist of JavaScript and XML. The JavaScript would validate the user input by checking the values in the XML file. The AJAX technology now can check user input even by checking it against the values stored in a database. This has allowed AJAX components to become versatile. While an XML file can store some database information, it cannot store trillions of pieces of data. Using a relational database with an AJAX component, you can now validate user input against even trillions of pieces of data that are stored in the database.

*Answer 6*

Model–View–Controller (MVC) is an architecture pattern where the model and the views are controlled by a third component called controller. A model is the business logic that does computations. A view is the outcome (of the model) in the form of displays on any user interface. In simple software architectures, there is only a model and a view, but in an MVC architecture, the controller plays an instrumental role. When a user triggers an event, the controller receives it. The controller then instructs the model to do some computation. Once the controller receives information about the computation, it may form a view. What kind of view will be formed can be defined in the controller. In effect, the controller controls the kind of output that can be displayed on the user screen. This view is then visible on the user interface.

*Answer 7*

A model in an MVC architecture is the business logic that does computations. When these computations are complete, the model sends this information to the controller.

*Answer 8*

A controller in an MVC architecture is the nodal point between a model and a view. The controller is like a traffic police. It regulates the information exchange between the model and the view. What information needs to be sent to the view or received from the model is totally controlled by the controller.

For example, suppose there is a shopping cart on a website. A user can select some items from the website and put them in the shopping cart. The user can keep browsing the website and keep adding items to the shopping cart. During browsing, the user navigates from one web page to another (of the same website) but the shopping cart information is never lost. When the user checks out after browsing that website, he or she is presented with the list of items that he or she selected during the entire browsing session on that website. The user can now buy the items that are in the shopping cart. In this example, you can see that the controller was determining what views need to be presented to the user. At the same time, the controller has been exchanging information with the model (shopping cart information) continuously, but the controller was not sending this information to the view. Only when the user clicks the Checkout button does the controller create a view that is visible on the user screen.

*Answer 9*

A view in an MVC architecture is the content that is composed by the controller and sent to the view. The view is then displayed on the user screen. Since the controller is controlling the information flow, no information comes directly to the view from the model. Thus, information always comes from the controller to the view.

*Answer 10*

A client-side script runs inside a web browser. Using JavaScript, personal information about a user can be stored on his or her computer in a small file that is known as a cookie. This happens when the user visits a website for the first time. If the user visits the website next time, then things such as personal preferences, automatic filling in of user information, and automatic login into the user account are possible by reading the information (by the JavaScript code) from the cookie.

Client-side scripts are also used to validate user inputs. For example, if a password field on a registration form on a web page is restricted to use up to eight characters, then there is no need for the business logic to do the validation. This validation can be done using the client-side scripts.

Some websites contain copyrighted images. Using client-side script, it is possible to disable the mouse right click on a web page so that the user cannot download any images from the web page.

*Answer 11*

A style sheet is a formatting language that formats the content on a web page. Style sheets provide formatting on web pages, in addition to the formatting already done by

HTML. For example, if you need to set the font color, font size, and so on at many places on one or more web pages, then you will need to set the values for these font properties at all such places. This is a labor-intensive work. Instead, if you define the values for these font properties inside a style sheet, then you can just call that style sheet and all the formatting will be done automatically.

*Answer 12*

HTML is an acronym for Hyper Text Markup Language. HTML is platform independent. Any web browser installed on any operating system can display HTML content. HTML provides static content on any web page. You can build any web page using HTML with all the user interface elements such as buttons, textboxes, and drop-down lists.

**Chapter 7**

*Answer 1*

Variable types include integer, character, string, date, and Boolean. An integer variable type can hold natural numbers. A character variable type can hold values for any English character. A string can hold values consisting of many characters. A date variable can hold date values. A Boolean variable type can hold values for "true" or "false." For example, in Java, we have "int" variable type for holding integer values and "String" variable type to hold values containing a string.

*Answer 2*

A variable can be defined at the class or method level in object-oriented programming languages. When a variable is defined at the class level and if the variable is declared public, it is accessible directly from any other class. If the variable is declared private, then it cannot be directly accessible to other classes. In those cases, access can be provided using accessor methods.

At the method level, variables are always defined as private. Thus, the variables defined inside a method are not accessible outside of that method.

*Answer 3*

A method can be declared public in a class. When a method needs to be accessible from other classes, that method should be declared public. For example, if a method does computations that are used by other methods inside other classes, then this method should be declared public.

*Answer 4*

In object-oriented programming, objects are used. These objects are created during runtime from the classes that are designed. These objects have their own behavior and attributes. Objects can also be thought of as data structures that can hold many types of data that are otherwise not possible using any of the variable types. Using these objects, an entire software product can be built. While programming, you can manipulate the objects so that the behavior of an object can be changed. For example, suppose you have an object that calculates the interest rates for a bank. The same object can be used to calculate the interest rates for fixed deposits (i.e., certificates of deposit) as well as for recurring deposits by changing the behavior of the object. This will allow you to reuse your code, thus making the programming work more productively.

*Answer 5*

Procedural programming uses procedures, functions, and global variables to create software programs. Values from one function to another are passed either using the parameters in a function or using a global variable. Use of global variables makes programming in procedural languages dangerous. The value of a global variable can be changed by another programming unit. There is no way whatsoever to put a control on this phenomenon. This phenomenon can lead to unintended behavior (i.e., software defect) in a software product.

*Answer 6*

A method is a unit of programming code that does some computation. After the computation, some computed value can be passed to another method using a return statement. Using the parameters, this return value can be used in another method. Methods that do not have a return value are used to do some computation and show the output on the user interface.

*Answer 7*

The state of an object is the value stored in a class variable of that object. When the value of a class variable (inside that object) is changed, it is said to be a change in the state of that object. When that object is destroyed, the state of that object is reset.

The behavior of an object is the ability to do some computation through the methods it contains. Depending on different input values passed to the object, the object will have different output values. For example, suppose an object has a method that computes the interest rates on deposits. This method takes the interest rate as a parameter and calculates the interest earned on a deposit. If you provide a 4% interest rate, then the interest will be different from that of a 5% interest rate.

*Answer 8*

Object-oriented programming allows building software products by reusing the source code. Reuse of source code enables developing the software products quickly. Reuse of source code also allows building a software product with fewer defects. The reuse concept comes in object-oriented programming languages through the implementation of polymorphism, inheritance, method overriding, and method overloading.

Object-oriented programming does not use global variables; therefore, the chances of defects entering the software products are lower.

*Answer 9*

Object-oriented programming languages are strongly typed languages. All the variables must be defined with their types (such as integer or float) before they can be assigned values. This makes learning an object-oriented programming language hard.

Object-oriented concepts are difficult to learn especially for people with a background in procedural languages.

Object-oriented programming languages do not work well when serialized or procedural access of information is needed. For example, when we need to access the data in a database, object-oriented programming languages will not work.

*Answer 10*

For web-based software products, managing the state information of web pages is difficult. Web-based software products use Hyper Text Transfer Protocol (HTTP). This protocol uses stateless objects. Whenever a web page is navigated or reloaded, information about the previous web page is lost. This makes it impossible to keep the user information persistent across the web pages (of a website) when the user navigates through these web pages.

To overcome this problem, session objects are used. Session objects keep user information. Whenever a web page navigation or reloading occurs, the user information is fetched from this session object.

*Answer 11*

A class is a template for creating objects. Once a software product runs, the objects are created based on the classes the software product is composed of. All the objects from the same class, when created, contain the behavior and the attributes defined in that class. If you need to have similar objects that have different behavior or attributes, you can create an abstract class and then inherit to create many similar classes.

*Answer 12*

A package is a folder structure that can hold many related classes. Similar to the folders in operating systems, the packages allow organizing and keeping the classes in a neat way.

*Answer 13*

Refactoring is the process of restructuring the source code. Refactoring becomes necessary when new features need to be added to the existing software product. In such cases, some of the existing classes or methods (in that product) do not allow the addition of new features. Sometimes, some methods or classes need to be moved so that the source code can be reused or managed better to add the new functionality. In some other cases, some methods or classes need to be renamed to make them more relevant in the context of the new reality.

In all these cases, the changes made in the source code are known as refactoring.

*Answer 14*

An object is a data structure that can hold the data in the manner required by a software component. An object can have relationships with other objects.

*Answer 15*

A sequence diagram depicts the invocation of events that trigger when a call is made for the first object. Messages are passed from one object to another in a sequence that is predetermined by the relationships defined for those objects. The messages are passed through the invocation of methods in those objects. Data pass from one object to another through these messages. A method belonging to an object returns a value when that method runs. This value is then passed on as input to another method belonging to another object in the sequence, and so on. Sequence diagrams depict the dynamic behavior (i.e., the behavior during runtime) of a software product.

*Answer 16*

A detailed design in software engineering is the lowest-level design for a software product. It shows the classes, their relationships, and the details of each class (methods, class variables, etc.). A detailed design is derived from a component design.

A class diagram depicts the software design in static condition. A sequence diagram depicts the software design in dynamic condition (i.e., when the software product runs). A statechart diagram shows all the states an object passes through from its initialization until it is destroyed.

*Answer 17*

A class diagram is the static representation of a software product when the detailed design of the product is made. A class diagram shows all the classes, their details, and their internal structure as well as the relationships between these classes.

*Answer 18*

A statechart diagram depicts the life cycle of an object from its initialization to its destruction. An object may undergo many changes at different stages of its life. All these stages of the object are depicted in a statechart diagram.

*Answer 19*

Database programming involves manipulating or creating the data in a database using a programming language. The actual operation of data manipulation or data creation in a database is done using Structured Query Language (SQL). However, a programming language is needed to connect or disconnect to the database. Similarly, the data that need to be written in the database are passed from a program (written in a programming language) to SQL. When some data are fetched from the database by the SQL statements, then these data are passed to the program. These data can then be used by the program to do some computation.

The programming that is used to perform all these operations on the databases is known as database programming.

*Answer 20*

A Model–View–Controller (MVC) architecture is used when a model (business logic) and a view (user interface) are not directly connected to each other. A controller interacts with the model and gets the updates. If the controller finds it suitable, only then will it create a view that will be shown on the user screen. Thus, the controller may get the updates from the model many times but may construct a view only once. This kind of architecture is useful when the responses against user inputs are saved and later, when appropriate, an aggregate response is constructed.

**Chapter 8**

*Answer 1*

A database is a structure where the data can be saved permanently. A database is similar to a file on an operating system but with a difference. The data saved in a file are difficult to query. For example, the contents saved in a Microsoft Word file cannot be queried. You can make searches inside the file but you cannot create queries that can provide you with structured information. In contrast, in a database, the data are saved in a structured manner. When you query these data, you can get the data in a structured format. For example, if you have data saved from an order management system in a database, then you can create queries to find out how many orders came from a specific customer. You can also find out a list of all the orders made by a particular customer in a given month. In fact, you can create queries to find out any information that is stored in the database.

When the data are saved in a file, the file treats the complete data as a single piece. In contrast, the data in a database are saved in atomic form. You can combine these atomic data to form meaningful information. The data are combined in many ways depending on the kind of combination stated in a query statement.

*Answer 2*

Permanent data are crucial for many purposes. When a customer purchases some goods from a vendor, the vendor must keep the sales data permanently as a record. The sales data can be used to make sales tax payments to the government later. The sales data can also be used to provide customer service. A sales manager needs the sales data to show his or her performance to the management. A marketing manager needs the sales data to analyze how a particular product is selling and, based on the sales, to prepare future strategies to increase the sales.

For payrolls, permanent data about employees and their salary are important. For a manufacturing company, permanent data about the production of goods are important. In fact, permanent data are crucial to run businesses, governments, organizations, and so on.

Permanent data are always stored in a database. Thus, databases are extremely important.

*Answer 3*

Databases are powerful monsters. They derive their power from their ability to store the data at the most atomic level and in enormous amounts (more than trillions of pieces of data). If a database is designed perfectly, then all these huge data will be free from ambiguity or duplication. These huge and pure data can be used to power the search engines that can in turn provide powerful reports.

Without these capabilities of a database, it is hard to imagine how businesses and governments could run or how any kind of scientific or medicinal research could be carried out.

*Answer 4*

NoSQL databases are primarily file systems where the data are stored in a very structured manner. They are a hybrid between a traditional relational database and a file system. An example of a NoSQL database is an XML file. An XML file can store the data in a structured manner. For relational databases, SQL is used to query the database. In the case of XML, a query language called XPath is used instead of SQL.

*Answer 5*

A relational database consists of many tables. Each table in turn consists of many columns. All the data are stored inside these tables and columns. The tables are created in such a manner that you can link the data belonging to two or more tables by using the primary and foreign keys. Using this relationship among the tables, it is possible to store the related data in two or more tables. Later, when you need to get these data, you can create queries to fetch the related data from these tables.

This relational aspect is the success factor behind relational databases. The term *relational database* was derived from this aspect.

*Answer 6*

Referential integrity ensures that the relationship among the pieces of data saved in two or more related tables is intact. Referential integrity ensures that no references among the tables could be breached. For example, if you delete a record in a master table, then all the records that exist in the child tables and are related to this record should also be deleted automatically. There should be no orphan records in the child tables. When a new record is created in a child table, a related record must already exist in the master table.

*Answer 7*

Referential integrity can be implemented using a referential clause in the SQL statement that creates the reference among the related tables. When a record needs to be deleted in the master table, how referential integrity should work can be defined in this clause. For example, the clause can state that the child records should also be deleted or the master record should not be deleted at all. Similarly, if a master record is updated, then how the child records should be updated accordingly can be stated in this clause.

*Answer 8*

An orphan record in a child table has no corresponding record in the master table. If a record in the master table is deleted and there is no referential integrity, then the related records in the child table will not be deleted. This will result in orphan records. Similarly, when a record in the master table is updated and this record has related records in the child table, those records in the child table will not be updated. This, again, will result in orphan records.

*Answer 9*

A primary key can be set on one or more columns in a table. A primary key ensures that the data in that (those) column(s) are unique and that there are no NULL values in that (those) column(s).

A primary key is used to create relationships with other tables. The primary key of the master table is used to make a join with another table that has a foreign key.

*Answer 10*

A foreign key is used to join a child table with a master table. A foreign key can consists of one or more columns in the child table. All the records in the child table must have a master record in the master table.

*Answer 11*

A sequence is a component in a database and is used to generate contiguous numbers to populate a column in a table (in the database). The numbers generated by a sequence are always unique; therefore, the sequence is suitable to be used to populate the data in a primary key column.

When a record is inserted in a table, the next value function of the sequence can be used to fetch the next number in the sequence. This next number can be used to populate the data in the primary key column of the table. The sequence generates the number next to the last number it has generated in the previous use of the next value function.

*Answer 12*

An index is created to search the records quickly. An index can be created for a table using the data dictionary language. When an index is created for a table, the index will create a list of search terms. Whenever a search is made on that table, an entry in the index is looked up and, based on this index entry, the proper records in that table are found. There will be no need to search the entire table for the keyword when an index is created for that table. This is why searching a table for a keyword becomes extremely fast when an index is created for that table.

An index for a table works in the same way that an index works for a book. Based on the keyword and page number, anybody can search a book easily, without searching each page of the book.

*Answer 13*

If an index is created for a table and there is an update, delete, or insert operation on that table, then the index needs to be updated. This will make these operations slow.

If a table is updated frequently, it will create a problem because the transactions will become slow and impact the users of the software product. Imagine a table getting updated every second and the problems an index will create here. For a large business house, thousands of transactions can happen hourly. These transactions will create thousands of records in the transaction tables. For each transaction, a record needs to be created in the transaction tables. Because of the presence of indexes, the creation of records in these tables will take time. This will result in the software product running very slowly and affect the users who will be doing those transactions.

Because of this problem, indexes should not be created in transaction tables.

*Answer 14*

Data Dictionary Language (DDL) is used to create database objects. It is also used to delete these objects. Once an object is created, it cannot be modified. It can only be deleted using the drop DDL statement.

You can create or drop database objects such as schemas, tables, primary keys, foreign keys, indexes, procedures, and triggers.

*Answer 15*

Data Manipulation Language (DML) is used to insert, update, or delete the data in the database tables. DML is also used to query a database to find the data in any database table. DML is implemented using Structured Query Language (SQL). You can use SQL statements to search, insert, update, or delete the data in any database table.

*Answer 16*

A stored procedure is a function that is created in a database to do a Data Definition or Data Manipulation operation in that database. A stored procedure is saved with a name. The stored procedure can be invoked using its name from the program to perform some operations in the database. Since the stored procedures are saved inside the database, they can perform very fast operations in the database.

*Answer 17*

In a database, two types of data can be stored: master data and transaction data. Master data relate to the entities about which the transaction data will be generated. For example, if you have a database to store the data about sales of a retail store chain, then the master data may include the store names, merchandise items, the price of these merchandise items, and store operation times.

*Answer 18*

Transaction data are always related to the data that are generated for each transaction. Transaction data for a retail store chain may include the time of sales, quantity of sales, counter number, total amount of sales, and merchandise sold.

Transaction data are stored in transaction tables. Transaction tables are always the child tables for the master tables that hold the master data.

*Answer 19*

Most databases are used to store the master data and the transaction data. Making a distinction between the master data and the transaction data is important when a database is designed. Master data remain constant and do not change. The volume of master data is small. In contrast, transaction data can change and large volumes of transaction data are created on a daily basis.

When tables are created to store the master data, you need not pay much attention to the amount of data to be stored. However, you must pay attention while creating the tables to store the transaction data. If a single column in a transaction table contains duplicate data (i.e., the data that are already stored in the master tables), the amount of these unwanted data will be huge. The duplicate data will unnecessarily consume the hard disk space and also slow down the transactions.

**Chapter 9**

*Answer 1*

Software testing is performed at unit, integration, system, and deployment levels. Unit testing is done at the class level. Integration testing is done to test the interface of a class to make sure that the class integrates with other classes. System testing is done when the complete system is built and the system's functionality needs to be tested against the system design. Deployment level testing is also known as user acceptance testing. It is done after the software product is deployed at the customer site.

*Answer 2*

Black box testing is done at the system level. Black box testing does not use the source code. It runs the executable binary code of the software product for testing. Since this testing treats the software product as a black box (i.e., no idea about the internals of the software product), it is known as black box testing. Black box testing is done for the functional and nonfunctional software requirement specifications.

*Answer 3*

White box testing is done against the source code of the software product. Since this testing involves knowing the internal parts of the software product, it is known as white box testing. Unit testing and integration testing are examples of white box testing. In unit testing and integration testing, the source code is run to test a component of the software product.

*Answer 4*

Unit testing is done to check some business logic implemented in a class. The business logic is implemented through the methods inside a class. For example, suppose a class is built to carry out interest calculations for the deposits made in a bank. The class takes different interest rates through parameters and returns the computed interest amounts.

To test this class, we can pass various interest rates to it and find out if it is doing the computations correctly.

*Answer 5*

An integration test is done to check the interfaces of a class so that it can be integrated with other classes. There could be integration problems associated with a class's parameters. For example, a class needs two parameters but it passed just one parameter from another class. This will result in integration failure. Another integration problem could be related to passing a parameter with a wrong data type. Again, it will create integration problems.

Integration testing is done to eliminate all these problems.

*Answer 6*

Testing tools are used for unit and integration testing to automate all the testing effort. These tools also save the test cases. When regression testing is done, these saved test cases can be used. In these testing tools, you pinpoint the class (unit) that needs to be tested. Then, you provide the test values for testing these classes.

These tools can also simulate (mock) the environment under which the testing needs to be done. For example, if a method in a class will connect with a database to manipulate or search the database, then the test tool can simulate a database to test this class.

*Answer 7*

Verification testing is done for the artifacts that cannot run (execute). Examples of verification testing include the requirement specifications review, software design review, and code inspections. Verification testing is mostly done in a formal environment. Generally, meetings are arranged where the reviewer will review and find the defects, if any.

*Answer 8*

Validation is done by running the source code or executable binary code of a software product. Examples of validation testing include unit testing, integration testing, system testing, and user acceptance testing.

*Answer 9*

Functional testing is done to test the business logic of a software product. Functional testing is always performed against the executable binary code of the software product. Functional testing uses the requirement specifications related to the functionality of the software product. In the functional requirement specifications, business logic is mentioned. This business logic is broken down into manageable pieces and test cases are written. These test cases are then used to perform functional testing.

*Answer 10*

Nonfunctional testing is donc for environmental factors that are not part of the business logic but adversely affect the user of a software product. Some of these factors include performance, security, and usability. Some other nonfunctional factors do not affect the user but may affect some other parties. One such factor is the maintainability of the software product. If the software product is built in such a way that it is difficult to do maintenance, then it creates a problem during maintenance time.

Testing all these environmental factors ensures that the software product can be used effectively by the users and also maintained by the maintenance people.

*Answer 11*

The first step in testing is the creation of test cases. These test cases are run against a unit or the system. If a test case fails, then it means that there is a software defect. This software defect is reported to the software developers. They fix it and ask the tester to test it again. If the test case again fails, then the defect will again be fixed by the developer. If the defect is fixed, then the defect is closed. All the test cases are run and all the defects that are found will be fixed.

*Answer 12*

A test case consists of preconditions, postconditions, testing data, and the test description. This test case is run against the target (i.e., system under test) to test whether the system is working correctly. If the test case passes when a test case is run, then this test case is marked as passed. When the test case fails, it is marked as failed. A defect is also logged against that failed test case. The defect will then be fixed by a software developer.

*Answer 13*

*Continuous integration* is the term used for very frequently integrating some pieces of source code to the software build. In fact, when a software developer writes a new class and once the tests find out that it is working fine, the developer immediately checks in (integrates) this class to the software build. Thus, the software build gets updated continuously. This type of working is known as continuous integration.

The benefit of this continuous integration is that a new release of the software product can be made anytime. The software build is always ready. The other benefit is that no extra effort is needed to integrate different pieces of software product later on.

*Answer 14*

In a continuous integration environment, a software developer keeps the updated copy of the software build on his or her local computer. If a class passes all the unit tests, then the developer will check that class in the local software build on his or her computer. If the build fails, it means that there are integration problems. The developer will look into the source code of that class and fix the problems if there are any. He or she will try to make a build again with the changed source code. If the build is successful, then the developer will check it in his or her code on the central software build (i.e., the server software build). Even if the local software build was successful, the integration could still fail on the server build. This is because the local build on the developer's computer may not be up to date (i.e., synchronized) with the server build. The developer has to look into the failure messages issued (after the build failure) and find out the cause of the problem and fix it.

*Answer 15*

Alpha testing is associated with an alpha release of a software product. Alpha release is used by some software vendors to get their software product tested quickly and free of cost. They release their untested software product to the customers for free. The customers use the product and find the software defects (if any) and report them to the software vendor. The software vendor fixes those defects.

This kind of testing done by the customers for an alpha release of a software product is known as alpha testing.

*Answer 16*

Beta testing is associated with the beta release of a software product. It can happen that, during alpha testing, a large number of defects were found. The vendor may fix these problems and want to get the software product tested again. For this purpose, beta release is used by software vendors to get their software product tested quickly and free of cost one more time. They make a beta release of their software product to the customers for free. The customers use the product and find the software defects (if any) and report them to the software vendor. The software vendor fixes those defects.

This kind of testing done by the customers for the beta release of a software product is known as beta testing.

**Chapter 10**

*Answer 1*

Once a software product is developed and tested and is found to be working fine, it needs to be released so that it can be implemented (i.e., deployed) at the customer's site. Release activities for a software product involve creating the technical and user manuals, providing user training, implementing the software product at the customer site, and so on.

*Answer 2*

On agile projects, only two to three product features are added during an iteration. An iteration can last from 1 to 4 weeks. At the end of each iteration, a minor release of the software product is done. The marketing team of the software vendor decides the major releases of the software product. A major release of the software product could encompass three or more minor releases.

*Answer 3*

A software vendor may want to get its software product tested quickly. Even if the vendor employs a large testing team, testing may still take some time. Moreover, if the software product is tested by the customers themselves, then the testing will be more thorough and the software vendor will also get an idea as to whether the customers perceive something as a defect. For these reasons, the software vendor may make an alpha release of its software product to its customers free of cost. The customers will use the product and report to the vendor if they find any software defects. There could be thousands of customers who receive this alpha release. Thus, testing done by the customers will be fast and thorough. This is the idea behind an alpha release.

*Answer 4*

Software projects that use agile methodologies are built incrementally. The incremental building of a software product allows the software vendor to bring the product to the market as early as possible. It also allows the software vendor to decide the product features on the basis of the latest demand from the market. If the current market demand is for some particular product features, then the vendor will immediately tell its project team to build those features in the next iteration. After the completion of that iteration, the software product will have the required product features ready to be released in the market. Thus, the software vendor can plan the releases of its software product exactly as per the market demand.

*Answer 5*

A software release checklist could contain items such as user manuals, technical manuals, user training, version of the software product to be implemented (i.e., deployed) at the customer site, and configuration items such as required files.

A checklist will ensure that everything to be done for the software release is completed by the project team.

*Answer 6*

Commercial Off-the-Shelf (COTS) and Software as a Service (SaaS) products can be implemented where there is already an existing legacy system in place. In such cases, the COTS system will replace the legacy system. On such implementations of COTS or SaaS products, software migration is also involved. This migration is about the data to be migrated from the legacy system to the newly implemented COTS or SaaS system. These data are important to the customer because they will be used for reporting purposes. In some cases, some transactions may not be complete on the legacy system and these open transaction data also need to be migrated to the new system. A data migration strategy thus becomes necessary.

**Chapter 11**

*Answer 1*

Once users start using a software product (in a production environment), they may encounter some software defects. This happens because, even after thorough software testing during software development, some defects may escape to the production environment. These defects in the production environment need to be fixed. Finding and fixing these defects is known as software maintenance.

In some cases, a software product that is already in the production environment needs some enhancements. Again, these enhancements are carried out during software maintenance.

*Answer 2*

Software maintenance requires a strategy so that maintenance can be carried out without problems. If the software product is hosted outside a firewall and is accessible to a maintenance team, then each software defect can be fixed immediately. However, if the software product is hosted behind a firewall, then fixing the defects immediately is not possible. In such cases, periodic maintenance can be done.

*Answer 3*

Reverse engineering is used when the source code of an existing software product is not available and it still needs to be modified. In such cases, a project team studies its user interfaces and tries to understand its business logic by running the product. Once the business logic of the software product is understood, the project team will create a design and then develop a software product that is very similar to the existing software product.

*Answer 4*

When a software product is deployed at the customer's premises, it is called a production instance of the software product. Users work on this instance and the transaction data are generated as a result.

When software maintenance needs to be done on the production instance, these customer data need to be taken care of.

*Answer 5*

Commercial Off-the-Shelf (COTS) and Software as a Service (SaaS) software products are already developed. When a customer wants an implementation of any of these products (for those customers' specific needs), the software vendor will evaluate the customer requirements. Some of the customer requirements may not be met by the existing features of the product. In those cases, the software vendor may develop new features and integrate them into the software product. This is known as the customer-specific version of the software product.

*Answer 6*

There are many similar phases between a software maintenance project and a software development project. At the same time, there are also differences. A software maintenance project starts when a maintenance request is received. The request will include the list of software defects that need to be fixed. The project team will need to classify all these defects on the basis of the parts (i.e., locations) of the software product where these defects have been found. Then, the location in the source code will be identified and need to be fixed. Each piece of code that needs to be fixed will be analyzed to determine the cause of the defect. On the basis of the analysis, a software design will be created to fix the defect. The software design will be implemented. Once all these defects have been fixed, a system test will be performed. After a satisfactory user acceptance testing, the solution will be deployed.

**Chapter 12**

*Answer 1*

Configuration management is used on software projects to manage software builds. A software product typically consists of many files. These files can contain the source code or the binary code (i.e., machine language). In a software build, these files are kept in such a way that they integrate well with each other. During runtime, the source code of a file that is executing will need the source code of other files because of the calls between the methods. If a required file (that is called) is missing or is not in its proper place, then the software product will not be able to run.

Configuration management ensures that the software builds are in good health by checking the integration requirements of all the files that are needed to create the software build.

*Answer 2*

During a software development project, project team members work on various kinds of project tasks and the end result of these tasks is in the form of files or documents. For example, a software designer will create the design of various parts of the software product. He or she will save these designs in some files. During the project, it is possible that many versions of the software product need to be created. This means that the software designer will also need to create many versions of the software design. He or she will save all these different versions in various files.

It is important to ensure that the right version of the software design is used for developing the software product. Otherwise, a wrong version of the software product will be developed if a wrong version of the software design file is selected. Managing various versions of the same artifact is done using the version control software. The version control software ensures that all the versions of an artifact are managed well so that there is no mixing or chance of using a wrong version.

*Answer 3*

A software product consists of many files containing the machine code (i.e., binary code). The software product is developed using a programming language such as C++. The source code (i.e., the code written in a language such as C++) is spread across many files. These files need to be arranged in such a manner that when a call is made for a piece of source code, it is available at the right place. For example, assume that source code A makes a call for a piece of source code available in file B. If file B is either missing or not available at the right place, then when source code A is executed, a compiler error will be raised.

A software build is a repository of all these files needed for a software product. In the software build, all these source code files are arranged in such a way that the required file is always available at the right place.

*Answer 4*

A CVS system manages various versions of artifacts of a software project. It also manages the configuration of software builds. On a software project, many software developers do their work and create many artifacts from time to time. A central place is needed to keep all these artifacts. The software build, which is developed during the project, needs all the files containing the source code developed by all these developers. A central place is needed for the software build as well, so that all the developers can submit their source code files on this central software build. For these reasons, a central CVS system is needed for software projects.

*Answer 5*

During software development, a software engineer creates (or modifies) a file (say, File X) and saves it on his or her local computer. Later, he or she checks in (i.e., submits) File X on the central CVS system. Sometimes, there is confusion as to whether File X is the same as (or different from) the file that is already in the central CVS system (if a file on the central CVS system has the same name and is at the same location where File X is being checked in by the software engineer, then the confusion starts). In such cases, a file comparison can be done to see the differences in the contents between File X and that file on the central CVS system. If the software engineer still decides to check in File X on the central CVS system, then synchronization will happen between File X and that file on the central CVS system. After synchronization, both files will have the same content.

*Answer 6*

A file system of an operating system has the ability to do version management among files. However, it can manage the files on one computer alone. It has limited ability to manage the files that are residing on many computers. A file system of an operating system has no configuration management capabilities.

## Chapter 13

*Answer 1*

Software project management is all about managing the following resources of a software project: schedules, budgets, suppliers, customers, and the project team. Software project management is more important for large projects. On large software projects, the large size brings many complexities into the project. Bringing a clear overview into project management becomes very difficult on such projects because many tasks are dependent on other tasks, and many tasks run in parallel. Determining whether a task is running as per the plan or is lagging behind its schedule becomes very difficult. The same is true for the project budget. Only through utilizing some project management techniques is it possible to manage such large software projects.

Software project management is relatively easier on small software projects. This is because there will not be many tasks running in parallel and the number of dependencies between the tasks will also be low. An overview of the project will also be easier on such projects.

*Answer 2*

CPM/PERT is actually an assimilation of two separate project scheduling techniques. CPM stands for Critical Path Method. PERT is a French acronym for Program Evaluation and Review Technique. While arranging, the tasks on a project should be in such a way that the project schedule is the shortest. To achieve this goal, the tasks are arranged in a line that is known as critical path. The task that has the earliest baseline start date is placed first on this line. The task with a baseline start date just after the first task is placed next on this line, and so on. Whenever a critical task is found, some buffer time is also allotted on this line for that critical task. We keep arranging all the tasks on this line until they are completed. Care is also taken about the dependency requirements. For instance, if task x should not start before task y finishes, then task x will be placed on the line only after the baseline end date of task y.

Once all the tasks are arranged this way (in a line), the baseline start date of the first task in the line will become the baseline start date for the project. The baseline end date of the last task in the line will become the baseline end date for the project.

The duration of the project will be from the baseline start date to the baseline end date of that project.

The CPM/PERT method can be successfully used to plan Waterfall-based software projects because these types of software projects are plan driven; therefore, all the project tasks can easily be plotted using the CPM/PERT method, and the complete software project plan can be drawn.

*Answer 3*

Earned value management (EVM) is a project tracking technique. EVM relies on the project execution data for the schedule and the budget, to create graphs that show how the project is progressing. By comparing the data for the baseline schedule and budget, you can find the schedule and cost deviation. Schedule deviation is the difference between the actual schedule progress and the planned schedule progress (for a task on a given date). Cost deviation is the difference between the actual cost incurred and the baseline budget (for a task on a given date). The baseline budget for a task on a given date can be calculated using the total budget for that task and the task schedule progress for the given date. For example, assume that the total schedule for a task is 10 days. On the fourth day, the task progress was measured. The schedule budget for the task on this day will be (4/10) × (total schedule budget for the task).

The formula to calculate the schedule deviation is

((Actual schedule progress – planned schedule progress)/(planned schedule progress))%.

The formula to calculate the cost deviation is

((Actual cost – planned budget)/(planned budget))%.

Software projects that are based on the Waterfall methodology can use EVM. This is particularly true when the detailed project data are available to plot EVM. On software projects where the project data are not easily available, the EVM method cannot be used effectively.

*Answer 4*

A Gantt chart is a graph that is plotted to measure the schedule progress for a project. A Gantt chart shows the baseline start dates and baseline end dates for all the tasks in a given project. The chart also shows the dependency among these tasks. Once the project starts executing, then the actual progress of all these tasks is again plotted on the same chart. Now, each of these tasks has baseline dates (for start and end) and actual dates (for start and end). Since the project schedule is in the form of a graph, it is easier to track the project progress as well as task progress.

Software projects that are based on the Waterfall methodology can use Gantt charts. The project can be easily tracked based on the baseline and actual dates.

*Answer 5*

A project consists of many tasks. Planning, monitoring, and controlling a project become easier if the project tasks are easy to manage. If the project tasks are not broken down into some manageable parts, then monitoring and controlling the project will become difficult. For example, suppose there are two software design modules that need to be developed. If the entire software design is taken as a single task, then it will be difficult to monitor and control the individual module designs. Thus, instead of one design task, we should have two design tasks: one task for each module design.

Analyzing all the project tasks and breaking them down into manageable tasks is done using a work breakdown structure.

*Answer 6*

A project schedule is the depiction of all the project tasks and their individual schedules inside the project schedule. A project schedule will have its own start and end dates. Each project task will also have its own start and end dates. Using CPM/PERT, it is possible to draw a critical path for the entire project. Using Gantt charts, Goldratt's critical chain method, or any other appropriate method, it is possible to track a project schedule.

*Answer 7*

Project planning is an activity where a project's schedule, budget, resource requirements, and scope are defined. Before a project can start, it is important to chalk out the details about all these requirements. A project manager must find out all these details so that the resources or budget for the project can be arranged for its duration. If project planning is not done properly, then the project will face resource or budget problems. A well-defined project plan will ensure that no serious problems arise once the project execution starts and that the project execution is carried out smoothly.

*Answer 8*

When a project starts executing, problems can occur on a day-to-day basis. A project can face resource problems because some project team members can leave the project midway. A project can face budget problems if the project spending is over the budget or the budget is simply not allocated properly. There could also be problems associated with product quality because the number of defects is more than expected.

To deal with such problems, project monitoring is done by the project manager. If any problems are encountered, then the project is controlled. For example, if the project is running behind schedule, then more resources can be procured so that the remaining project tasks can be completed more quickly. If it is found that the budget expenditure exceeded the planned budget, then a tighter control over the budget expenses can be applied.

*Answer 9*

On software projects, a supplier is an outsourcing partner. Some part of product development work can be outsourced to this supplier. The relationship with the supplier is controlled through an agreement. The supplier should get the work done as per this agreement. The most important aspect about supplier management is the timely delivery of the pieces of work assigned to the supplier.

*Answer 10*

On Waterfall projects, a customer is never on the project site. The project manager needs to send project status reports to the customer periodically. The customer expects the project to be carried out as per the schedule. If there are any delays, then the project manager must provide an explanation to the customer. The project manager should also provide remedial measures so that the project gets back on track. To ensure that the customer feels confident in the project team's capability to deliver the software product as per his or her expectations, a prototype should be developed and feedback should be obtained from the customer before the project work starts.

*Answer 11*

Velocity is a measurement of a software project team's ability to deliver a specific number of story points in an iteration or sprint. Each project team on any agile project finds out its velocity to design, implement, and test the software features. There is no universal definition for a standard velocity (to deliver the product features). Only after a project team works on two or three iterations of a project is it possible to measure the velocity of that project team.

*Answer 12*

Story points indicate the complexity of a product feature. A complex product feature will have more story points. Conversely, a simple product feature will have fewer story points. There is no universal definition for story points. Each project team determines its own way of finding the number of story points for each product feature.

*Answer 13*

A timebox is an iteration in an agile project. The timebox derives its name from the fact that the project team has to deliver (some product features) in an iteration that has a fixed time frame. For example, an iteration may be 10 days long. During these 10 days, the project team has to design, implement, and test all the product features that are planned for that iteration. Since the time and resources are fixed in an iteration, the product scope is left open ended. The product owner assigns importance to each product feature that needs to be developed. The most important product features need to be developed first. The least important features need to be developed last. In this arrangement, if time does not permit, the least important features can be left to be developed in the next iteration.

*Answer 14*

Project scheduling for traditional software projects is done by considering factors such as fixed scope, fixed resources, and fixed budget. In contrast, the agile projects have no fixed scope. Since the scope is not fixed, no fixed work breakdown structure can be defined for an agile project. Since there is no fixed breakdown structure for agile projects, traditional scheduling techniques such as Gantt charts, CPM/PERT, and earned value management do not work for them.

# Index

Page numbers followed by f and t indicate figures and tables, respectively.

D

S

X